ONLY THE WING

ONLY THE WING

REIMAR HORTEN'S EPIC QUEST TO STABILIZE AND CONTROL THE ALL-WING AIRCRAFT

With a New Introduction

RUSSELL E. LEE

A Smithsonian Contribution to Knowledge

Smithsonian
Scholarly Press
WASHINGTON, D.C.
2020

Published by
SMITHSONIAN INSTITUTION SCHOLARLY PRESS
P.O. Box 37012, MRC 957
Washington, D.C. 20013-7012
https://scholarlypress.si.edu

Originally published, in association with Rowman & Littlefield Publishing Group, 2011.
Reissue with new introduction 2020.

Cover image: The Horten 229 V3, front view. Photo by Ben Sullivan, 2013, courtesy of Smithsonian Institution.

Library of Congress Cataloging-in-Publication Data

Lee, Russell, 1956–
Only the wing : Reimar Horten's epic quest to stabilize and control the all-wing aircraft / Russell E. Lee.
p. cm.
Includes bibliographical references and index.
1. Horten, Reimar, 1915-1994. 2. Aeronautical engineers—Germany—Biography. 3. Airplanes, Tailless—History. I. Title.
TL540.H677L44 2011
629.1092—dc22
[B] 2010044472

ISBN 978-1-935623-03-8 (cloth)
ISBN 978-1-944466-38-1 (pbk)

Printed in the United States of America

⊗ The paper used in this publication meets the minimum requirements of the American National Standard for Permanence of Paper for Printed Library Materials Z39.48–1992.

To Natalie and Ian

Contents

Preface		ix
Acknowledgments		xi
Introduction		1
Chapter 1	From Flying Models to Piloting a Wing	23
Chapter 2	Freedom of Flight	49
Chapter 3	Competition Wings	75
Chapter 4	Jet Wings	111
Chapter 5	End and Beginning	141
Chapter 6	Winglets and More Wings	169
Appendix 1	Horten Aircraft Type Data	195
Appendix 2	Horten Aircraft Performance Characteristics	209
Appendix 3	Horten Aircraft Airframe Data	221
Notes		233
Glossary of Terms, Phrases, and Abbreviations		261
References		275
Index		285
About the Author		295

Preface

IN JANUARY 1994, four Horten sailplanes belonging to the Smithsonian National Air and Space Museum (NASM) arrived at the Museum für Verkehr und Technik (MVT) in Berlin, Germany. MVT had borrowed the wings from NASM after agreeing to undertake the long and difficult work required to restore and preserve the aircraft. The loan marked a milestone in efforts that began as early as 1978 when Jan Scott, soaring enthusiast and president of the Vintage Sailplane Association in the USA, wrote to NASM suggesting the museum lend one or more of the artifacts to Germany for restoration. As the curator responsible for the sailplane collection, I drafted guidelines that explained how to restore and preserve the Horten aircraft, prepared a loan agreement, located historical information, made the documents and information available to the specialists in Berlin, and monitored their progress via correspondence and telephone. I also visited Berlin to inspect the work and confer with MVT staff. This book grew out of the research I conducted to support this project.

It became apparent during my investigations that understanding how Horten worked, and what forces—both internal and external—guided him, were areas of inquiry fully as worthy of study as determining what he had accomplished. When the lack of an advanced academic degree and experience working with the aviation industry were not blocking the official support he might have received from the government and military, then his austere working conditions were slowing construction and restricting attempts to flight-test the wings. Political instability and social upheaval related to the Nazi seizure of power also disrupted his work, yet repeatedly he overcame these obstacles and ultimately refined the control and performance of all-wing aircraft to a degree that no one else had achieved.

Like most artists, the creative process often consumes aircraft designers to the exclusion of much else, including occasional pauses to note what influenced them to follow a particular line of development. Such is the case with Reimar Horten. He coauthored with brother Walter a single technical paper, "Stabilitätsbetrachtungen an Pfeilflügeln" (Considerations on the Stability of Arrow Wings), that appeared in spring 1943 in Berlin, choosing instead to describe a good portion of his work in articles written for popular magazines, a venue that precluded detailed explanation of complex aerodynamic concepts. The designer never worked with a large firm that might have preserved his working correspondence in a company archive. He dispersed his personal correspondence between two countries separated by language, culture, and geography, making this evidence difficult to collect and interpret. Almost no government records of his work exist because officials sanctioned few of his projects and technicians tested only one small model of a Horten design in a wind tunnel in Argentina.

Studying his wings in wind tunnels or by flying them after World War II ended could have provided accurate and independent assessments of performance and helped to determine whether the average pilot could fly them under practical conditions encountered routinely or whether only expert pilots should fly the wings under ideal conditions. Unfortunately, Allied soldiers and slave labor imported by the Nazis from the countries they had occupied destroyed most of the airframes built in Europe as the war ended. By 2010, only a few Horten aircrafts remained intact, yet their scarcity has not stifled enthusiasm for Reimar's work.

Acknowledgments

I COULD NOT HAVE spent more than fifteen years researching and writing this book without the support of my colleagues at the National Air and Space Museum (NASM). Director Jack Dailey, Deputy Director Don Lopez, and Ted Maxwell, Associate Director of the Collections and Research Department, encouraged me to write the Horten story. Tom Crouch backed the MVT's initial proposal to borrow and restore the Horten sailplanes, Dom Pisano read the manuscript and recommended improvements, and Peter Jakab funded an intern who translated original and published sources written in German. John Anderson took time to explain aerodynamic concepts that I found difficult to understand, while Howard Wolko generously shared his own research on German aircraft designers. Anita Mason and Collette Williams swiftly guided me through the travel bureaucracy when I travelled to Germany to oversee the Horten sailplane restorations and to research the Horten story.

Phil Edwards provided sources to important records on all-wing aircraft held in the National Archives and gave me a complete copy of "The Development and Airborne Testing of the PALE Seat." Larry Wilson teased out obscure references to Horten's work as he inventoried more than five thousand reels of documents in the Captured German and Japanese Technical Documents on Microfilm collection preserved at the Paul E. Garber Facility in Suitland, Maryland. Dan Hagedorn suggested Record Group 342, the Sarah Clarke Collection, held at the National Archives, and helped me to contact historians in Argentina familiar with Horten's work in that country. Lars McLamore introduced me to Barbara Harding, whose father, Horten test pilot Erwin Ziller, had perished while flying the Horten *H IX V2* jet fighter. Jeremy Kinney, Roger Launius, and Michael Neufeld gave advice and encouragement.

In addition to support from staff, NASM volunteers and interns contributed their time and considerable expertise. Geoff Steele was examining U.S. Office of Naval Intelligence records at the National Archives when vandals broke into his car and robbed him. Steele served as intermediary between David Myhra and NASM when Myhra donated the collection of interviews he recorded with Reimar and Walter Horten. Frank Carr helped to transcribe these interviews, and Tom Culbert explored files at the National Archives and records from the U.S. Army's Air Disarmament Division held at Maxwell Air Force Base, Alabama. John Fitzpatrick dug into the NASM library holdings and met with the test pilots, Eric Brown and Rudi Opitz. Both men had flown a Horten *H IV a* sailplane. Lars Urban interned in the Aeronautics Division at NASM from July to September 2003, tirelessly translating German documents and offering valuable feedback on Horten's work.

In Berlin, the staff of the MVT (renamed the Deutsches Technikmuseum Berlin in 1997 and abbreviated DTM) gave generously of their time and expertise. Holger Steinle arranged visits to Germany so that I could monitor the restoration work, and he once drove me to the village of Gut Tierstein in southern Germany to search for buried treasure: buried milk tins said by reputable sources to contain original drawings of Horten aircraft. Michael Hundertmark spent an entire day driving me from Berlin to the Wasserkuppe in south central Germany, the epicenter of German soaring history and the site where pilots flew Horten sailplanes on many occasions. Hundertmark and his family also generously shared their home with me.

On short notice, author and historian Peter Selinger offered photographs and rare memorabilia and answered many questions, and aerospace engineer Reinhold Stadler explained aerodynamic principles and interpreted Horten's theories on the technology of all-wing aircraft. Stadler provided drawings and other references, and he shared with me his important collection of Walter Horten memorabilia, including a training aid that Walter had developed to teach deflection shooting to German fighter pilots. Edward Uden made available items from his personal collection, including photographs taken during the war of the Horten *H III h* belonging to NASM and restored by DTM. Stadler and Uden deciphered the secrets to Reimar's distribution of lift and then demonstrated how he used the idea to improve the flying qualities of Horten wings.

Albion Bowers, Deputy Director of Research at the National Aeronautics and Space Administration Dryden Flight Research Center in California, carefully considered the work of Stadler, Uden, and others, and then posted his conclusions via the Internet to the community of Horten enthusiasts. He was especially engaging on the Internet discussion list created and managed by all-wing aircraft enthusiast Doug Bullard. Bowers generously provided much information for this book, sending me copies of obscure documents and answering many questions. Karl Nickel worked for Horten during and after World War II and married his sister, Gunilde. The Nickels allowed me into their home in Freiburg, Germany, in October 2005, and set aside the better part of five days to explain Reimar Horten's work from their unique perspectives. Stadler, Bowers, and the Nickels also provided invaluable comments on the manuscript. I must emphasize that much of what we know about Horten's work is due to the willingness of this small group to share their knowledge.

My research also relied on the work of amateur historians whose expertise and dedication to the truth rivaled the work of professionals. During the early 1980s, David Myhra began traveling to Argentina at his own expense to meet with and interview Reimar Horten. He also interviewed Walter Horten in Germany and by 1988, Myhra had collected more than sixty hours of conversations rich in information and anecdotes covering a wide range of subjects. This is the only oral history of the Horten brothers ever recorded. David Myhra very generously donated all sixty of his original tapes to NASM in 1998. Chris Bryant spent many hours looking for obscure documents at the Public Records Office in England before finding a report about the aerodynamic aspects of Horten wings. Bryant also helped me contact people in England who had flown the Horten aircraft.

Philip Wills had served on the team of Allied intelligence specialists that recovered two Horten gliders in 1945. Wills's son, Chris, sent me his own carefully compiled and annotated list of Horten aircraft plus videotapes of German gliding activity during the period between the world wars. Test pilot Dezsö George-Falvy, who flew a Horten *H IV a* at Mississippi State University during the 1950s, lent me slides and answered many questions, as did Reimar Horten's son, Diego, who also offered reports and articles describing his father's career in Argentina and his work on the PUL-9 and PUL-10 sport wings. Fernando Siarez illuminated hitherto unknown aspects of Reimar's career in Argentina and laid out the political context that profoundly influenced Horten's work. Jan Scott inspected the NASM collection of Horten aircraft several times during the 1970s and 1980s, and once the staff at the Deutsches Technikmuseum Berlin began the restoration work, Scott offered his collection of reports, articles, correspondence, and photographs. To all these people I owe my deepest gratitude. Of course, I am solely responsible for any errors in *Only the Wing*.

All-wing aircraft will never usurp the popularity of conventional aircraft, which perform so many important jobs safely and economically, but I hope that *Only the Wing* encourages others to make all-wing aircraft more practical and appealing by embracing the Horten lift distribution as the elegant means to achieving greater structural efficiency, positive control during stalls, and favorable yaw while maneuvering.

Russell E. Lee
April 2010

Introduction to the 2020 Paperback Edition

ALMOST A DECADE has passed since I finished writing the first edition of *Only the Wing*. Scientists, engineers, and museum professionals have continued to study Reimar Horten and his all-wing aircraft. Their findings have added to our understanding of the work and inspired enthusiasts to create new content related to the Horten aircraft and renewed public interest in the Horten story. In California, scientists proved the characteristics and benefits of the Horten bell-shape lift distribution (BSLD). At the Smithsonian National Air and Space Museum, conservators and treatment specialists began the first detailed examinations of the Horten 229 V3 all-wing prototype jet.[1] Their findings have answered questions but also posed new ones.

NASA Studies Horten Aerodynamics

At the National Aeronautics and Space Administration (NASA) Neil A. Armstrong Flight Research Center in California, Chief Scientist Albion Bowers began studying the aerodynamic characteristics of Horten aircraft in 1994. Bowers deciphered the mysteries of Horten aerodynamics and freely shared his findings. The worldwide community of aviation enthusiasts and historians owes a great deal to Bowers' intellectual curiosity and generosity. In 2013, specialists and interns built, tested, and evaluated an all-wing flying model called the Prandtl-D Ship 1. (Prandtl is the acronym for Preliminary Research Aerodynamic Design to Lower Drag.) They recorded the first scientific data derived from actual flight tests on June 26. During many flights that followed, they collected data that has improved our understanding of Reimar Horten's important contributions to the theory that Ludwig Prandtl first described. Bowers explained the key findings this way:

> The flying model Prandtl-D Ship 1 returned the first scientific data proving that proverse [or favorable] yaw is possible. Proverse yaw is the

> motion in which both roll and yaw happen in the intended direction on the initiation of a turn. In [conventional] aircraft it was the invention of the vertical tail [and rudder] that gave yaw control to overcome adverse yaw.[2]

Adverse yaw has prevented more intuitive control of aircraft since the beginning of heavier-than-air flight. It happens when a pilot rolls an aircraft in one direction to begin a turn, and the aircraft responds by yawing in the opposite direction. Most aircraft are equipped with a vertical tail and a rudder that allows pilots to overcome the adverse yaw. Favorable yaw allows an aircraft to roll and yaw in the same direction and makes controlling aircraft more intuitive for pilots. Aircraft efficiency could be increased if the tail and rudder could be reduced of eliminated.

To understand how the BSLD came about, Bowers also studied its evolutionary history as an idea. He discovered that aerodynamicist Ludwig Prandtl showed, in a paper dated June 11, 1933, how he had

> derived a wing spanload [lift distribution] to minimize induced drag for a given wing weight, not realizing his solution was also optimal in providing a method for proverse [favorable] yaw as well. . . . Reimar Horten proposed what was necessary to produce proverse yaw, but he did not prove it. . . . Further, [NACA aeronautical engineer] Robert T. Jones derived a similar solution [in 1950] to that of Prandtl, not realizing Prandtl had already solved the problem. . . . This [Prandtl-D Ship 1] is the first aircraft to document and demonstrate, with data, that all three of these men were right and more. And that the solution that all three of them sought were exactly what birds use to turn and maneuver.[3]

Bowers found that, in addition to Prandtl, Horten, and Jones, Sathy Viswanathan and Armin Klein also performed important work to reduce aerodynamic drag, but they missed the implications of the BSLD for eliminating adverse yaw. NASA aeronautical engineer Richard T. Whitcomb calculated how to design the winglets seen on many aircraft. Though winglets measurably reduce drag by creating some forward thrust at each wingtip, as does the bell-shape distribution, Whitcomb, too, missed the potential to counter adverse yaw.[4]

Bowers' belief that Reimar Horten missed the true potential of the BSLD marks an ironic twist in the story. "Horten," he wrote, "never understood what was happening. In his final works, [published in the magazine] *Soaring* [in]

1981, he wrote of the spanload and states that there is a negative angle of attack at the wingtips. Yet the wings still create no negative lift [to counter adverse yaw]. So clearly there is a disconnect in what he writes."[5] Although Reimar might not have fully understood the possibilities, he may have accidentally experienced favorable yaw as a result of the bell distribution (see Chapter 2).

Following on the Prandtl-D Ship 1, two more flying models followed: in 2015, the Prandtl-D 3 and, in 2017, the Prandtl-D 3c (c meaning carbon-fiber construction) to replace the Prandtl-D 3 damaged in a landing mishap.[6] NASA announced in September 2015 that chief scientist Al Bowers and NASA Langley senior research engineer Sue Grafton were collaborating to evaluate the BSLD using the wind tunnel at the NASA Langley Research Center in Virginia. Technicians built a model for the tunnel designed to validate the data derived from the two Prandtl-D flying models tested in California. Flight tests tell only part of the story. "We can only surmise the aero[dynamic] characteristics in a very piecemeal approximate way," said Bowers. "The wind tunnel data show all the individual pieces that go into that data set . . . comparing the results from the wind tunnel data set to the flight data is the real proof" that the behavior in flight is truly understood. Grafton remarked on the behavior of the tunnel air interacting with the model. "I don't think I have ever before seen these exact flow patterns that we saw with this test," she said. "How the vortices started and changed, where they went and their makeup. It seemed to look a lot different than the typical airplane wing that I have seen the vortices [form] on before."[7]

Bowers described how the past few years of tests combined to tell a coherent story in 2018.

> The work of 2013 [Prandtl-D Ship 1 test results], and 2015 [Prandtl-D 3 test results], was only explained by the final set of data taken in the summer of 2018. A very special instrumentation box was created allowing the pressure distribution over the wing to be taken. The instrumentation box and its software was designed and built by a team of seven students to whom I will be forever indebted . . . (Noah Edwards, Lydia Hantsche, Abigail Waddell, Deborah Jackson, Rachel Suitor, Chris Jensen, Victoria Hawkins). The data from that box explained in minute detail the aerodynamics of Prandtl's 1933 solution.[8]

The Electronic Pressure Measurement (EPM) device allowed the team to take 100 pressure readings per second simultaneously at 96 different locations on the wing. The team used this data to assemble a detailed map of the distribution of lift and drag as the wing flew. "In every pressure system we did before," Bowers explained,

> [W]e [recorded] the pressures sequentially, that is we took one pressure, then the next, then the next, etc. We did this with 32 pressures at a time. It would take a few seconds to sequence through. The EPM was the first mass simultaneous pressure system I am aware of. Because of this, we could make movies of the pressures (this had never been done previously).[9]

Bowers credited Serbian aeronautical engineer Marko Stamenovic with writing computer code that makes it easier for aerodynamicists to design straight wings with a bell lift distribution. Reimar Horten swept the wings of all his aircraft, and the flying models built to explore the BSLD had swept wings too. The straight wings developed by Stamenovic demonstrate favorable yaw, which Frank Bullot of Australia demonstrated when he built and flew an all-wing flying model designed by Stamenovic.[10]

So unprecedented is Bowers' extraordinary body of work that distinguished scientific journals refused to publish it. NASA recognized its importance and, in 2016, published "On Wings of the Minimum Induced Drag: Spanload Implications for Aircraft and Birds" in its Technical Paper series.[11] The authors argue that the bell lift distribution allows birds to maneuver without a vertical tail and to fly in formation with their wingtips overlapping with other birds, even though the birds have narrow wingtips that should be prone to stalling. Applying the theory to aircraft would require a

> properly designed aileron, on a bell spanload wing [that] could produce just the right amount of [favorable] yaw such that an auxiliary yaw [rudder] device would be entirely unnecessary for coordinated turning flight. A design without an auxiliary yaw device means there would be neither added drag nor complexity from such a device. It should be noted that this does not mean for all aircraft designs employing a bell spanload that an auxiliary yaw device would not be needed.[12]

Even conventional aircraft could benefit from "the improved aerodynamics and minimum structure approach."[13] It is exciting to imagine how designers might soon apply the bell lift distribution to more mainstream aircraft. In 2019, NASA took an important step to preserve the legacy of Bowers and his team when the agency transferred the historic Prandtl-D Ship 1 that had generated the first scientific data to the Smithsonian's National Air and Space Museum (NASM).

Horten 229 V3 Museum Artifact Treatment, Study, and Display at the Smithsonian's National Air and Space Museum

In December 2010, NASM Chief Curator Peter Jakab proposed, and Museum Director John R. Dailey agreed, to move the *Horten 229 V3* and several other aircraft out of storage at the Paul E. Garber Facility in Suitland, Maryland, and into the new workshop at the Steven F. Udvar-Hazy Center in Chantilly, Virginia. NASM Chief Conservator Malcolm Collum saw the move as an opportunity to examine and conserve the *Ho 229 V3,* and work began the following year. Conservator Lauren Horelick led a skilled team, which included Anna Weiss, Peter McElhinney, and Karl Heinzel, to first carefully clean the three major components (center section and two outer wing panels) and then study them to determine condition and treatment needs. Stands were built to support these parts and protect them during road transport. Museum staff moved the components to the Mary Baker Engen Restoration Shop at the Udvar-Hazy Center in January 2015. For the first time, visitors could view the wing, through a window wall on the second floor, as conservation treatment continued.

On web pages posted to the NASM website, Horelick described the construction materials used to build the *Ho 229 V3*, and she explained how a lack of technical specifications for the one-off Horten jet led her team to compare the physical evidence on the artifact to wartime documents about materials in general use during the war, in order to develop a contextual framework to better understand the team's research results. She explained the location history of the aircraft up to the time that treatment began, and she defined *preservation* and how it differs from *restoration*. Horelick listed in a bibliography section forty reports and articles published in three categories—wood, metal, and paint—which aided the team's research and analysis of the *Ho 229 V3*. More than 70 images of the aircraft, including many views of the preservation work, are also posted on the website.

As often happens when staff can closely study artifacts, new information was revealed. Burn marks discovered on the plywood belly panels fitted beneath both Junkers Jumo 004 turbojet engines suggest that U.S. Army Air Forces personnel may have operated them sometime after the aircraft arrived at Freeman Field, Indiana. U.S. Army Air Forces leaders considered flying the *Horten 229 V3*. "It is requested that this office be notified when the Horten 9 Jet Glider [sic] will be completed and ready for flight."[14] A response two days later noted that on January 25, 1946, "a visual inspection of the Horten 9 was conducted. There was complete concurrence in the fact that reconstruction is not feasible and is impossible without a re-design of the basic aircraft."[15] Did the Army actually operate the engines to determine their mechanical state? Perhaps, but if so, efforts to prepare the jet to fly appear to have ended quickly when it became apparent how much work was needed.

The project provided the NASM conservation staff an opportunity to use the techniques of modern materials science to evaluate one of the most interesting claims made by Reimar Horten about the *Ho 229*: that he took steps to make the jet wing stealthy and more difficult to detect with radar. Reimar wrote in 1983 that when German industry began to mass produce the *Ho 229*, he would mix sawdust and charcoal with the glue that would join the layers of wood forming the jet's exterior surfaces. He would do this, he said, to shield the "whole airplane" from radar as "the charcoal should absorb the electrical waves. Under this shield, then also the tubular steel [airframe] and the engines [would be] invisible [to radar]."[16] A production model never existed, but Horelick and her team wanted to study the NASM *Ho 229 V3* for any evidence of the charcoal-infused glue.

The team conducted detailed analyses of tiny samples of the wood that they had carefully removed from various locations on the center section. In addition to publishing the results in peer-reviewed scholarly papers, they presented their findings on the NASM website[17].

Summary of Methodology and Findings

A preliminary assessment of the wood samples showed that crafts persons had used two different glues to join the wood, but the team determined with research in the scientific literature that only one, urea formaldehyde, could hold charcoal. Between each of the 5-plys, or layers, of plywood assembled with another type of glue (phenol formaldehyde), the team found layers of urea formaldehyde joining the stacks of 5-ply layers.

Next, the team devised a systematic approach to sampling the adhesives. Minute areas of glue worthy of study were removed, catalogued, and photographed using a digital microscope capable of making high-resolution images. The team went through many steps to isolate and concentrate potential charcoal particles and place them on a glass slide for study using X-ray diffraction (XRD) and Fourier-Transfer Infrared Spectroscopy (FTIR). (Figure 1)

The team analyzed each sample three times with equipment housed and maintained at the Smithsonian's Museum Conservation Institute (MCI), located in Suitland, Maryland. To study how the black particles were distributed within the glue, they first embedded thin sections (30–50 microns wide, about .0012–.002 in.) of glue in clear polyester resin. After the resin dried, a

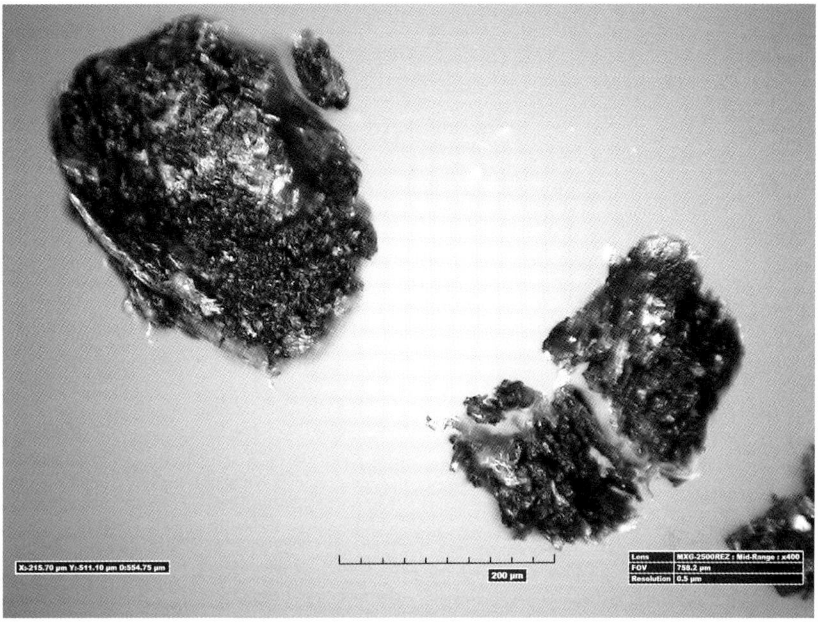

Figure 1. Photomicrograph of tiny black particles extracted from the adhesive matrix. These samples were further characterized with XRD and FTIR. Image: Lauren Horelick, 2014. Courtesy of Smithsonian Institution.

team member sliced it with a diamond blade saw and polished it before using an optical microscope to study isolated black particles, each about 200 microns in length (about .01 in.). Several particles are visible in Figure 1.

Assisted by MCI staff, the NASM team used the XRD and FTIR instruments to characterize the black particles. No conclusive evidence was found that charcoal was used in the glue to reduce the reflected radar signal.[18]

Northrop Team's Stealth Analysis

Horelick's team was not the first to probe the *Ho 229 V3* for evidence of stealth properties. In September 2008, a team from the Northrop Grumman Corporation studied the aircraft during a visit to the Paul E. Garber Facility in Suitland, Maryland. Before Northrop bought Grumman in 1994, the company had built 21 B-2 Spirit "stealth" bombers for the U.S. Air Force. The company's expertise in the technologies that make aircraft difficult to detect using radar and by other means is recognized worldwide. The Northrop team measured the leading edge of the *Ho 229 V3* using a handheld reflectometer device called the Next Generation Sensor (NGS) to illuminate a five-inch diameter area with electromagnetic radar energy and to measure the characteristics of the energy returned to the NGS.[19] A few months later, Northrop Grumman staff built a full-size replica *Ho 229* out of wood and mounted it on a stand at their outdoor range in California to conduct a second series of tests. They beamed radar energy at the replica at "electromagnetic frequencies equivalent to the allied radar systems of World War II. Imaging techniques were utilized to understand the main RCS [radar cross section] scattering sources of the *Ho 229 V3*."[20]

The Northrop Grumman team presented a paper describing their evaluation processes and findings two years later at the 10th American Institute of Aeronautics and Astronautics Aviation Technology, Integration, and Operations (ATIO) Conference held September 13–15, 2010, in Fort Worth, Texas. Dobrenz and Spadoni explained that, "during our inspection of the Horten 229 at the Smithsonian museum, it appeared that a material similar to carbon black or charcoal was mixed in with the glue between the thin layers of the leading edge shape." After examining the aircraft, the group analyzed sections of the leading edges of the jet's center section using a portable radar reflectometer called the Next Generation Sensor. They compared the Horten data to a plywood sample about ¾ in. (2 cm) thick, presumably a "control sample," against which the Horten data could be compared. The findings "indicates that the design using the carbon black type material produced a poor absorber" of radar energy.[21] An inability of the material to absorb radar energy suggests that it would reflect a strong signal and have weak stealth characteristics.

The results of the second test series confirmed that the aerodynamic shape of the jet would have lowered the radar cross-section compared to conventional World War II combat aircraft fitted with fuselages and tails.[22] They concluded that

> it is evident that the Horten brothers made some attempt to incorporate radar absorbing materials. Even though the material did not perform as an RF [radio frequency or radar energy] absorber, the increase in the dielectric constant of the outer mold line did help shield the inside of the platform from RF energy.[23]

These science-and-technology-based studies by the conservators at NASM and the Northrop specialists produced weak evidence at best for special substances added to reduce the *Ho 229 V3* radar cross-section. The reader will find more background to Reimar's claim for using stealth technology in Chapter 6.

Horten Enthusiasts and Popular Culture

Encouraged by the work of Lauren Horelick's conservation team to prepare the *Ho 229 V3* for public display, the staff of Zoukei-Mura, Inc., studied the museum aircraft then designed and manufactured an extremely accurate plastic model in 1⁄32 scale with a 20¾ in. (53 cm) wingspan. The design staff added the complete interior of the *Ho 229 V3* including cockpit, engines, fuel tanks, landing gear, and the airframe. The company released the model kit in 2014 to great acclaim from the hobbyist community. Zoukei-Mura published separately a 112-page, illustrated, soft-cover book titled *Concept Note SWS No. VII, Horten Ho 229*, written to guide enthusiasts in building the model and to supply historical context to the full-size original jet.

Two years later, Marek Rys and Marek Murawski published *Horten Ho 229, Monographs Special Edition in 3D* (Lublin, Poland: Kagero, 2016). The book stands out for the many pages of highly detailed and full-color digital renderings of every major part of the aircraft. Like *Concept Note SWS No. VII*, this book aims to inspire enthusiasts to build scale models of the *Ho 229*. In 2018, after spending 350 hours on construction, George Johnson completed the Zoukei-Mura kit of the *Ho 229 V3*. George built the model to the highest standards and in competition later that year with some of the best model builders in the USA, his model earned second place in its category at the 2018 International Plastic Modelers Society annual show and Best Aircraft, Best in Category, and Best in Show at the Oregon Model Society show in September 2017. George generously donated his winning model *Ho 229 V3* to NASM in April 2019. The museum plans to display it near the original Horten wing.

Reimar began his experiments in the late 1920s. From flying models equipped with conventional fuselages and empennages, he gradually removed these components between many test flights until only the wing remained. Sensing that the all-wing aircraft was the purest form of mechanical flight, the course of his life's work was soon set—to design all-wing aircraft unencumbered by components that did not produce lift. His approach had become multidisciplinary by the time his first piloted wing flew in 1933. Aerodynamics, structures, propulsion, and accommodation all factored into his approach to all-wing flight. Fearing that the established aviation industry would consider his ideas too radical, he avoided working for a major aircraft manufacturer to maintain his design independence. Thus freed, Reimar advanced all-wing technology beyond any rival, but he failed to consider the practical requirements of successful designs. Except for modest successes among the community of sport sailplane pilots, he failed to successfully market his other all-wing designs.

Defining All-Wing Aircraft

What I have chosen throughout the book to call the conventional configuration consists of a fuselage supporting main lifting surfaces toward the front and a tail at the rear that combines a vertical stabilizer and rudder with a horizontal stabilizer and elevator, also called an *empennage*. Nearly all aircraft have a conventional configuration. It had become omnipresent before World War I began in 1914. Yet, by the end of World War II, enough all-wing aircraft had flown with consistently limited success or outright failure to convince one writer in 1944 that building such aircraft amounted to "the useless repetition of a futile idea."[24] The configuration failed to attract sustained private, civil, commercial, or military business until well after World War II and then only for specialized military applications requiring low-observable characteristics. Not only were all-wing aircraft more difficult to design with stability and control comparable to conventional aircraft, but they were also more difficult to develop into practical flying machines capable of repeated and sustained flights while being economically competitive with conventional aircraft. Engines enclosed inside wings were harder to access for maintenance. Providing reasonable access and accommodation for pilot and crew while maintaining the crew's ability to see out of the wing to navigate was challenging. Maneuvering an all-wing aircraft on the ground was more difficult due to the constricted placement of the landing gear.[25] The conventional aircraft with fuselage could mount a nosewheel or tailwheel and use the extra leverage from the longer moment arm provided by the fuselage to give a pilot more stability and control when turning on the ground.

The deceptive simplicity and potential performance advantages of the all-wing airplane nonetheless continued to attract designers. Serge Krauss identified more than 350 projects built throughout the twentieth century that resulted in flyable all-wing aircraft.[26] Persistent interest in all-wing aircraft sprang from the notion that deleting the non-lifting components, such as the fuselage and tail, would lessen drag, reduce weight, and make an aircraft more efficient by increasing speed and lengthening range. Fewer parts were needed, which theoretically would lower manufacturing costs and speed production. After World War II, some designers perceived another advantage. With fewer components protruding at different angles, the all-wing would reflect less radar energy and therefore could be more easily concealed, rendered more "stealthy" than conventional aircraft of roughly equivalent mass.[27] Before they could reap any of these benefits, designers had to overcome a daunting challenge: duplicate the solid aerodynamic stability and control provided by the tail of a conventional aircraft working through a powerful fulcrum, the fuselage.

The German word for all-wing aircraft is *Nurflügel,* literally meaning "only wing" as in "only the wing."[28] Reimar Horten often used the English term *all-wing.* I use *all-wing* or simply *the wing* throughout this book. The terms *flying wing* and *tailless* airplane appear frequently in the USA and Britain, but I think these terms are misleading or carelessly applied to almost any aircraft missing parts of the fuselage or tail. Note that between the all-wing and conventional configurations, there are further variations. The semi-tailless aircraft, for example, derives aerodynamic stability and control from some portion of a conventional tail projecting behind the wing. There is also the tailless aircraft (but not all-wing) without a fuselage or tail employing some form of stability or control that does not generate lift, for example vertical stabilizers and rudders mounted at the wing tips.

I have included in the text the English translations in parentheses after each German and Spanish term, and some translations appear in the glossary as well. German aviation historian Peter F. Selinger suggested that the most correct and authentic method for designating Horten aircraft designed was Reimar's own system. I have therefore abbreviated Horten with an *H* followed by the Roman numeral he assigned to each of the aircraft he designed in Germany; for example, *H II.* Reimar used lowercase suffixes beginning with *a* to identify the various subtypes of a particular design; for example, the *H III d* and *H IV a.* Some of the aircraft that Reimar designed in Argentina bear the designation *I. Ae.*, an abbreviation for the organization that built them, the Instituto Aerotécnico Córdoba (Aeronautical Engineering Institute of Córdoba). I italicized all the designations and type names for aircraft designed by Reimar Horten in Argentina, such as *I. Ae. 34 a Clen Antú.* The types names were a feature unique to aircraft designed in Argentina.

Military authorities assigned *Versuch* (attempt, test, or experiment) numbers to several Horten aircraft during World War II, for example, the two Horten sailplanes designated the *H VI V1* and the *H VI V2.*

Historiography

Oscar Ursinus began writing about Horten gliders in 1938 in his twice-monthly journal *Flugsport.* Shortly after World War I, when Germany took up sport gliding, Ursinus became the unofficial master of ceremonies at the annual gliding contests held on the Wasserkuppe.

Periodicals in Britain and Europe published the occasional article about Horten aircraft after the war. With cowriter Peter F. Selinger, Reimar published his autobiography in 1983. *Nurflügel: Die Geschichte der Horten-Flugzeuge 1933–1960* (Herbert Weishaupt, 1983) remains a valuable reference source despite newer accounts that have appeared as recently as 2006. Reimar and coauthor Peter F. Selinger begin the book by surveying the early all-wing aircraft that influenced Horten's work. They follow with descriptions of each of Reimar's all-wing aircraft accompanied by photographs and drawings, and they explain how brother Walter helped to build and test the wings when the demands of his own busy career as a fighter pilot in the Luftwaffe (German Air Force) would allow. Because each chapter of *Nurflügel* begins with a summary in English, the text flows somewhat unevenly. Reimar tries at the end of the book to explain how his lift distribution improved the handling and performance of the wings he designed, but he does so in a brief, untranslated appendix peppered with graphs and formulas that only a professional aerodynamicist fluent in German could comprehend. Not included is a history of the bell-shaped lift distribution. That would not be written until NASA Chief Scientist Albion Bowers began to untangle the full story beginning in the mid-1990s.

Ten years after *Nurflügel*, Karl Nickel and Michael Wohlfahrt published *Tailless Aircraft in Theory and Practice* (American Institute of Aeronautics and Astronautics, 1994). The authors supplement their test with many graphs, drawings, and photographs to explain both the pitfalls and best practices of designing aircraft that are missing some significant portion of a conventional tail. From his unique perspective as former employee and brother-in-law of Reimar Horten, Nickel presents unique insights into many Horten aircraft. Unfortunately, he dismisses Reimar's BSLD as flawed speculation. Nickel and Wohlfahrt wrote their book primarily for the designers and builders of flying model aircraft and piloted sport aircraft; however, it is widely read within the online community of Horten enthusiasts, a group that includes aerodynamicists, engineers, and mathematicians.

David Myhra mentions the BSLD in his 1998 biography, *The Horten Brothers and Their All-Wing Aircraft* (Schiffer Publishing), but he provides no explanation of its function or origin. His large-format book is noteworthy for the many photographs drawn from the collections of Reimar and Walter Horten. The text suffers factual, grammatical, and typographical errors, and citations to source material are missing for what otherwise are many interesting claims and anecdotes. Myhra brings no fresh perspective on the cultural forces that guide Reimar Horten; rather, he frames the designer's accomplishments within the standard mantra of so-called wonder-weapons that could have won the war for Germany if only the Nazis has fielded them sooner and in quantity.

Horten Ho 229, Spirit of Thuringia: The Horten All-Wing Jet Fighter (Classic Publications, 2006), written by Andrei Shepelev and Huib Ottens, focuses on Reimar's jet-propelled and all-wing military prototype developed and flown during World War II. Black-and-white line drawings accurately rendered using original sources, along with detailed 3-D color computer graphics, depict previously unknown details of the jet. The text covers old ground but in detail and with accuracy, and the authors treat some previous claims with much more skepticism, for example, Reimar's assertion that minimizing the jet wing's radar cross section was a specific design goal defined when that project began in 1942.

All these publications have led readers to believe that Horten's most important achievement was the jet wing *Ho 229 V2* and the *V3*. Their out-of-this-world appearance implied superweapon performance, which fed a powerful if-only mythology. I have tried to dispel myth with documentation from original sources and interviews with key actors. My research leads me to regard Reimar's key role in studying the BSLD as his most important accomplishment. I lay out here the history of this idea, from Ludwig Prandtl's concept paper published in 1933, through the murky circumstances leading Reimar to adapt it to his own special requirements, to the studies made by engineers and aerodynamicists in the 1990s. Their work has proved the idea valid, and it continues in 2020.

Social historians such as Richard Evans (*The Third Reich in Power*, Penguin Press, 2005), and Peter Fritzsche (*A Nation of Flyers: German Aviation and the Popular Imagination*, Harvard University Press, 1992) have provided vital context for understanding why Reimar cultivated close ties to the *Nationalsozialistische Deutsche Arbeiterpartei* (National Socialist political party, hereafter shortened to Nazi Party or Nazi) led by Adolf Hitler. Reimar's relationship with the Nazi Party facilitated his immigration to Argentina when Juan Perón welcomed German scientists and engineers into the country after World War II.

Neither Nickel, Wohlfahrt, nor Myhra compare the work of Reimar Horten with that of John K. "Jack" Northrop. This prolific designer worked in California and, from the 1930s to the 1950s, designed a series of all-wing aircraft from small twin-engine models to heavy bombers powered by jet engines. My reading of *Northrop Flying Wings: A History of Jack Northrop's Visionary Aircraft*, by Garry R. Paper and John M. Campbell (Schiffer, 1995) and my research in Record Group RG 342 allowed me to correlate the two designers' careers.[29] I learned that the weak yaw control and stability in many all-wing and semi-tailless aircraft also caused the U.S. Air Force in the 1950s to cancel multi-million-dollar contracts for Northrop bombers. In addition to records in the National Archives, here I cite a diverse assortment of other original primary and secondary sources from the USA, England, Germany, and Argentina. Many of the foreign language documents I read in translation, but others I translated myself. Individuals whom I interviewed spoke English.

Origins of All-Wing Aircraft

All-wing aircraft epitomize the science of streamlining, whose origins date to the eighteenth century.[30] Designer, engineer, and art historian Donald Bush defined *streamlining* and specified the natural forms from which it sprang:

> All forms are simplified by the design principle of *absorption* [italics in original], the merging of one substance into another with transitional curves, and *reductivism*, the elimination of extraneous details. The resultant forms are intended to penetrate air or water with the least resistance. They . . . tend to resemble the organic forms in nature. Indeed, natural forms, like fish and birds, suggested low-resistance forms to the scientists who pioneered hydrodynamics and aerodynamics.[31]

English aeronautics pioneer Sir George Cayley, in 1809, visualized how the dolphin's slender shape, which contributed so much to the animal's speed and grace in the water, could also promote low-resistance aerodynamics. As the nineteenth century ended, a product of nature demonstrated that the all-wing configuration could be highly stable in flight. The seeds of the climbing gourd vine, *Zanonia macrocarpa* (also called *Alsomitra macrocarpa,* the Javan cucumber), is native to the Sunda Islands in the Maylay Archipelago. Each seed is less than 1⁄16 in. (1 mm) thick and lies inside a membranous wing spanning up to 6 in. (15 cm). When a gourd seed pod dispenses hundreds of individual seeds in still air, they can glide in a steady descending spiral as wide as nearly 20 ft. (6 m). With a wind, the seeds can glide hundreds of meters.[32]

In his paper published in 1897, naturalist Dr. Friedrich Ahlborn of Hamburg, Germany, called attention to the seed's stable gliding qualities and its potential benefits to the design of aircraft. Franz Wels of Austria corresponded with Ahlborn and received several seeds from him in 1904, one year after the Wright Brothers made their famous flights at Kitty Hawk, North Carolina. Wels and Igo Etrich studied the seeds, and then began an epic series of experiments with flying models that grew by 1907 to encompass piloted gliders and powered aircraft. Between 1908 and 1909, Wilbur Wright flew with precise control in all three axes of flight as large crowds watched at air meets held in France, Italy, Germany, and England. Interest in the potential of aviation soared. Etrich's work soon focused on the *Taube* (Dove), a name derived from the shape of its wings and ease of control. A pilot flying a Taube completed the first extended cross-country flight in Austria in 1910. Etrich could not patent the Taube wing design and numerous manufacturers made them.[33]

Zanonia macrocarpa contributed to the work of a young British army lieutenant who built and flew all-wing aircraft. Struck down with typhoid fever in 1900 during the South African War, John W. Dunne had time during his convalescence to indulge an interest in soaring birds. While testing flying models, Dunne apparently discovered the effects of designing permanent twist, or washout, into wings to improve stability and control in flight. Dunne also examined, and no doubt observed in flight, the gourd seeds from the Sunda Islands. Even a cursory examination of them would have revealed to Dunne that each seed's wing was twisted with washout across its outer sections. He shared his findings with the author H. G. Wells before returning to active duty. News of the Wright Brothers experiments at Kitty Hawk led Dunne to resume his work in 1903. Wells advised Dunne that a stable aircraft might appeal to the military.[34]

Dunne's work attracted the attention of Sir Hiram Maxim and earned support from Col. John E. Capper, head of the Royal Army's Balloon School, who arranged working space and government support in 1906 that allowed Dunne to build airplanes. Based on his extensive tests with flying models and encouraged by Wells, Dunne devised swept-wing, tailless biplanes of "arrowhead" configuration that flew with "automatic stability." Dunne used "elevons," control surfaces that combined the functions of elevators and ailerons, to control the aircraft in pitch and roll. Jack Northrop would not coin the word *elevon* until 1940, but the concept of combining pitch and roll control in this way predated Dunne's work. The aeronautical theorist Richard Harte had suggested the idea in 1870, and Dunne appears to have been the first to use elevons on several of his all-wing aircraft that flew in 1906 and 1908.[35] He used tip rudders suspended between the wings and because they provided no lift, the Dunne aircraft must be categorized as tailless but not all-wing.

Bad luck and technical limitations of that era prevented Dunne from winning a production contract in England, nonetheless, the beginning of World I in 1914 goaded the U.S. Navy and Army to buy modified versions of Dunne airplanes built by W. Starling Burgess. The Navy conducted its first trials for aerial gunnery using a pair of Burgess-Dunne AH-7 biplanes. A Navy pilot ascended in one to a world record altitude of 3,040 m (10,000 ft.), but both services soon abandoned the tailless configuration.[36]

Dunne must be credited with building the first successful tailless aircraft with swept wings to increase longitudinal stability and with introducing twist to the outer sections of the wing to maintain stability and control at low airspeeds.[37] Other designers used different techniques to stabilize and control all-wing aircraft, but Harte, Wels, Etrich, and Dunne developed ideas that Reimar Horten drew upon to build his first piloted wing in 1933. However, Reimar drew his initial inspiration and confidence to begin experimenting with the all-wing layout from the German designers Hugo Junkers and Alexander Lippisch, who had to some degree inspired Reimar's chief rival, American designer Jack Northrop.

Rivals

Reimar tracked developments in aeronautics around the world through the pages of popular aeronautical magazines of the 1930s and 1940s such as *Flight*, *Flying* and *Aeroplane*.[38] Surely Jack Northrop perused some of the same publications. The two men were aware of each other's work. Like Horten, Northrop worked for years on the all-wing concept. He completed a series of aircraft that flew, ranging from small test airframes up to large, multiengine bombers. Reciprocating engines powered the first bombers, but Northrop later modified the design to use jet engines. Horten focused more on unpowered wings, and the only Horten design comparable in size to Northrop's bombers did not fly until the late 1950s. What his wing's lacked in size, they made up for in aerodynamic refinement; Horten's proficiency in aerodynamic control of all-wing aircraft exceeded that of Northrop.

Reimar was a year old in 1916 when Jack Northrop got his first job drafting plans to build conventional airplanes at age twenty-one. Northrop's interest in all-wing aircraft blossomed in the early 1920s. With support and encouragement from barnstormer and shop superintendent Anthony Stadlman, he progressed step-by-step. After designing a streamlined monoplane with a conventional configuration and an exceptionally smooth surface with no external wire supports or struts to cause drag while he was working at the Douglas Aircraft Company, Northrop jumped to the Lockheed Aircraft Company in 1927 to build it. It was called the Vega and became popular with the likes of Wiley Post, Amelia Earhart, and Frank Hawks, who set numerous records for altitude, speed, and distance flying. The 128 Vegas built by

Lockheed from 1927 to 1934 exceeded the sixty-seven airplanes that Horten built during his entire career.[39]

Soon Northrop took another step toward the all-wing layout. Lockheed was not ready, however, so Northrop quit in March 1928 and formed his own small company, seeking the freedom to design that Reimar also cherished.[40] The first aircraft Northrop designed under the new company banner was the Avion Model 1. Finished late in 1928, the thick wing of the Model 1enclosed a cockpit and engine, but "we didn't dare go the whole way and eliminate the tail," so Northrop mounted a conventional empennage on two booms attached to the wing (Figure 2).[41] The idea to enlarge the entire wing to accommodate the crew, engines, and cargo echoed the work of German industrialist and designer Hugo Junkers. Northrop conceived and executed the Model 1 to study how well the wing could hold major components before he took the significant next step and removed the tail entirely. The aircraft flew in 1929 but the onset of the Depression grounded it permanently in 1930.[42]

Figure 2. Jack Northrop designed the Avion Model 1 in 1928 to experiment with minimizing the conventional fuselage and using the wing to enclose the cockpit and engine (National Air and Space Museum, Smithsonian Institution, NASM 00078785).

Northrop sold the Avion Model 1 and spend the next nine years designing successful conventional civil and military aircraft. When President Franklin Roosevelt called on American industry in November 1938 to build ten thousand military aircraft as war loomed in Europe, Northrop saw the opportunity to develop all-wing aircraft on a grand scale. Immediately after forming the Northrop Aircraft Corporation in 1939, Northrop discussed the recent history of such aircraft with his new chief of design, Walter Cerny, who later recalled that Northrop acknowledged the work of Alexander Lippisch in Germany but not Reimar Horten. Cerny also stated that Northrop and his design team had obtained a *Zanonia macrocarpa* seed from Australia, watched it fly, and then tried unsuccessfully to make the seed tumble out of control.[43] The design of a Northrop all-wing aircraft soon began to take shape. The director of the California Institute of Technology (Caltech) Daniel Guggenheim School of Aeronautics was Theodore von Kármán and he "found the [all-] wing fascinating."[44] Von Kármán helped Northrop and his design staff resolve various problems, and probably facilitated access to the wind tunnels at Caltech and Pasadena Junior College. Horten repeatedly claimed he was denied access to wind tunnels in Germany, and I explain how he overcame this handicap in later chapters of this book.[45]

Seven years after Reimar and Walter Horten flew their first all-wing aircraft, the Horten *H I* glider, in July 1933, Jack Northrop flew his first all-wing aircraft, the twin-engine Northrop N-1M in July 1940. The Northrop wing was more complex mechanically than the *H I* and more closely resembled the twin-engine Horten *H V b* that flew in 1937, although the Horten wing exceeded the Northrop wing's span by 4 m (14 ft.).[46] Greater wing area gave the *H V b* less wing loading at 38 kg/m^2 (7.8 lb/ft^2) compared to 58.6 kg m^2 (12 lb/ft^2) lifted by the N-1M. The Horten weighed 45 kg (100 lb) less but cruised a bit faster on engines of comparable horsepower.[47] Horten built the *H V b* in one solid piece from wing tip to wing tip, whereas Northrop designed the outer sections of the N-1M to pivot fore and aft when mechanics adjusted the wing on the ground. This allowed Northrop to use one airframe to test various sweep angles and positions of the center of gravity and center of pressure.

The N-1M was not the first all-wing aircraft as Northrop claimed in a May 1941 letter to General Henry H. Arnold, who commanded the U.S. Army Air Corps. This was Northrop's largest potential customer for military all-wing aircraft, so the hyperbole is understandable.[48] Reimar and Walter Horten believed that a photograph appeared in the *New York Times* in 1938 showing Walter flying an *H III* wing, and that it led the U.S. Army to contract with Jack Northrop for an all-wing bomber, but my search of major U.S. newspapers found neither a photograph nor a written account of Walter's flight.[49]

The N-1M was purely experimental, but it gave Northrop's design team confidence to build and fly larger wings. In September 1942, Northrop proposed to the U.S. Army Air Corps his first all-wing fighter. He designed it for propulsion using rocket engines, and it led a year later to a contract for the XP-79. The aircraft used a pair of vertical stabilizers when it finally flew in 1945. Another fighter design made almost entirely of magnesium and called the XP-56 flew in 1943. It, too, was semi-tailless. A new and larger all-wing aircraft with swept leading and trailing edges flew in December 1942. It was designated the N-9M and marked the beginning of a series of experimental wings that would culminate in the XB-35 heavy bomber. The N-9M spanned 18 m (60 ft.), and two pusher engines propelled it. The span was greater and the aspect ratio higher than the N-1M. Northrop built four, and they probably were the most successful wings the firm constructed. The wings were relatively easy to build and fly, and had performance adequate to their role, which was to prove that the general approach taken to design the heavy bomber was a sound one. Many pilots flew the N-9M wings to become familiar with the all-wing configuration. The third N-9M built flew fifty times before November 1944. One N-9M survived until 1981 when the Planes of Fame Museum in Chino, California, began restoring the airplane. It flew again thirteen years later in 1994 and continued to be flown at air shows until destroyed in an accident in 2019.[50]

Whereas the N-9M was a relatively trouble-free design to build and fly, other Northrop all-wing aircraft were problematic. Reaching the goal to develop an all-wing aircraft that pilots could repeatedly and safely fly was already very challenging, but Northrop significantly increased the difficulty by adding exotic and poorly understood technology. Reimar and Walter overreached as well and nearly died when they attempted to fly the *H V a* in 1937. They had constructed most of the wing using an experimental composite material, and it failed during the first take-off. When Jack Northrop proposed the XP-79 fighter in 1942, he included plans for a prone-pilot cockpit, a welded magnesium airframe, and rocket propulsion. Northrop fell so far behind in meeting his contract goals that he had to transfer design analysis and drafting work to the U.S. Army Air Material Command Engineering Division at Wright Field in Dayton, Ohio. Two years passed before the plan to use rocket engines to power the XP-79 was declared unachievable and abandoned in favor of jet turbine engines. Another year passed before the wing finally took off under its own power and then crashed, killing the test pilot.[51]

Rather than use readily available engines to propel the all-wing XB-35 heavy bomber, which spanned 52 m (172 ft.), Northrop chose an engine still under development. He envisioned four of these engines buried inside the wing and driving counter-rotating propellers via drive shafts in a pusher configuration. Designers and engineers had to design a pressurized crew cabin; work out crew ergonomics inside the cockpit; determine how to install, aim, and control defensive gun turrets; and design the landing gear. Unfortunately, the critical engineering work slowed after Northrop lost staff to the wartime draft. Northrop also could not control the delivery of government-furnished-equipment. Vital components such as engines, propellers, gearboxes, autopilots, and auxiliary power units were delayed many months, in part because war priorities dictated allocations to production aircraft already in service, not to new types in development such as the XB-35.[52]

U.S. Army Air Forces leadership nearly cancelled the project, but in August 1944, they proposed converting the bomber to accommodate jet engines. Northrop finally flew the propeller-driven XB-35 two years later, but another two years passed while the design team attempted to resolve numerous technical problems. A year after the U.S. Army Air Forces was separated from the Army, reorganized, and renamed the U.S. Air Force, planners cancelled the XB-35 in 1948 to concentrate on the jet-propelled all-wing bomber called the YB-49 (Figure 3). They also began to assign to operational bomber squadrons the Convair B-36 Peacemaker, a bomber consisting of a conventional fuselage and tail. Although the YB-49 promised to fly faster, the propeller-driven B-36 hauled more bombs, could fly farther, and was ready for operational missions. This was important because the Cold War with the Soviet Union was heating up and Air Force leadership was unwilling to be without a nuclear-capable bomber. Northrop probably needed several more years to make the YB-49 ready for operations.

Yet no amount of time and effort could have overcome a fundamental characteristic of all-wing aircraft: weak yaw control. Like Reimar Horten, Northrop found it especially difficult to control all-wing aircraft about their yaw axis. The four N-9M wings handled better in flight than the other Northrop wings and yet, "rudder [yaw] control was always a problem," said Northrop aerodynamicist Irving L. Ashkenas.[53] U.S. Army Air Forces Capt. Glen W. Edwards assessed the N-9M after an hour-long familiarization flight in early May 1946:

> the airplane flew surprisingly well, was more stable and handled far better than most would expect. It would take a few hours practice to make good take-offs and get the proper coordination on turns. But the technique could be mastered without too much difficulty.[54]

Edwards was an Army test pilot who appears to have easily transitioned to flying the all-wing configuration. His statement nonetheless implies that

Figure 3. The two all-wing, jet-propelled YB-49 bombers built by Northrop in 1947 and 1948 were the largest and heaviest all-wing aircraft ever flown until the Northrop Aircraft Corporation flew the B-2 bomber in 1989. The YB-49 weighed 95,850 kg (213,000 lb.) fully loaded with fuel, bombs, and flight crew and spanned 52.3 m (172 ft.). None of the Horten wings that flew approached the size and complexity of the YB-49 (National Air and Space Museum, NASM A-4957-A, Smithsonian Institution).

Northrop had not overcome the adverse yaw that compromises the "proper coordination on turns" of many aircraft but is particularly troublesome to all-wing aircraft without tails. Yaw control becomes critical to stabilizing an aircraft when it stalls. The first N-9M fell out of control and crashed during stall tests in 1943.[55]

The YB-49 program suffered a heavy blow in June 1948 when the wing crashed, killing the crew. No specific cause was determined, but evidence pointed to a loss of control and structural failure. Had Northrop improved the bomber enough to make it safer to fly, he still faced the difficult challenge to make the aircraft more stable. In the words of Carl Reichert, chief of the design branch of the Aircraft Laboratory in the U.S. Air Force Materiel Command, "The low order of longitudinal and directional stability will render all-wing configuration incapable of precision bombing."[56] The last flight of a Northrop all-wing bomber occurred on April 26, 1951, and two years after that, the Air Force cancelled the program. While Northrop struggled to fix the jet wing, Boeing had introduced into service in 1951 a new jet bomber with swept wings fitted to a fuselage equipped with a tail. The B-47 Stratojet was nearly 160 km/h (100 mph) faster than the YB-49 and designed to carry nuclear weapons. In development when the B-47 entered service was a much larger bomber with the same conventional configuration, the Boeing B-52 Stratofortress.[57] Jack Northrop had also tried in the late 1940s to interest the commercial airlines in an all-wing passenger and cargo aircraft. He even made a promotional film, but his overtures were ignored. Disappointed at failing to win production contracts for all-wing aircraft, Northrop retired in November 1952. His bombers remained the largest and heaviest all-wing aircraft to fly until the Northrop Corporation flew the B-2 stealth bomber in 1989.[58]

Like Jack Northrop, Reimar Horten excelled in meeting the challenge to build and fly all-wing aircraft but then failed to meet the end user needs for practical machines. How and why this happened is the story you are about to read.

CHAPTER 1

From Flying Models to Piloting a Wing

REIMAR HORTEN was born in Bonn, Germany, on March 12, 1915, to Max and Elizabeth Horten. Two brothers preceded him, Wolfram in 1912 and Walter in 1913. A sister named Gunilde was born in 1921.[1] Both parents were strong in the liberal arts. Elizabeth Horten had studied teaching and attended history, geography, and English language classes at Oxford University. She also worked as a private tutor for the family of the German ambassador to Spain around the turn of the century. Max Horten was forty-two when Reimar was born, and he was a respected university professor who had devoted his professional life to understanding and writing about Islamic philosophy. He had lived in the Middle East from 1898 to 1900 and often traveled with Bedouin tribes to study their various dialects. His work earned him doctorate degrees in theology, philosophy, and oriental languages (specializing in Arabic) by the time he married in 1910. During his career, Max wrote some thirty books, eight in one focused three-year effort.[2]

The Hortens owned considerable property before World War I, including a railroad in Mexico, but they lost most of their holdings soon after the war ended. Large cash sums that Max had lent to finance the war effort in exchange for government bonds had turned worthless following Germany's defeat. The hyperinflation that seized the economy in 1922 to 1923 consumed what remained of his investments, forcing Max to find a state job to support his family and qualify for a pension.[3] In 1929, he moved to Breslau, more than 800 km (500 miles) east of Bonn, to work at the university as a librarian. He did not return permanently to Bonn until 1935, yet he and Elizabeth continued to support their children's various interests.[4] Raising them in a relatively relaxed environment allowed Max to compensate for his own disciplined upbringing. Reimar and Walter in particular benefited from their parent's tolerance and encouragement as they investigated aeronautics through study, practice, and experimentation.

Walter Horten remembered in 1979 how his father encouraged the children to "study the forces of nature, particularly wind and weather, to see the parallels between nature and pure philosophy."[5] Reimar claimed that he and Walter studied the gliding flight of the *Zanonia macrocarpa* seed and observed seagulls, raptors, and other soaring birds. They grew more enthusiastic about aeronautics after Max and Elizabeth bought tools to build flying model aircraft. All four children eventually learned to fly, but the boys settled on aviation careers while Gunilde gravitated toward physics. Reimar's siblings supported his early work when their own careers and finances allowed. Wolfram gave money to construct models and buy flying lessons, and he helped build the first piloted wing, the Horten *H I*; however, Walter became his brother's chief collaborator. During the early years, he also helped build the *H I* and discussed flying wing theory and practice, particularly during construction of the *H II*.[6]

Reimar learned to fly sailplanes in 1927 and powered aircraft soon thereafter, but he admitted he was not a good pilot. When he appeared at the airfield, the mechanics "told me [that] when I flew, I created more work for them every time!"[7] Pilots must commit time and energy to maintain flying proficiency, but Reimar focused obsessively on his research. He neglected schoolwork and earned only the "minimum grades to avoid failing."[8] His parents nonetheless maintained their strong support, and his father never pushed his son to match his own academic accomplishments. Reimar set aside "no time for girls at all. In fact, I didn't do any dating, didn't attend dances or [other] social functions."[9] The designer tried to exclude all but "schooling, gluing [aircraft parts], and thinking."

In contrast to Reimar, Walter (Figure 4) was extroverted, gregarious, and athletic, and he enjoyed sculling and swimming. His social calendar was usually full, which occasionally upset Reimar, who preferred that Walter focus on aeronautics, but their disparate interests never caused a serious break. Perhaps because of gender and family rank, but certainly also due to Walter's preference for final design, construction, and flying the wings, along with the adroit networking and organizing skills he developed later, the two brothers forged a strong bond and worked closely together until forced to separate after the war. Walter began flying gliders in 1927 and passed an early milestone in his flying career in 1930 when he made a solo flight in a powered aircraft at the Bonn-Hangelar Airfield. He earned his civil license to pilot powered planes in 1934 and then joined the army in April. The following year he transferred to the Luftwaffe.[10]

National Pastime

By prohibiting most forms of powered flight, the Treaty of Versailles drove a national enthusiasm for soaring flight that surged through Germany during the 1920s and swept all three Horten brothers along with it. Successful sailplane competitions and public demonstrations held at the movement's epicenter, the Wasserkuppe peak in the Rhön Mountains, "stimulated a national

Figure 4. Luftwaffe Leutnant (Lieutenant) Walter Horten in 1940 when he was twenty-six years old (Rolf Schroedter via Gunilde Horten and Peter F. Selinger).

grassroots gliding movement," wrote historian Peter Fritzsche.[11] From 1927 to 1931, membership in the Jungflieger groups, subsets of the Deutschen Luftfahrtverband (German Aeronautical Association, the DLV), "more than tripled to nearly 10,000, mostly but not exclusively young male aviators between the ages of eighteen and twenty-six."[12] The movement spawned nearly forty gliding schools by 1931, and during the remainder of the decade, "an enormous literary production of songs, stories, and plays recounted the happy weeks spent at gliding schools."[13]

In the midst of this surging passion for soaring flight, a neighbor introduced Reimar, then eleven years old, and Walter, age thirteen, to kites in 1926. They learned how to vary the curve shape, or camber, of their airfoils to enhance the stability and increase the lift of their kites, but soon the boys yearned for new challenges, and their interests turned to aircraft.[14] The brothers joined the Lower Rhine Flying Club, Bonn District, manned by a group of young men who had started building a primary training glider. The brothers began attending workshop classes to learn aircraft construction two evenings a week. In 1927, they accomplished solo flights in the club glider. Their flying lessons continued on weekends and by year's end, both boys had earned their A Certificates to become the club's youngest pilots. Earning the A Certificate required them to demonstrate their knowledge of basic airmanship and to carry out their first solo takeoff and landing.[15] Like other young enthusiasts, they began to construct and fly model airplanes, too. In those days, the air sports movement regarded this activity almost as highly as piloting full-size aircraft.

To supplement their practical training, the Horten brothers studied aeronautical theory in Jungflieger courses sponsored by the DLV and taught at the Städtische Berufsschule (city vocational school) in Bonn.[16] The curriculum included piloting, aerodynamics, engineering, structures, and propulsion. It was here that Reimar acquired the mathematical and analytical skills necessary to begin calculating wing lift distributions for flying models. He later focused these same skills on designing piloted all-wing aircraft. Reimar believed the coursework was as beneficial to his career as the engineering seminars that he attended much later at the University of Bonn.

Reimar supplemented his class work with considerable outside reading, favoring the series on aircraft design and engineering edited by the famous aerodynamicist and father of fluid dynamics, Ludwig Prandtl, and published in 1927 by the Aerodynamische Versuchsanstalt, the Aerodynamic Experimental Center at Göttingen, Germany, or AVA. Reimar also studied the technical papers published by the Deutsche Versuchsanstalt für Luftfahrt e. V. (or DVL), the German Aviation Experimental Institute located at the Adlershof facility in Berlin. Horten monitored the popular German periodical *Flugsport* that catered to sport aviation in Europe, and he tracked the broader international developments as the monthly British publication *Flight* reported them.[17]

Around 1927, Reimar began to imagine how an airplane might operate more efficiently without a fuselage and tail. Carefully studying the seminal work that Germans had already done in this field, he found that on February 1, 1910, the Kaiserliches Patentamt (Imperial Patent Office) had granted Hugo Junkers patent 253788 for "a hollow wing form."[18] The patent proposed a wing broad and tall enough along much of its span to contain the items normally carried perpendicular to the wingspan inside a conventional fuselage—cockpit, passenger cabin, baggage area, and power plants. Several drawings that accompanied the patent show how Junkers also envisioned using the thick wing to enclose the external struts and wires that joined the wings, fuselage, and tail of many conventional aircraft of that era, thus substantially reducing drag.

Historian Günter Schmitt argues that Junkers did not propose the type of controllable all-wing designs that Reimar developed more than two decades later, but the perception arose that he did because the drawings accompanying his patent were widely misinterpreted. Regardless of Junkers's true intentions, the 1910 patent almost certainly inspired Horten to seriously consider the all-wing configuration. Reimar wrote early in his career about the profound and nascent influence that Junkers had on the development of all-wing airplanes.[19]

Junkers tested several ideas related to the all-wing configuration when his firm built and began to fly the G.38 transport aircraft in 1929, then the largest civil, land-based airplane in the world. The giant's wings spanned 44 m (144 ft.), and a man of average height could stand at their thickest point (2 m/6 ft.) along the root. In this space, Junkers and his chief designer, Ernst Zindel, placed the four engines, their fuel, most of the baggage and freight, and a considerable portion of the passenger cabin. They joined the wing to a fuselage, which contained the cockpit and additional passenger accommodations, and supported an empennage, or tail, equipped with elevator and rudder. The G.38 was not even semitailless, but it nonetheless impressed young Reimar, who later recalled admiring it as "the right way to go" to make aircraft more efficient.[20] He and Walter watched the behemoth fly and examined it on the ground at Bonn-Hangelar Airfield in 1930.[21]

Reimar never met Junkers, but a man who had worked for him, Franz Wilhelm Schmitz, became an early and important mentor guiding Reimar's work. From 1922 until 1941, Schmitz taught business, auto mechanics, and flight physics at the Bonn vocational school. Prior to teaching, he had been a member of Junkers's pool of airplane designers and helped to design the Junkers F.13, the first successful airliner constructed entirely of metal and equipped with a cantilever wing.[22] Both Horten brothers attended his classes. Gunilde and her husband, Karl Nickel, a mathematician who went to work for Reimar in 1942, remembered that Schmitz taught the boys "how to design, construct, and build models and [piloted] airplanes."[23] The Nickels recalled that F. W. Schmitz must have taught Reimar how to calculate the aerodynamic stresses of flight. Nickel understood Reimar's work piecemeal, having helped during the war to analyze performance and lift distributions and to figure stress analysis and the like, and he occasionally contributed his own ideas, yet he seems not to have assimilated Reimar's complete design strategy.[24]

Schmitz tried to convince the brothers that the future belonged to tailless aircraft. He occasionally visited the Horten family and sometimes played piano while Elizabeth sang. The former Junkers designer mentored other youths to embark on successful careers as aeronautical engineers and pilots, including the German aerobatic flying champion, Liesel Bach.[25] The designer Fritz Gosslau, who had worked at the Klemm factory designing light airplanes during the late 1920s, visited the Bonn-Hangelar Airfield and lectured on aircraft design in 1927 when Reimar was twelve. Gosslau met young Reimar in 1932 and later followed him to Argentina after the war. Reimar admired his design and piloting skills, but Gosslau's willingness to share knowledge and his progressive attitudes led Horten to declare him "one of the first men to have a great influence on me and [help] me get my start in aircraft design."[26]

In 1930, Junkers proposed a tailless aircraft with wing sweep, which became a key element in the Horten approach to all-wing design, but something else fired their initial enthusiasm for the configuration. A year earlier, the brothers had witnessed Günther Groenhoff flying the swept wing Storch I. Alexander Lippisch designed this tailless sailplane and equipped it with nonlifting rudders mounted at the wing tips and used to control side-to-side, or yawing, motion. They watched in astonishment as Groenhoff performed aerobatics while flying a powered Lippisch wing called the Delta I in 1931. These flights gave the brothers confidence that a small sporting wing could be a practical alternative to developing a large and costly multiengine transport such as the Junkers G.38.[27]

Model Trials

Lippisch had begun to design conventional and semitailless sailplanes and powered aircraft in 1921. His work culminated during World War II with the Messerschmitt Me 163 interceptor propelled by a liquid-fuel rocket engine.

Lippisch had followed the methodology employed by many young independent designers of his day to develop their ideas. He built flying models and tested them before "proceeding to the full-scale glider, and finally to the building of the powered airplane. The use of wind tunnel measurements was not feasible for financial reasons."[28]

In 1927, members of the Bonn flying club asked Reimar and Walter to compete in a contest for flying models planned to coincide with the Düsseldorf Aviation Exposition. Walter submitted a powered design, but Reimar offered a glider spanning 1 m 80 cm, or 5 ft. 10 in., with a fuselage and tail, and made using thin plywood and parts of a cigar box. The contest judges believed it too unorthodox to compete, obviously the "work of a child," as Reimar put it, but mentor Franz Schmitz interceded and convinced the judges to relent.[29]

Reimar won a prize for flying 270 m (886 ft.) against about a dozen other young modelers. The victory gave the teenager confidence in his design skills and motivated him to excel. Walter's model did not fly as well, a sign their career paths were already diverging. Reimar was starting to exercise his gift for designing aircraft, and although competent enough as a pilot, he never had the hand-eye coordination required to excel at flying. Walter was less intellectual and was fast maturing into a skilled pilot, yet different interests and temperaments did not kept the two apart. "Walter was always with me," Reimar said, ". . . we [worked constantly] together."[30]

The boys shared the challenge and exhilaration of competing together in the national competition for model sailplanes held annually on the Wasserkuppe. Located in central Germany about 330 km (200 miles) southwest of Berlin, this tallest mountain (950 m, 3,116 ft.) in the Rhön chain was still the spiritual home of the German sport soaring movement in 2010. Then, as now, trees dotted the grassy slopes, but rain and mist concealed the crest for most of the year except late summer, hence the name *Wasserkuppe*, meaning "waterhilltop." The summit covered about 5 sq km (2 sq. miles). When the prevailing westerly winds were strong, pilots launched their sailplanes from slopes, some bisected by gorges choked with fir trees, and soared over a vast plain.[31]

Gliding enthusiasts discovered the mountain before World War I, but it remained obscure until after the first official contest in 1920. The Germans erected permanent buildings, and by 1926, the camp "was serviced with electricity and water and boasted a police station, a hotel, restaurants, a theater, and a post office" to accommodate the throngs of crowds covering the slopes on most contest days. "Tens of thousands of visitors made the pilgrimage . . . [and on] one typical Sunday in August 1931 there were twenty thousand sightseers on the Wasserkuppe."[32]

The brothers' first visit came in 1927, but they probably did not fly a model in the national contest until 1929. The most challenging event was the test of maximum flight distance and duration. After the young men had tossed their models off the slopes, judges measured the results as the models soared over both steep and nearly flat terrain that varied in texture from grass to rocks to dense thickets of tall evergreen trees. This uneven surface generated turbulence, and those models stable enough to resist upsets flew far.

Reimar observed that during many flights, a wind gust pushed a model off course by sharply rolling its wing. During a steep roll, the wing lost lift and dropped its nose, falling away into a short dive that Reimar called a sideslip, before returning to level flight on a new and often wildly divergent course, or crashing into the ground.[33] Reimar understood that

> sideslip was the principle thing I wanted to avoid . . . That, to me, was important. I didn't want my models to make curves or turns while flying but instead to fly perfectly straight. If it would curve, [the model] would enter a dive into the earth and this was not good . . . I wanted it to be self-correcting when disturbed [by rough air].[34]

Soon the designer was constructing and testing models specifically to evaluate the design features that promoted stability. His experiments not only confirmed the need for more roll stability but also disclosed how a gust could hit the tail of a conventional model from different angles to push the aircraft far off track. The position of the empennage at the end of a powerful fulcrum, the fuselage, magnified this phenomenon.

Reimar aimed to develop models that could win on the Wasserkuppe, but this goal also dovetailed with the coursework taught in the Jungflieger program and the hands-on experience that both brothers were gaining in the Bonn flying club. The programs pushed their model work to mature rapidly into a deliberate and systematic effort designed to test and evaluate specific aerodynamic design elements:

> The construction of airplane models enabled us to investigate fundamental problems of flight mechanics. We usually built a series of three glider models [at a time], with only a stick as the fuselage and reinforced plywood pieces as stabilizers, since the wing was [becoming] our main interest.[35]

Reimar and Walter often conducted a series of flight tests that began with a conventional model and progressed to removing portions of the tail until none remained. "We had different models," explained Reimar, "some with tails and some without, as early as 1929."[36] A few models had permanently lost their vertical fins and rudders by 1930, but Reimar still clung to the horizontal stabilizer.[37] "We just couldn't remove the tail section [completely] and immediately have a success," he said, ". . . it involved sort of an evolutionary process and required many test flights of models . . . we thought that it would be better to fly it with tail and know the [results] and then fly it without tail and see the difference . . . Each test gave us new information."[38]

To maintain their progress, the brothers fed the test program a prodigious quantity of models, and Reimar recalled often working "through the night until dawn" to finish building them.[39] Aeronautics so consumed him that in 1929 he flunked chemistry and English and had to repeat the school year. He and Walter flew a model as often as ten to twenty times a day, three days every week, for as long as twenty to thirty weeks. They carried out modifications and repairs at night. Reimar guessed that several models accumulated as much as two hundred hours flight time. Max sometimes accompanied his sons on outings to test the models. He bought them a carpenter's bench and a set of tools, and he and Elizabeth burdened the boys with few chores. They were equally generous about where the work took place, saying to Reimar, "the house is for you."[40]

After Max moved to Breslau in 1929 to work as a librarian, the two teenagers found plenty of space in the Bonn house to work. The three-story dwelling was warm and dry—ideal conditions for constructing wooden aircraft—and staying there cost the boys nothing. The house concealed their work at a time when Reimar wished to remain obscure until he could prove his concepts, although neighbors sometimes dropped by to help. Elizabeth's sister somehow cleaned the house, but history has not recorded how she dealt with the clouds of sawdust and puddles of glue that inevitably accompany the construction of wooden aircraft.[41]

In order to ensure that the tests yielded accurate data, Reimar developed a precise routine for launching each model, even using consistently the same hill near Bonn as his launch site. For the first flight, Reimar simply tossed the model into the air by hand and observed the basic flight characteristics. This simple test sometimes revealed a dead-end design without further effort. If this short hop was successful, he moved to a more powerful launch method using a rubber band 5 m (16 ft.) long to propel the model into the air. Reimar staked one end of his catapult firmly to the ground, hooked the free end to the model, and always stretched the rubber 3 m (11 ft.) before releasing it.[42] As he and Walter reduced the size of the tail and fuselage, they had to compensate for the stability these structures had given their models.

Stability around the pitch axis was largely a matter of correctly locating the center of gravity, the point at which the wing balanced fore and aft and side to side. They mastered this problem before addressing stability about the yaw axis.[43] With the rudder and fin becoming steadily smaller, Reimar concentrated on the wing's directional stability:[44]

> If the glider was to maintain its course, it was essential that after tilting of the wing from a gust and the subsequent sliding [away from original course], the slide-roll moment would be used to right the model. Therefore it had to have a positive slide-roll moment [to roll and yaw back on course] and, at the same time, great roll damping [resistance to rolling too far the other way], so that the righting was non-periodic. This was achieved by sweepback and appropriate dihedral wing shape.[45]

To understand the stabilizing and self-correcting effect of wing sweepback, it helps to visualize a wind gust pushing a model's right wing tip up and the left tip down. As the model began to slide, or fall, toward the ground, the airflow meets the left side of the wing at nearly a right angle to the leading edge because of the sweep angle, and lift increases over this wing half. The sweep of the right wing angles the leading edge away from the airflow and reduces lift. The resulting lift imbalance rolls the model back to level flight.

Dihedral increases stability through a process resembling wing sweep, but the stabilizing force acts against the underside of the wing rather than its leading edge. Recalling a gust that drops the left wing when it hits the right, the left wing meets the airflow at an angle reduced by dihedral, which tends to return the wing to level.[46]

As they had done to refine pitch stability, the brothers tested repeatedly, first to determine the proper degree of wing sweep and then to balance it with the correct amount of dihedral. If they overdid the sweep and dihedral to right the model, it could easily roll beyond level flight and even continue to roll back and forth if struck by another gust, which the brothers saw on several occasions. "Excess dihedral shape would become evident by wobbling, which could increase up to a complete roll," Reimar explained.[47]

He wanted the model to return smoothly to level flight but acknowledged that some roll instability was inevitable without active control, which was not yet possible for technical and financial reasons. This would come later when the brothers designed a piloted wing. Reimar and Walter determined that a positive dihedral of about five degrees in the center section, with the wing tips set flat or angled down slightly, worked best. The arrangement resembled the wing shapes of soaring birds, but Reimar saw examples in the work of other designers, too. "It should be dihedral positive in the center," he said, "like a bird [and] wing tips negative . . . or zero [angle] . . . [as seen on the] Minimoa . . . and Fafnir," two well-known sailplanes designed for competition soaring by Martin Schempp, Wolf Hirth, and Alexander Lippisch.[48]

Now that Reimar and Walter understood the aerodynamic requirements necessary for stability, it was time to compete. In May 1930, they soared a model 535 m (1,750 ft.) across the slopes of the Wasserkuppe. The aircraft was not tailless, but it flew far enough and remained airborne long enough to encourage them to continue experimenting.[49] The brothers built and flew a larger model (span 3.6 m, or 11 ft. 8 in.) in 1931, based on the previous year's design.[50] They finished about one test model every month through early 1932 by foregoing a high-quality finish and expending no effort to repair or replace the damaged and destroyed models.

As spring gave way to summer and the Rhön contest approached, Reimar and Walter focused their best efforts to build a model incorporating their latest findings. They carefully trimmed every wooden component, glued smooth joints and clamped them with care, and neatly applied the final coats of weatherproof varnish to the best grade of plywood from Finland, purchased through a Berlin supplier for five marks a square meter (10.8 ft.^{2}), rather than buying the standard grade that sold for a mark-and-a-half.[51] This model had no fuselage or tail, alighted on a skid landing gear, and weighed 10 kg (22 lbs.). The brothers finished constructing it in August 1932.

With some two hundred other contestants looking on, Reimar launched the model and watched it fly eight to ten minutes and cover 4 to 5 km (2.5 to 3.1 miles) before encountering disturbed air that spun the aircraft around and sent it sailing back toward the mountain. It landed less than 600 m (1,970 ft.) from the launch point in the rocky terrain that covers much of the east side of the peak, shattering one wing. The flight won first place for duration but ranked very low for distance.

The flight also helped push Reimar, with six years of model testing behind him, to step beyond the limitations of flying models to construct a wing large enough to carry an occupant who had sufficient control to resist turbulence, keep the aircraft on course, and avoid mishaps.[52] Such a vehicle would allow Reimar to expand the scope of his work to include not just stability but also active and precise control in all three axes of flight. Owning their own aircraft would also let the brothers build their flying experience more quickly. Distracted by the demands of school and the model test program, they had each eked out less than an hour of flying time since earning their A certificates five years earlier, not nearly enough if they hoped to take up careers in aviation.[53]

First Piloted Wing

Reimar was seventeen and Walter nineteen when they began building their first piloted wing, the Horten *H I*, in February 1933; however, the construction began not simply because the brothers had come of age physically and intellectually when their work reached a logical turning point.[54] The momentous, and tragic, political and social upheaval entailed by Hitler's rise to power presented the brothers an important opportunity. Existence of the Luftwaffe remained secret, but during April 1933, Hitler's Nazi government openly began to centralize the nation's management, design, and manufacture of aircraft by creating the Reichsluftfahrtministerium (German Air Ministry), or RLM. At the same time, the Nazis started to hire and educate a large and modern workforce and to organize two million workers to build new airfields and factories; in short, laying "the foundations for a modern and powerful aircraft industry."[55] The Nazis also dissolved several of the sport flying organizations including the old DLV, and then reorganized them into one organization called the Deutscher Luftsportverband (German Airsports League), still identified by the old DLV initials but operating within a wholly different political context.[56]

Participation in the industrial transformation came at the price of official membership in the Nazi Party. On February 28, 1933, a day after fire gutted the Reichstag (seat of the German Parliament), the new Hitler government pushed through legislation giving it the power to jail political opponents on minimal pretexts. The police and the party's own security forces arrested at least 100,000, mostly Communists, but also many people who had simply disagreed with Hitler's policies. More than five hundred people died in custody that year.[57] Beatings and other physical abuse often preceded arrest. The violence that continued into the first half of 1933 was "comprehensive in scope, affecting anyone who expressed dissent in public, [and] nonconformists of every kind."[58] As the only designer in the world at that time experimenting with all-wing aircraft, Reimar Horten epitomized the nonconformist. Despite repeated denials, official party membership cards, complete with photographic identification, prove that Reimar joined the Nazi Party at the beginning of April 1933.[59] Party membership provided the safe haven necessary to continue Reimar's work.

Walter remembered the audacious confidence with which his brother undertook the *H I* project. "Simply the fanatical belief," he said, "in the idea that the most efficient airplane design was one that generated the greatest amount of overall lift."[60] Yet Walter urged a conservative approach to developing the wing. A crash was possible, and with it might come negative publicity, which could destroy their credibility during a precarious time in their budding aeronautical careers. He advised building and testing a conventional tailed aircraft first before proceeding with an all-wing design, but Reimar believed they could control the publicity if a crash occurred. He pointed out that even with generous parental support, they probably lacked the resources to build piloted versions of both a tailed and an all-wing type. His arguments carried the day.[61]

Reimar first envisioned enlarging one of his model airplane designs to accommodate a pilot, but the naïve notion quickly succumbed to the acute disparities in cost between models and piloted airplanes, and the greater complexity of adding active control to a model design. Reimar wisely patterned the *H I* span, sweep, and taper on a semitailless aircraft with an established flying record, the Lippisch Delta I.[62] He attempted to plot the distribution of lift over the wing, but soon asked F. W. Schmitz to help with the calculations.

When Reimar began drafting the construction drawings, he wrote to Alexander Lippisch for advice on selecting airfoil profiles. Horten claims that Lippisch did not respond but when he visited Bonn to give a lecture, Reimar heard enough to create the profile himself. Karl Nickel says that Reimar copied the shape from the Delta I sailplane. Regardless of origin, it was a symmetrical profile across the entire span, ranging in thickness from 20 percent of the chord length at the wing root to 10 percent at the tip, following Junkers's approach to designing the F.13 and G.38 powered airplanes. Schmitz probably encouraged Reimar to adopt this arrangement.[63]

Horten also designed twist, or washout, into the outer sections of the wing. Following Reimar's specifications, craftspersons attached each internal rib to the central wing spar at a slightly reduced angle until the last ribs installed at the wing tips were oriented at 7° negative angle of attack compared to the center section. Washout was necessary to prevent stalling at one or the other wing tip, a condition usually followed by a spinning dive from which recovery was not possible because the aircraft lacked a fuselage and tail and the extra measure of control these components provided.[64]

Reimar juggled different parameters as he calculated the wing's overall dimensions: wing area required to support the sailplane's estimated flying weight, volume and weight of accommodations for the pilot, and maximum wing thickness at the root. Reimar estimated the maximum flying weight at about 210 kg (464 lb.), including a pilot weighing 90 kg (200 lb.).[65] He meant to soar competitively against other sailplanes at this weight, so he calculated that the wing loading must be near 10 kg/m^2 (2 lb./ft.^{2}). To reach this loading, the wing area had to be 21 m^2 (226 ft.^{2}).

With the area figure in hand, Horten could refine the overall shape, reckoning that the root had to be about 3 m (9 ft. 10 in.) long to hold a pilot, seat, and instrument panel and allow space for control fixtures and linkages. Space inside the wing was tight, and Reimer placed the seat close behind the wing spar so that the pilot's arms had to pass right through the spar framework to manipulate the controls.

Reimar's decision to set the wing root thickness at 20 percent of chord length yielded a depth of 60 cm (almost 2 ft.) to the wing, enough space to enclose a seated pilot's legs, torso, and upper body, but not the head. To cover it, the canopy had to stand another 30 cm (nearly 1 ft.) above the wing, or half the root thickness. This was Horten's first attempt to merge cockpit and wing smoothly, and the result was not as graceful as it would become later in his career. He designed a keel for the *H I* that conformed to the bottom curve of the wing to support the seat and controls and to support the landing skid. The sides of the canopy and keel met the wing at nearly right angles, and these corners roiled the airflow, causing modest drag, but otherwise, the wing was quite smooth. Reimar would devote considerable effort throughout his career to refining the junction where canopy and landing gear met the top and bottom surfaces of the wing.[66]

Horten also claimed to have sized the *H I* wing tips according to Osborne Reynolds's theory, proven, he said, in wind tunnel tests performed at the AVA, which argued against a tip chord greater than one-half meter due to a high Reynolds number. Although Karl Nickel questioned how widely known the theory was at that time and how well Reimar understood it, Reynolds had published his formula late in the nineteenth century, and it had become widely disseminated by 1930. A Reynolds number combined into a single value several important characteristics of a moving fluid, for example airflow interacting with a solid body, such as a wing tip. It accounted for both the size of the body and the speed of the fluid, and then compared these values to the combined density and viscosity of the fluid. With this number in hand, a designer could determine such useful qualities as the tendency of the airflow to transition from smooth to turbulent flow, a critical consideration in the design of wing tips that resisted stalling.[67]

Using Only the Wing for Stability and Control

Reimar's trials with models had been largely about stability, but the challenge now was to add precise control while maintaining stability. The *H I* pilot must be able to navigate a precise course, right the wing after sudden turbulence, and round out a descent mere inches from the ground before touching down smoothly for a landing—all while flying an aircraft without a conventional tail mounted on a powerful fulcrum, the fuselage. To allow only the wing to perform all this work, Reimar used three sets of aerodynamic controls. For control about the pitch axis, he hinged a pair of elevators to the wing trailing edge immediately behind the cockpit, in keeping with the scheme that Lippisch had employed on the Delta I. Moving away from the center section and farther out along the wing, Horten installed ailerons that extended right to the wing tips to control roll. The pilot moved a control stick to actuate both elevators and ailerons.

To yaw the wing, slow it down, and adjust overall angle and rate of descent, Reimar abandoned the tip rudders used on the Lippisch Delta I, and near the wing tips and ahead of each aileron, he installed control surfaces called drag rudders that combined the actions of the dive brakes and rudder fitted to most conventional sailplanes. The pilot depressed the left or right pedal to start a turn, or toed both pedals simultaneously, activating both drag rudders at once to break the wing's speed or adjust descent rate and angle. When not used, each control lay within a recess that conformed smoothly to the wing surface and caused no drag.[68] Elevator, ailerons, and drag rudders all contributed to lift, hence the *H I* was a pure all-wing aircraft.

One of the most difficult obstacles facing Reimar was adverse yaw, caused when a wing yaws opposite to the direction a pilot is trying to roll during turns. Horten made the problem harder to solve by insisting that all control surfaces lie flush against the wing, generating no drag, or retarding force, until the pilot activated them. The designers of semitailless aircraft, Lippisch among them, had relied on some form of conventional rudder to control the adverse yaw; however, these surfaces violated Reimar's embargo on diluting the purity of his all-wing designs with components that produced drag but not lift. The few designers bold enough to attempt control with all-lifting surfaces had failed.[69]

Fredrich Wenk had designed an all-wing sailplane with lifting control surfaces that flew briefly at the Wasserkuppe in 1921 before crashing, and in 1924, soviet designer B. I. Tscheranovsky flew an experimental all-wing design lacking dihedral and wing tip washout. It flew poorly, and the examples that followed employed nonlifting control surfaces. Alexander Soldenhoff experimented in Switzerland and Germany from 1927 to 1931 with results similar to Tscheranovsky's.[70]

In 1929, Reimar had helped a group of university students to build a conventional sailplane designed by Lippisch called the *Professor* in a workshop at the

Akaflieg Bonn. That experience helped prepare Horten to build the *H I*; nonetheless, the task of managing his own project demanded discipline and ingenuity.[71]

> By eating only peeled potatoes and blood sausage could we afford to buy essential material such as plywood, wood strips, glue, sheet metal and steel tubing which cost 320 Reich Marks. The merchants of Bonn sometimes donated material to us. In the end, we painted the glider black to mask the inexpensive variety of plywood used to cover it.[72]

Reimar was breaking the law by using the less expensive, substandard wood, but the material cost less than one-third the amount charged for the higher-quality grades approved for aircraft use. Sailplanes with performance similar to Horten's wing cost about one thousand marks, but Reimar reckoned that he and Walter reduced this figure significantly by cutting corners and doing the work themselves whenever possible.

The brothers needed more hands for some tasks, so rather than shun public contact as they did during the model test program, the brothers invited members of the Bonn flying club to help build the wing. Elizabeth accepted their presence without comment until they began staying late into the night. "My mother very often was angry for [she] told them to go home early so that we could sleep. But we were working the night and could not stop." Reimar "learned to sleep with eyes open in school," but his studies continued to suffer.[73]

Engineer Rolf Schrödter also contributed to the project. A combat pilot during World War I and one of the founding members of the Bonn flying club, Schrödter owned the Imperia Motorcycle Company in Godesberg. He gave the brothers building materials, but his greatest contribution was intellectual. Reimar said that Schrödter taught them to understand "why designers did what they did" and to see aircraft designs with a "critical" eye. As they had done during the model trials, the Horten family supported the *H I* project. When Reimar was shocked with a bill for seven hundred marks, Max stepped in to pay it and then set up his son's own bank account. Brother Wolfram contributed cash and labor whenever he could take a break from his career as a merchant seaman.[74] Max and Elizabeth again allowed their sons to turn the family dwelling in Bonn into a workshop. They were relieved to know where the boys were, what they were doing, and with whom they associated, as the streets turned dangerous during the violent transition to National Socialism in early 1933.[75]

After working for about 1,200 hours from February to May, Reimar and Walter finished the sailplane (Figure 5). Their friends were thrilled, according

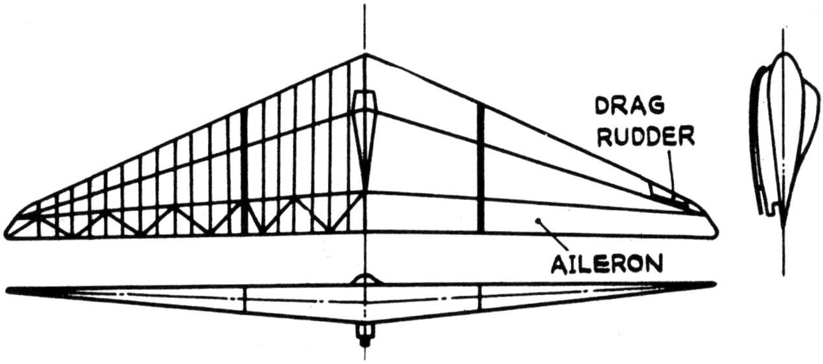

Figure 5. Horten *H I* three-view drawing Reimar used elevators inboard of each aileron. (D. C. Appleyard and M. A. Biot, "The Horten Tailless Aircraft," May 1945, CIOS Target Number 25/157, published by the Combined Intelligence Objectives Sub-Committee G-2 Division, SHAEF [rear] APO 413, 41), 1, 27.

to Walter, and the excitement increased in anticipation of the first flight. Work commenced immediately to move the wings and center section out of the house and across the Rhine to the Bonn-Hangelar Airfield. It was necessary to remove a door from its hinges and open the window above it to create enough space to pass these large components out of the house and onto a waiting wagon. Horses drew the load through the crowded streets and across the river to the airfield, and by day's end, the glider was assembled and almost ready to fly.[76]

During the next month, Reimar made brief visits to the airfield to tidy up the glider. He turned most of his attention to neglected schoolwork, but his thoughts must have strayed to the first flight. It would mark the beginning of a new dimension in the evolution of his aerodynamic ideas: manned tests flown by the designer himself. Reimar was about to test a completely new and untried configuration, one that incorporated several exotic elements that previous designers had tried but failed to use. To take to the air in such a bold experiment was risky.

Reimar decided to make his first flights in the new wing short, controlled hops limited in altitude by the launch apparatus, which consisted of about twelve fellow flying enthusiasts pulling two ends of a bungee cord hooked under the sailplane's nose to catapult it into the air. This method was preferred for most primary glider flights, particularly during training, because it was inexpensive and lacked the velocity to carry an inexperienced pilot to an unsafe altitude. Reimar would climb no higher until he better understood the wing's flight characteristics. Still, the thought of exploring unknown territory must have been sobering, for he risked not just wood, glue, and some month's labor but also his own safety. Reimar had earned his C Certificate in gliding flight by the summer of 1933. He was a careful flyer and treated the *H I* with the respect due an untried airplane, but had he been tempted to try something foolish, members of the Bonn flying club would have reminded him to exercise good judgment.

Horten made the first modest hops in July 1933. These flights rose no higher than 2 to 3 m, about 6 to 10 ft., and carried the aircraft between 50 and 100 m (165 and 330 ft.) across the airfield. Horten immediately found the sailplane difficult to control. If he pulled the control stick back, the wing pitched up as if to climb but settled to the ground. In short, the angle of attack increased but not the lift, so refining control in the pitch axis was the first task on the road to taming the spirited wing. According to Walter, his brother was lucky to survive the first flights in the *H I* due to the "dangerous flight characteristics."[77]

Reimar correctly diagnosed the problem as too much weight near the wing's nose, which pushed the center of gravity (CG) too far forward, unbalancing the sailplane. Horten was familiar with the problem because Lippisch had encountered it when he first flew the Delta I. When preparing the *H I* to fly, Reimar had anticipated the need to adjust the CG by installing ballast, or "trimmer weights," as he called them, in a line running fore and aft inside the wing center section. Starting forward and working aft, Horten removed a small amount of weight and then flew the sailplane to measure the results, repeating the process to move gradually the CG aft until it fell at a point 1.55 m (5 ft.) behind the nose. By this process, Horten "nearly doubled the flight distance," but he remained disappointed with the elevator control and determined to improve its response. He did not fully comprehend in 1933 the cause, but he came to believe that positioning this surface at the middle of the wing was a mistake that he would correct in future designs.[78]

Reimar was soon ready to fly the wing higher than his launch team could comfortably loft it, so he began flying behind an automobile to altitudes as high as 30 m (almost 100 feet), and nearly the full length of the airfield. Further experiments with the center of gravity set further aft would have been dangerous, so the boy turned his attention to the two ailerons. By simply tightening the turnbuckles that linked each surface to a control cable, Horten could raise the trailing edges and dramatically alter the wing's flight characteristics in just a few minutes. "A mere 4° upward deflection rendered the elevator 'normal' without reversing the action" of pulling the stick back to ascend and pushing it forward to descend.[79]

Trimming the wing to fly at a more nose-up attitude forced the ailerons to share the elevator's workload to raise the nose and utterly transformed the wing's handling characteristics with a relatively easy modification. Reimar still rated the elevator "rather feeble in action," but he could now take off, climb, and land using normal control inputs. It was time to try turns and explore the wing's behavior about the yaw and roll axes, which meant stepping further into territory not previously mapped during his experiments with models.[80]

Adverse Yaw Encountered

As a car towed Reimar across the airfield, he made gentle curves right and left, and immediately encountered one of the most vexing aerodynamic quirks in the all-wing's bag of nasty tricks: adverse yaw, which occurs when a pilot rolls an aircraft to start a turn. During a right turn, for example, as the left aileron moves down and deflects the airflow to raise the left wing tip, the wing's camber increases, thereby increasing lift, along with the inevitable, but in this case unwanted by-product, drag. Simultaneous to the actions taking place on the left, the right aileron moves up, deflecting airflow to lower the right wing tip, decreasing camber, and thereby reducing lift and drag. This drag imbalance yaws the airplane left and away from the intended direction of the turn (Figure 6).

Many conventional aircraft suffer varying degrees of adverse yaw, but the lack of a fuselage and tail makes the problem considerably more difficult for the designers of all-wing aircraft to overcome. Designers of conventional airplanes mount a rudder at the end of the fuselage for extra leverage to overcome adverse yaw. Pilots deflect the rudder by depressing pedals with their feet to execute smooth curving turns that cause the airplane to skid neither left nor right. Pilots call these coordinated turns because they require manipulating the rudder and ailerons in subtle harmony.

Indeed, it was not enough simply to manage one control governing movement about a single axis, for a single change usually had secondary effects on other controls governing movement in different axes. Developing precise control and predictable handling characteristics depended on managing these complex interactions. Reimar found that during shallow turns, the *H I* drag rudders were almost powerless against adverse yaw. Turning was possible, but the wing rolled and yawed drunkenly about the sky, rough and uncoordinated, and when he tested the braking effect by deploying both drag rudders

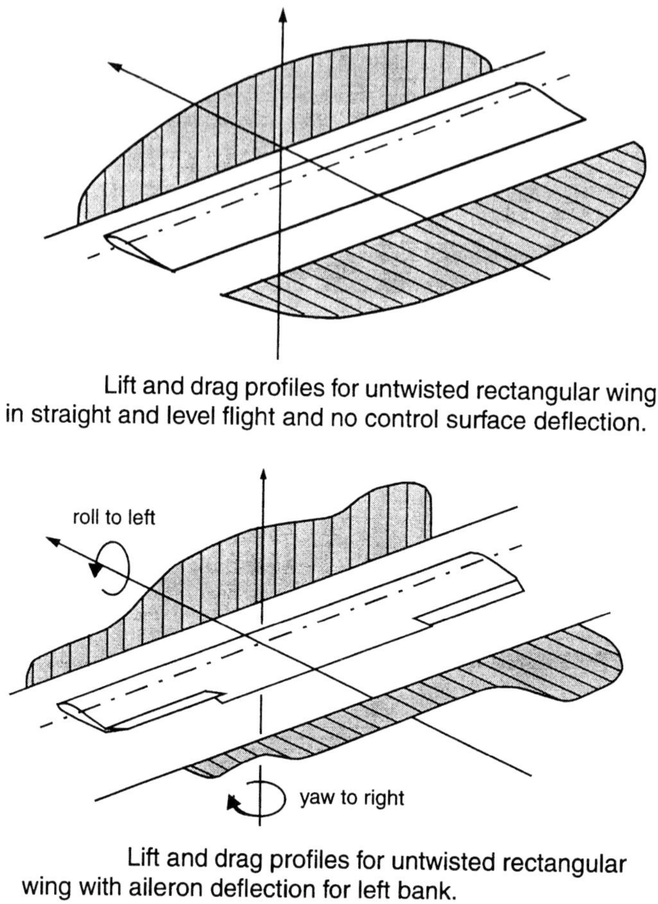

Figure 6. The shaded areas in this depiction of adverse yaw mark the distribution and amount of lift and drag for a simple rectangular wing in straight and level flight (top), and banking left to turn (bottom) (Bill & Bunny Kuhlman/B2Streamlines).

simultaneously, the wing pitched nose down and the elevator control was barely adequate to pull out of the resulting dive, regardless of how far he pulled the control stick. Reimar tried to cure both adverse yaw and pitch-down by installing another set of drag rudders closer to the wing tips. This change tripled the braking force, but the sailplane continued to misbehave.[81]

Horten decided to try a different approach, so the ailerons that controlled roll became the next devices to command his attention. Since the pitch control had improved so dramatically simply by adjusting the control cable turnbuckles to raise both ailerons, raising them even more might make the aircraft more tractable in yaw. He thought that enlarging the surface area of each aileron might help, too. This change would also satisfy the technical committee whose members ensured that all sailplanes flown in the national contest conformed to strict specifications. The *H I* met all their requirements except one, an obsolete measure of the pressure loading carried by a portion of the wing. Extending each control surface a mere 20 cm (8 in.) would satisfy the requirement.

Following these modifications, when Reimar rolled into turns during his next flight, the airflow became turbulent beneath the aileron deflected up, which drove the wing tip down on the inside radius of a turn. Horten probably felt the turbulence vibrating the airframe and control stick. This disturbed air increased the drag enough that Horten could finally make coordinated turns.

Flight control was improving enough that Horten decided to begin the next phase of testing by towing the *H I* aloft behind a powered airplane to observe the wing's behavior at higher altitudes and faster speeds. The aircraft showed promise but remained difficult to fly. It was time to delegate the test flying to Walter, who was considerably more experienced in the air. After a tow airplane hauled him up to 1,200 m, almost 4,000 ft., to begin his first flight in the *H I*, he gently curved the wing left and right to feel out the controls and then approached to land.

Reimar watched as Walter began to round out the sailplane's trajectory, slowing his speed in preparation for touchdown. Suddenly the approach soured, and Walter sailed 150 m, more than 320 ft., beyond the spot where he should have touched the grass. He finally pulled the elevator to full-up position, and the wing dove its nose into the ground. Walter was not injured and damage to the wing was minimal, but what had happened? The brothers forgot that Walter weighed slightly more than Reimar, enough to make the sailplane nose heavy once again. The elevator action had reversed as Walter approached to land, just as it had during Reimar's first launch by bungee cord months earlier. Reimar made the next flight and confirmed normal control. From that day forward, the brothers vowed to adjust the trim weights and reset the center of gravity whenever a different pilot flew the wing.

Powered airplanes towed the wing high enough to try steep extended turns, but the brothers were disappointed when the wing refused to recover smoothly. It was possible to roll using the ailerons into a turn, but when trying to stop the turn and roll back to level flight, the wing dropped into a diving spiral. Deflecting the drag rudders in any combination had no effect. Only after Walter centered the stick to neutralize the control surfaces could he roll the wing out of the spiral and into a straight dive. Recovery to level flight was then a simple matter of hauling back the stick. Looking back on his flying career, Walter described the *H I* as the poorest handling Horten aircraft that he flew.[82]

The cost and scarcity of tow airplanes at Bonn-Hangelar Airfield slowed efforts to improve handling, but soon, Reimar discovered that small weight increases worsened the spiral dive. A heavier wing required more elevator deflection to hold up the nose. Reimar observed that as Walter deflected the ailerons to recover from a turn, a wide gap opened between the elevator and the aileron, causing turbulence that made the wing tip partially stall and drop into the dive. He rebuilt and tested the ailerons several times, altering washout and other characteristics, before arriving at a crude fix. By cutting off the outer end of each aileron and widening the wing tip trailing edge to fill the resulting gap, Reimar increased the washout angle and reduced the stall speed of the wing tip. The modification also moved the gap closer to the center section, made the aileron smaller, and thereby weakened the turbulence considerably. The brothers finished the modifications in a few days, then waited three months for a tow airplane to test the changes in flight. When one became available at last, Walter took the wing aloft and executed turns "flawlessly" with no tendency to spiral dive.[83]

It was now early spring 1934—the start of a new flying season. The Bonn chapter of the DLV approached the brothers and asked them to fly the *H I* during the June air show at Bonn-Hangelar Airfield. They were quite pleased to do so, but first, a DLV official had to witness a flight before the organization would certify the wing safe for a public demonstration. The boys accomplished this handily. Flying the *H I* a few weeks later during the air show marked an important milestone for Reimar. In the span of a year, he had refined the all-wing configuration until it was safe to demonstrate publicly, without compromising its aerodynamic efficiency by using auxiliary control surfaces that produced drag but not lift.

To be sure, other designers had flown pure tailless configurations before him, but no one had established such complete control. For a nineteen-year-old with so little experience designing piloted aircraft and flying them, the achievement was all the more remarkable. Reimar's ability to doff his lab coat in one moment, and don his flight suit in the next, stepping from designer crafting an experimental aircraft into the role of test pilot carrying it out, had been a key factor in his success. The years of methodical research with flying models had provided a solid aerodynamic foundation and had given Reimar confidence to adapt the Lippisch Delta I into the milestone *H I*. The Delta I's inherent stability had freed Horten to concentrate on improving the control of the pure-wing configuration. To acknowledge Lippisch's contribution, Reimar christened his wing the *D-Hangwind*. *D* was for Deutschland (Germany) and *Hangwind* was for the nickname given to Lippisch during the 1920s for his sailplane piloting skills.[84]

Flight to the Wasserkuppe

Flying the wing for the home crowd at Bonn-Hangelar had been fun; however, the boys undertook more serious flying less than two months later. They had planned before the air show to attend the fifteenth Rhön competition, which began on July 22, 1934, and ended August 5. Reimar had to miss the first week because of school, but by air towing the *H I* to the Wasserkuppe, they would arrive with plenty of time remaining to fly. Reimar would pilot the wing while Walter towed him behind a two-seat Klemm sport aircraft. They faced a flight of several hours, but in good weather, and had planned an early takeoff, but a petty bureaucratic squabble held them up and forced the brothers to fly through a storm front.

After plodding through rain, fog, lightning, and turbulence, Reimar jettisoned the towrope with the Wasserkuppe in sight. Fatigue overcame him, and he undershot his landing approach. The wing touched down among rocks, shattering the landing skid, and damaging both wing tips. Repairs could take days, but Horten was caught up in the moment. "Thousands of people" were watching as he touched down, and officials had to restrain the curious crowd, lest they further damage the aircraft.[85]

The crowd's interest stemmed in part from the well-publicized problems with tailless aircraft encountered by Lippisch and others. Reimar and the wing had to perform well for the group of seasoned designers and pilots at the contest. These veterans had campaigned for years at the Wasserkuppe and watched a variety of crackpot flying contraptions appear with great fanfare only to fade quickly away, or fail spectacularly before their eyes, often with fatal consequences. The widely respected designer Hans Jacobs argued that no one could design a tailless aircraft to fly well. Reimar could not persuade Jacobs that he had taken the configuration to a new level of stability and control. Only seeing the wing fly would convince him and other skeptics. Reimar also wanted to earn state sponsorship from the Nazi officials managing both the meet and sailplane research funds. Soaring over the throngs gathered on the peak would prove the concept worth developing further. Walter was the better pilot, and the best choice to make these important flights, but he was unavailable.[86]

A good distance flight could really impress the critics but would require Reimar to land away from the mountain. The wing was too big to retrieve with a standard sailplane trailer, and he might damage the aircraft by alighting on rough terrain. His best strategy was to fly directly over the Wasserkuppe, find a good thermal already occupied by a few pilots circling in conventional sailplanes, and then to climb well in the *H I*. During contests, as many as ten to twenty sailplane pilots often worked the same thermal, tightly spiraling their machines inside the rising column of warm air as they clawed for altitude.

As the meet neared its end, Reimar had been unable to repair the *H I*, and his frustration was acute. Finally he flew on Sunday, August 5, 1934, the last contest day. His twenty-one-minute, noncompetitive demonstration flight was less spectacular than Reimar had wanted, but it impressed the contest officials. They awarded him a cash prize of six hundred marks for innovative design, a portion of which he paid to the Bonn flying club for outstanding towing fees.[87]

The meet was over, but Reimar remained on the mountain another week waiting for the right conditions to fly again. Wolfram Horten arrived to help prepare the wing for another flight, but just as Reimar was about to launch, the local Nazi official overseeing the Wasserkuppe soaring school, Jupp Schmitt, challenged Reimar to an informal contest. Schmitt would soar a conventional sailplane of average performance called the Super Falke and try to out-climb Horten in the *H I*. Reimar recalled no difficulty climbing faster than Schmitt, but any satisfaction he felt vanished upon landing when Schmitt ordered the *D-Hangwind* removed before the next day ended.

Reimar needed to obtain a tow airplane and quickly. He called the Bonn flying club, and another club in Essen, but neither group could help. Reimar then contacted the headquarters of the German Research Institute for Gliding (DFS) at Darmstadt and spoke directly to Lippisch, but he also declined. Walter arrived on leave the next day, and the three brothers conducted impromptu structural testing on the wing before removing the useful hardware. Walter volunteered to put a match to the remains, and Reimar watched fire consume the *H I*, his first piloted wing. Reimar and Walter had spent one thousand hours or more building the *H I* at a cost of at least 320 marks but had flown the wing less than seven hours.[88]

When asked years later to recall the flights he enjoyed most during his career, Reimar said that "flying and soaring the [*H I* at] the Wasserkuppe . . . to look for when the trees are moving and the wind was [blowing] . . . to remain airborne over the mountain and then to climb a bit. Our flying in the *D-Hangwind* had been wonderful."[89] His emotional attachment to the wing, in part, fueled the feeling that Lippisch had betrayed him by not helping save the aircraft. Reimar claimed that without his permission, Lippisch had inspected the *H I* the previous winter, and it had impressed him enough that during several visits to the Horten family in Bonn he had spoken to Max about Reimar joining the staff of the glider research institute after he finished high school.

Reimar skewed Lippisch's refusal to help into jealous retaliation for young Horten's competence in all-wing design, yet Karl and Gunilde Nickel stated that Lippisch simply lacked the resources to tow it to another airfield. In a pattern that dogged both his professional and personal relationships for most of his career, Reimar held a personal grudge toward Lippisch for years thereafter.[90]

CHAPTER 2

Freedom of Flight

REIMAR HORTEN began thinking about the *H II* as early as fall 1933. By the following summer, he longed to shed the expense and inconvenience of using a tow airplane to tug the *H I* aloft and to increase the tempo of his experimental flights. The fiery destruction of his first wing sharpened this desire for "freedom of flight."[1] By installing an engine, Reimar could take off without a tow, climb to altitude, switch off the motor, and then soar. A powered sailplane would allow Reimar to more rapidly evaluate fresh ideas and introduce improvements. The *H I* had been an experimental prototype with idiosyncrasies he tolerated to further his research. Horten could now build on that work and begin to refine his second wing for more practical, everyday use.

Reimar improved the *H II* pilot's compartment by repositioning the main spar. The top edge of this heavy component had passed through the cockpit of the *H I* within a few centimeters of the pilot's chin, not an ideal location during a hard landing or crash. To simplify the control system, increase pitch authority, and prevent the control reversal that had troubled the *H I*, Reimar eliminated the elevator. This change also made it easier to narrow the center section width, which he had found too broad to easily move down the street.[2]

For control in pitch, Reimar modified the control system to deflect the ailerons simultaneously as well as differentially, combining the actions of elevator and aileron into dual-purpose control surfaces called elevons. He did this by cleverly engineering two pushrods and a bell crank hinge between the control stick and each elevon in a manner Karl Nickel described as "brilliant" for its simplicity and effectiveness, particularly compared to the complicated arrangement that Lippisch had used from 1933–1940 in his series of tailless airplanes.[3]

Reimar did not make the *H I* easy to disassemble into smaller components; however, this requirement became important as he designed the *H II*, so that the second wing was easier to move on the ground and more practical in everyday use. Graceful in the air, sailplanes became cumbersome and awkward as soon as they touched down. As a result, soaring was a group activity because it required a crew to manhandle the aircraft. Ground crews often had to journey hundreds of kilometers to retrieve a sailplane following a cross-country flight. Designers therefore configured their sailplanes to break down into three or four components—wings, fuselage, horizontal stabilizer and elevator, and sometimes the rudder—for manageable loading onto a trailer built for the purpose and towed behind an automobile.

Reimar had no fuselage or tail to contend with, but to accommodate a cockpit, and for aerodynamic reasons, his early wings had about twice the root chord (wing width, forward leading edge to rear trailing edge) of conventional sailplanes. He divided the *H II* into a center section and two wing panels and modified a trailer to hold the pieces. To manage the bulky center section, Horten had to install lifting handles and handholds into the leading and trailing edges. The ground crew pushed the handles flush into the wing before the pilot took off to cut drag in flight.[4]

To make the *H II* a more practical aircraft, it was important that Reimar improve yaw control beyond good handling qualities consistent with conventional airplanes. Pilots had to believe they could stall a Horten wing and then resume normal flight without losing control and entering a spin. This could happen at the end of a stall when a wing tip dropped sharply and the aircraft began rotating and falling rapidly toward the ground. Most conventional aircraft were designed to allow pilots to spin them and recover at will. The fuselage and tail were vital to good spin control, but Reimar had only the wing to work with. Since his drag rudders were too weak for the task, the key was to prevent spinning altogether. Horten accomplished this by placing the center of gravity well forward and by carefully adjusting the distribution of lift across the wing.

Reimar reshaped the *H II's* aerodynamic signature by distributing the lift not in the traditional and widely accepted elliptical form, as he had done with the *H I*, but in a pattern shaped like a parabola (Figure 7). The distribution of lift over the wings of many aircraft varies in flight and can often resemble a parabola; however, Horten's method of distribution was unique in its characteristics and functions. In a process not unlike an automobile mechanic adjusting valve timing, cylinder head configuration, combustion mixture, carburetor settings, and gear ratios to improve engine performance, airplane designers can "tune" lift distributions to produce certain flight characteristics by altering singly, or in various combinations, basic design elements such as wingspan, plan form, and aerodynamic and geometric twist.

Reimar altered all of these characteristics as he created the *H II*. Compared to the *H I*, he increased the span by almost 25 percent to 16.5 m (54 ft. 2 in.), broadened the chord by 14 percent, and boosted aspect ratio by 14 percent to 8.5. These modifications sharpened the taper ratio to 8.75, an extraordinarily high value at that time. Horten also increased the sweep angle of the leading edge by 22 percent to 29.5 degrees and this change, coupled with the increased span, gave pronounced sweep to the trailing edge (Figure 8). Using Junkers's formula for the G.38 passed to him by F. W. Schmitz, Horten thickened the *H II* wing at the root to 20 percent of the wing chord and at the tip to 10 percent. He also twisted the wing's geometry with 6° of washout, gradually increasing the *H II* wing's angle of attack from rib to rib, in a nonlinear progression beginning at the root and extending to the wing tip.[5] The *H II* was no tweaked copy of the Lippisch Delta I but rather a bold new design.[6]

Varying the shape of the airfoil profile from the wing root to the wing tip to impart aerodynamic twist was yet another modification that altered the lift distribution. Horten recalled using a reflex profile at the root of the *H II* and a symmetrical profile at the wing tip (Figure 9), but he did not specify where the transition between the profiles occurred.[7] Experiments conducted late in the nineteenth century had demonstrated that the reflex airfoil section added stability to a wing.[8] Reimar combined this airfoil architecture with a carefully considered center of gravity location to make the wing handle well during a stall. Of course, the distribution of lift would change whenever a pilot deflected the control surfaces, which led Reimar to test the wings many times to evaluate the effects of subtle shifts in lift during various maneuvers.

The Horten distribution provided another design benefit that had nothing to do with control. To understand it, we must go to its origin. In the November 1933 issue of *Zeitschrift für Flugtechnik und Motorluftschiffahrt* (*Magazine for Flight Technology and Air Transport*), Ludwig Prandtl published "Über Tragflügel kleinsten induzierten Widerstandes" (About Wings of Smallest Induced Drag). Reinhold Stadler and Al Bowers interpreted this short paper as solving a problem that had stifled sailplane design for some years: how could wing design give the maximum lift with the least penalties in drag and weight?

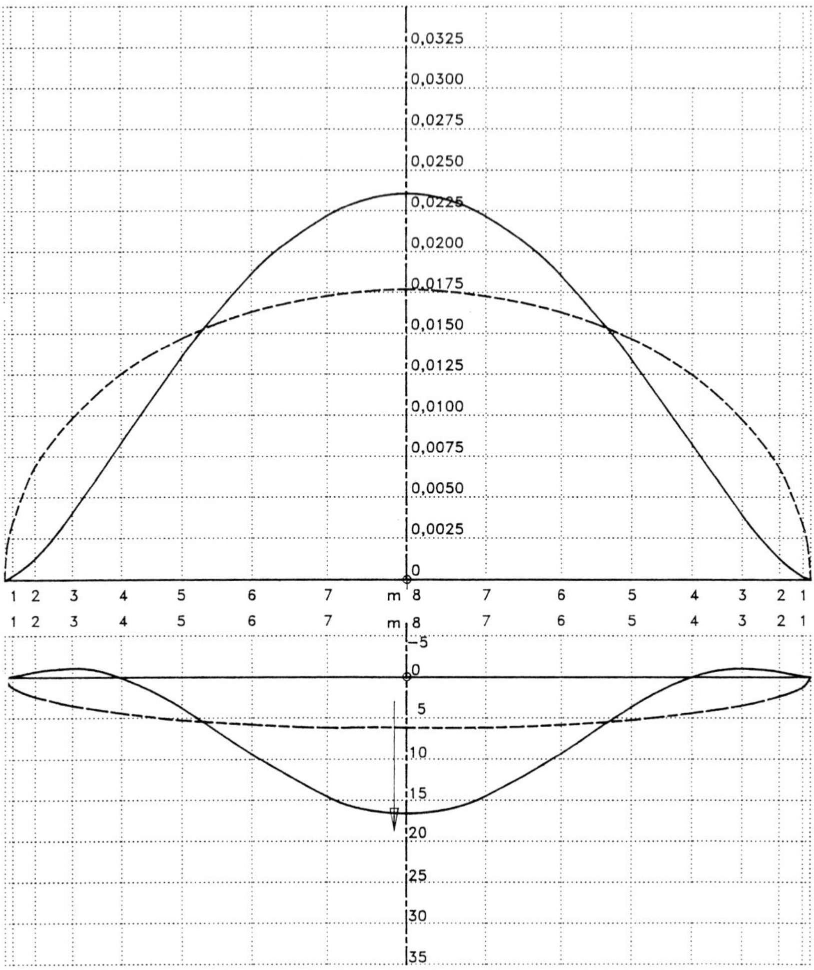

Figure 7. Curved lines track the distribution of lift (top) and drag (bottom) for a Horten bell form (solid lines) and an elliptical form (dashed lines). The bell makes considerably more lift at the center of the wing than at the wing tips, thus lessening the wing root bending moment. As the wing's geometry twists to form washout at the wing tips, the drag track rises above the 0-lift line, indicating a modest but measurable amount of lift, which becomes favorable yaw when the pilot deflects the elevons to roll into a turn (Edward Uden).

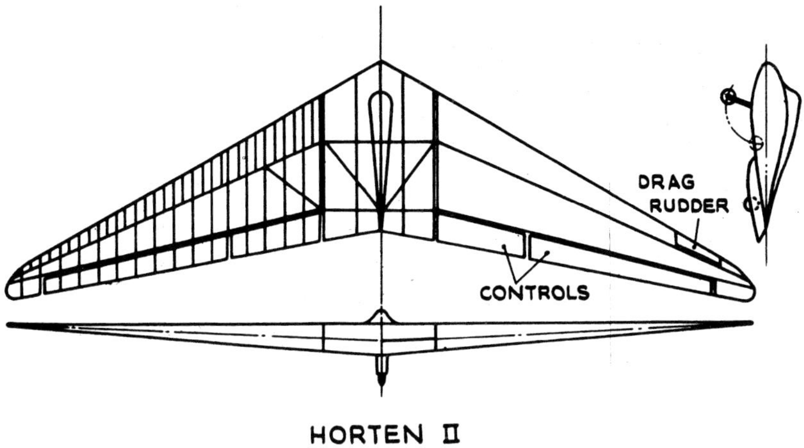

Figure 8. Horten *H II* three-view drawing (Appleyard and Biot, "The Horten Tailless Aircraft," 41).

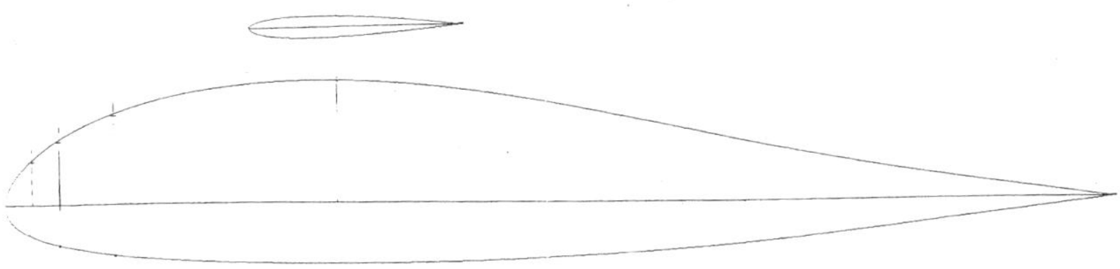

Figure 9. Reimar frequently employed a reflex profile (bottom) at the center section and a symmetrical profile (top) at the wing tip to stabilize the wing during a stall. The characteristic concave upper surface near the trailing edge of the reflex profile caused turbulence, loss of lift, and a stall before the uniformly smooth upper and lower surfaces of the symmetrical profile (Edward Uden).

In the late 1920s, sailplane designers had begun to push wing aspect ratios ever higher to boost lift while holding down induced drag. Wings became longer and thinner, and to keep them strong enough to withstand the forces of flying, designers had two choices: fit stout support struts beneath the wing roots, which caused more drag, or beef up the main spars inside the wings, which increased weight. As Al Bowers explained, an elliptical design does not "produce the minimum induced drag for a given spar size (wing root bending moment)."[9] With only the elliptical model of lift distribution to work from, designers groped to find the right balance of performance, drag, strength, and weight.[10]

At a single stroke, the parabolic lift distribution described by Prandtl in 1933 superseded the elliptical distribution as the optimum method for achieving the correct balance to design the most efficient wing. In other words, parabolic lift produced the least wing spar bending moment, the force of lift pushing the wing up at its tips and bending it at the root.[11] The spar is the single most massive component in a wing; therefore, a lower bending moment allowed designers to use smaller and lighter spars. They could remove the struts and lighten the wing spars because the parabolic form made the center section carry more lift, rather than the outer wing sections.

In addition to the structural benefits, applying Prandtl's theory led to significant aerodynamic gains. Provided the designer allowed the wingspan to vary while calculating the lift distribution,[12] then given two wings lifting the same amount, Reinhold Stadler deduced that the parabolic wing would exceed the span of an elliptical wing by 22 percent, but it would induce 11 percent less drag![13]

Despite these important advantages, nearly all designers ignored Prandtl's 1933 paper and continued to use the elliptical form, also introduced by Prandtl but much earlier in 1920, because it had become, in the words of Bill Kuhlman, "dogma in most aerodynamics texts."[14] Designers had put the elliptical form of lift to immediate use in all types of powered aircraft with wings of relatively low aspect ratio. The sailplane wings with higher aspect ratios that could benefit the most from the parabolic form did not begin to appear until late in the 1920s.

University of Southern California professor and certified engineer Hugh H. Hurt Jr. still championed the older lift distribution when the Aviation Training Division of the Office of the Chief of Naval Operations published his classic textbook, *Aerodynamics for Naval Aviators*, in 1960. The book remained in print nearly fifty years later as a selection in the Federal Aviation Administration Handbooks series. "In the aerodynamic sense," Hurt explained, "the elliptical wing is the most efficient plan form because the uniformity of lift coefficient and downwash incurs the least induced drag for a given aspect ratio. The merit of any wing plan form is then measured by the closeness with which the distribution of [both] lift coefficient and downwash approach that of the elliptical plan form."[15]

A month after Prandtl's 1933 paper appeared, Alexander Lippisch published "Bestimmung der Auftriebsverteilung längs der Spannweite" (Definition of the Lift Distribution along the Wingspan) in *Flugsport*, the first article of a three-part series published between December 1933 and February 1934. Lippisch aimed to offer designers an improved method for calculating lift distributions by simplifying the process without degrading its accuracy. He said that existing methods did not always account for variations in lift distribution due to different flight regimes, from slow flying to high-velocity maneuvers such as aerobatics. Lippisch claimed that his new formula enhanced flight safety by giving wings benign stall characteristics and good performance, without compromising maneuverability.

In his third article, Lippisch complimented Prandtl on his 1933 paper, and he depicted a simplified "triangular-shaped" lift distribution that resembled the parabolic form. Lippisch clearly specified that the triangular form required geometric and aerodynamic wing twist, the latter derived using a reflex airfoil section from the root to midspan and a symmetrical section at the wing tips.[16] Reimar claimed that Lippisch gave him the formula for calculating the parabolic lift distribution when Lippisch visited Bonn in 1934, but Horten never acknowledged Prandtl's role as the originator.[17]

Building the *H II*

The sight and smell of the charred remains of the *H I* must have been fresh in young Reimar's mind when he and Wolfram stopped on their way home from the Wasserkuppe in mid-August 1934 to order materials for constructing a new glider. The following month, he set to work laying out the *H II*. Reimar again chose to set up shop in the family residence. Several local businesses donated supplies and material, but the designer's reluctance to discuss the new project limited the scope of donations. Reimar invited members of the Bonn flying club and a few other enthusiasts to help, and Wolfram joined him during the first months, but Walter could stop by only infrequently due to the demands of his military career.[18]

Lippisch's simplified method made calculating the new *H II* lift distribution somewhat easier, but actual construction was difficult. Reimar and his helpers easily handled the wooden portions of the wing, but the center section had to be made of welded steel tubes to support an engine and to make the airframe strong enough to withstand aerobatics. Reimar had never worked with metal, and Dr. Rolf Schrödter stepped forward again to provide expertise and resources. Using specifications from Reimar as a guide, Schrödter tasked specialists in his motorcycle shop to cut the steel tubes and delicately bend them to form the wing's profile shape before welding the pieces together at precise angles.

Schrödter delivered the incomplete framework to the house in Bonn and sent along a welder to help finish it, but even with his friend's help, Reimar required nine months to complete the aircraft. He was shorthanded without Walter, and school matters often interrupted his work on the wing. The brothers had spent about one thousand hours building the *H I*, but the second wing consumed roughly five thousand hours due to its size and complexity.[19]

Compared to the *H I*, the *H II* required fittings to mount additional controls, extra instruments, and a fuel tank, all to accommodate an engine. Reimar planned to perform aerobatics, but the wing weighed almost double the *H I*, so he had to precisely determine the effects of this additional stress and design the wing to support itself, the power plant, and the pilot during extreme maneuvers. The wing's size and weight made a skid landing gear impractical, and Reimar designed a more complex arrangement consisting of two wheels mounted in tandem. The steel tube frame that held the engine also supported the landing gear. A pilot could retract the front wheel after taking off while the rear unit remained extended but sheathed inside a streamlined enclosure to minimize drag. A linkage allowed the pilot to steer the aft wheel using the drag rudder pedals.

It was important to design the gear to track straight ahead during the first moments of a takeoff, after the wing began moving, but before it had reached sufficient speed for aerodynamic control. Particularly during tows behind aircraft, the wing pilot had to reach flying speed as quickly as possible. Failure to track straight ahead risked ground looping the wing and stalling the tow airplane, which frequently lifted off first, yet the wing pilot had neither rudder nor steerable tail skid to correct for a swing to one side.

For control in the air, Reimar wanted elevons that extended almost to the wing tips, drag rudders set into the tops of the outer wings, and landing flaps mounted inboard near the center section. Horten made the elevons easy to adjust, as he had done with the ailerons on the *H I*. The designer had perceived loss of lift at the center section of his first wing, the result, he suspected, of turbulence caused by the cockpit enclosure projecting well above the smooth wing surface. To eliminate this protuberance on the *H II*, reduce drag as much as possible, and restore the lift, Reimar ignored the standard dogma on seating a pilot upright in the cockpit and struck out in a completely new direction by introducing the first supine (lying on the back) seat, or couch, installed in an airplane.[20]

Otto Lilienthal had written in 1889 of accommodating pilots in a "horizontal" position for "easier passage [less drag] through the air and the consequent saving of energy in flying"; however, he envisioned piloting prone, or stomach-down.[21] Taking their cue from Lilienthal, Orville and Wilbur Wright became the first pilots to fly prone in 1900 and continued to do so aboard all of their gliders and powered aircraft until they began carrying passengers in 1908. The Wrights learned of Lilienthal's theory on horizontal flying posture in Octave Chanute's book, *Progress in Flying Machines*, published in 1894.[22] Horten did not identify the origins of his idea to fly supine, but he worked out the placement and manipulation of the instruments and flight controls, and their positions relative to his eyes, hands, and feet, while lying on his bed.

Reimar's supine couch cradled the pilot's back nearly flat, head buried deep within the wing, but propped up enough to see forward. To improve the view, Horten covered nearly the entire forward half of the center section with clear plastic panels that gently curved to maintain the airfoil shape. His earlier work with the bedroom mockup made it easier to place the controls and instruments, but the space between the couch and the clear plastic wing skin measured a somewhat claustrophobic 0.7 m, or 28 in. This bold experimental leap in all-wing design soon collided with practical limitations unforeseen by the designer until he installed an engine and propeller and began to conduct powered test flights.[23]

Although he would not earn his PhD until after World War II, Reimar began working toward the degree early in 1935 after taking a course in differential equations and realizing that he "had the key that would open the door" to understanding lift distribution theory, as he later recalled.[24] Perhaps distracted by his coursework at the time, Reimar could not consistently recall when he first flew the *H II*, but this probably occurred sometime in May or June 1935 at Bonn-Hangelar Airfield. No engine was fitted at this time. Reimar said that after a few short hops, he carefully eased the wing into turns and found none of the control reversal that had marred the *H I*. Most remarkable, he claimed no need to use the drag rudders, turning solely with elevons because their deflection made favorable yaw (Figures 10a, 10b, and 10c), thanks to his lift distribution.[25]

Reimar never explained the origins of favorable yaw, but A. R. Weyl mentioned in 1945 many individuals who observed or deduced the effect of wing tip design on the yaw stability of tailless and all-wing aircraft as early as 1883, when French engineer Alexandre Goupil saw what he called "negative drag" (or favorable yaw) on a flying model.[26] Experiments conducted by Ed Uden and Al Bowers in the 1990s seemed to prove that no Horten wing that flew could have produced favorable yaw, therefore either his recollection was faulty or he lied to magnify his accomplishments in the waning days of his career. Reimar began referring to his version of Prandtl's parabolic distribution as a bell-shaped form in 1981, and henceforth, I will use Horten's terminology.[27]

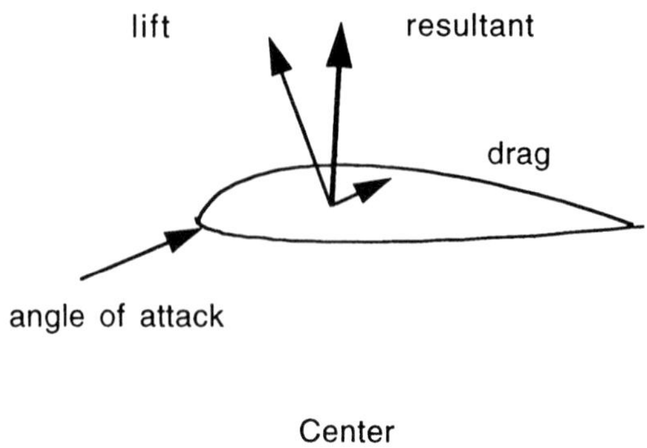

Figure 10a. The first step in understanding how the bell lift distribution makes favorable yaw is to visualize how the airflow interacts with the center of the wing in normal cruising flight, as illustrated here. This angle of attack produces lift, skewed by drag, leaving a resultant oriented along a vertical axis (Albion H. Bowers, NASA Dryden Flight Research Center).

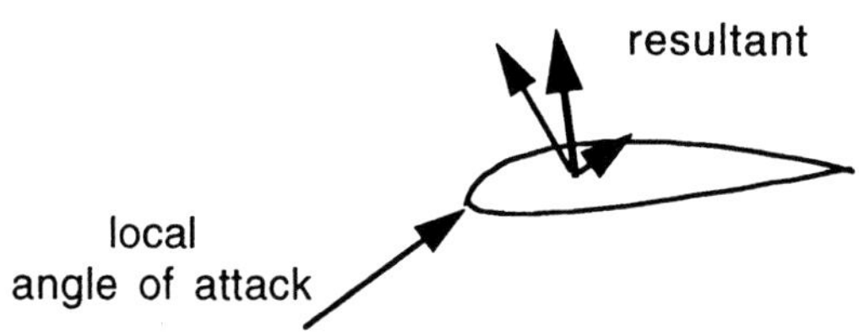

Figure 10b. Geometric washout at the wing tip reduces considerably the local angle of attack and tilts the resultant slightly forward (Albion H. Bowers, NASA Dryden Flight Research Center).

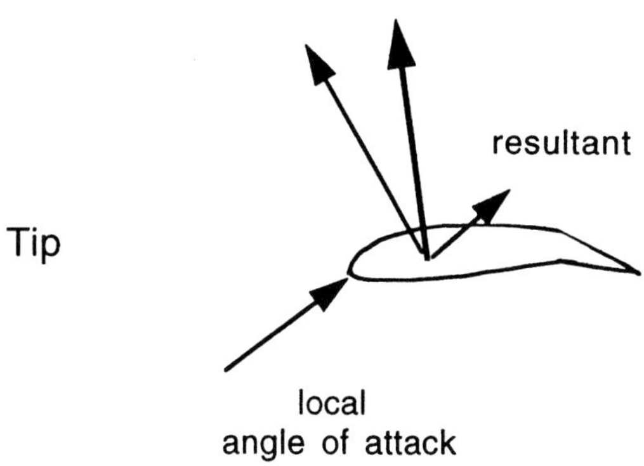

Figure 10c. When a pilot deflects the elevon down to roll the wing, camber increases, along with lift and drag, but there is enough geometric twist to produce a resultant sufficiently powerful to tug the wing tip toward the direction of the roll rather than away, as happens with adverse yaw. Correct placement and configuration of the elevon ensures more lift and less drag. The reverse occurs on the opposite wing tip, where deflection up weakens considerably the resultant (Albion H. Bowers, NASA Dryden Flight Research Center).

Reimar used the results of experiments with flying models to calculate the center of gravity and balance the *H II* to fly with no tendency to pitch up or down uncontrollably; however, the first flights seemed to show again the same loss of lift at the center section that he had detected when flying the *H I*. To compensate, Reimar placed about 50 kg (110 lb.) of ballast in the wing trailing edge, and wheel fairing, and continued flying. He found the undercarriage failed to keep the wing rolling straight ahead or to dampen sufficiently the pounding motion caused by irregular surfaces during the takeoff and landing. Reimar also noted the poor visibility from the supine couch when looking toward the rear during turns, yet he did not immediately change the cockpit. The top surfaces of the center section had to remain undisturbed to allow him to measure the odd loss of lift. It is worth noting that by the late 1980s, nearly all conventional sailplanes seated the pilot supine.[28] The wings did not block the pilot's view because designers were able to position the cockpits well ahead of them.

Reimar had completed about ten flights lasting perhaps one to two minutes each before the winch used to tow him aloft failed. He then turned to motorizing the *H II*. A friend lent him a four-cylinder Hirth HM 60 engine that produced about 48 kW (65 hp), and Reimar worked for three months to install it behind the cockpit. A drive shaft was required to harness the motor to a pusher propeller, and Dr. Schrödter constructed a smooth aluminum fairing to cover these components. The complete installation added enough weight at the back of the center section that Horten could remove the ballast added earlier to compensate for the loss of lift discovered during the first flights. Designers loathe superfluous weight as much as excess drag, so Reimar felt relieved to swap useless weight for an engine, which paid its way by furnishing power; however, he would have reinstalled the ballast if he removed the engine to compete as a sailplane at the Wasserkuppe.[29]

Losing lift across the center section of his wings, or the Mitteneffekt (middle effect) as he called it, haunted Reimar until well into World War II. He attempted to eliminate it using several different approaches, but all were unsuccessful. During the 1990s, Reinhold Stadler evaluated the structural and aerodynamic qualities of many Horten aircraft, and he determined that as late as 1944, Reimar had failed to consider accurately the effects of wing sweep when he calculated the lift distributions, due in part to limitations in the mathematical calculation method that he had used rather than an error in the designer's judgment. Nonetheless, he expended considerable effort working on a problem that did not actually exist.[30]

As he was more skilled and experienced with powered aircraft, Walter Horten made the first flight under power about September 1935. His flight was also the first of a powered aircraft equipped with a supine cockpit. When Reimar's turn came, he took off and climbed to 1,000 m (3,300 ft.), then cut the engine and soared the wing before landing about fifty minutes later. He was surprised and pleased to take off under power at 50 km/h (30 mph) and accelerate to 180 km/h (110 mph) in about 800 m (2,600 ft.). At that velocity, he could climb to 1,000 m (3,300 ft.) in two-and-a-half minutes. The wing proved strong and maneuverable enough to perform high-speed dives, so Reimar named it *Habicht* (Hawk) (Figure 11).[31]

After noting the performance at high speeds, Reimar flew slowly, the wing sputtering along with its nose toward the sky, and discovered that the supine seat was quite uncomfortable. Pilots of conventional airplanes sat upright, their feet below their heads, even with the nose of the wing pointing high during slow flight, but Reimar had violated this convention. The sensation reminded him of pulling up into a "looping" aerobatic maneuver. Besides the physical discomfort, the awkward position increased the danger of collision because Horten found it nearly impossible to twist his head around to scan the sky for other aircraft. This could become even more dangerous whenever a pilot circled in thermals crowded with other gliders.[32] Horten's bold experiment in cockpit design had failed, but his efforts were

Figure 11. On the Bonn-Hangelar Airfield in 1935, Walter and Reimar stand at the nose of the *H II m* fitted with a motor that turned a propeller mounted at the trailing edge of the wing (Reimar Horten via Peter F. Selinger).

not totally wasted. The work revealed a direction in cockpit design that other designers should avoid.

The uncomfortable cockpit did not deter Reimar from continuing the tests. With about fifteen hours of flying time on the wing, he stalled it at a forty-five degree angle with full engine power during a flight in 1937, and when his "legs seemed to be vertical in the sky," Horten suddenly felt a blast of heat from the engine laboring furiously behind his seat.[33] Smelling the hot fumes of something burning, he landed quickly and discovered that the engine had overheated and now required a complete overhaul. He immediately returned the damaged power plant to its owner and then traveled to the Hirth factory to request a new engine. The firm gave him a more powerful HM 60R engine that propelled the wing for more than one hundred additional hours of flying time. This motor was reliable enough that other flyers could evaluate a Horten wing in the air.

Via informal "hanger talk," letters to the editor of popular magazines such as *Flugsport*, and in the typed pages of official reports, pilots shared their views on performance and the practical aspects of flying and operating the *H II*. The Nazi leadership closely monitored the discussions and stood ready to support this unorthodox technology if it showed promise. Flying the glider was straightforward enough that even inexperienced pilots added it to their flying log books, and even Reimar Horten earned his glider rating in the *H II*, but he exaggerated by saying that a "baby" could fly the wing. As we shall see, other pilots assessed the handling characteristics differently.[34]

Reimar's work reached a milestone in 1937, even as Lippisch bemoaned the "defective flying qualities" of all-wing aircraft.[35] At Lippstadt in 1937, for the first time a Horten wing design entered low-rate serial production when Reimar and Walter managed a team of craftspersons who completed two more *H IIs*, plus a third the following year. These aircraft differed from the prototype by accommodating the pilot in a standard cockpit with upright seating, and in improvements to the landing gear. Using the aft wheel to steer the wing during takeoff had not worked well.

Building multiples of this wing, which Reimar had designated *H II L*, came about primarily because of support from a high-level RLM official, General Ernst Udet, whom Hitler had appointed in 1936, at Hermann Göring's suggestion, to lead the Luftwaffe Technical Development Office, where he was responsible for nurturing promising experimental programs. Udet had learned about Reimar's work from General Friedrich Christiansen, commander of the Nationalsozialistisches Fliegerkorps (National Socialist Flier Corps, or NSFK). Christiansen had known of Reimar's work since 1934, when he had awarded him the six-hundred-mark design prize at the Wasserkuppe for designing the *H I*.[36]

Christiansen had assumed his command in April 1937 after the Nazis dissolved the DLV and replaced it with the NSFK. The new organization soon absorbed all sport flying groups in order to create a pool of skilled pilots with the correct political attitudes from which the Nazis could "select their new blood for the aeronautical industry and Luftwaffe."[37] NSFK officials decided who could become sport pilots and enforced standards on the aircraft types used for training. Developing unorthodox new designs such as all-wing aircraft did not fit the NSFK agenda, and the organization would have forbidden flying clubs from teaming with Reimar. Udet's interest kept the designer working at a time when his career could have been vulnerable.

Major Oskar Dinort commanded Luftwaffe fighter group Jagdgeschwader (fighter wing) JG 134, based at Lippstadt near Cologne, about 25 km (15 miles) north of Bonn, when Walter Horten joined the unit in May 1936. Walter had just finished pilot training and asked Dinort for a posting near his brother, who had been working since February a few kilometers from Lippstadt at the Dynamit A. G. firm in Troisdorf, less than 10 km (6 miles) northeast of Bonn.[38] While working to restore the pilot's view from the *H II* by covering the center section with an experimental clear plastic, Reimar had met a chemist at Dynamit named Leysieffer. There is no record of the chemist's background or first name, but he became intrigued with Reimar's work after seeing the *H II* fly and asked the young designer to join the firm to help develop aircraft components made from new synthetic materials.

Dinort was an enthusiastic champion of soaring flight who had set an endurance world record for sailplanes in 1929. He was probably keen to help Reimar design and build all-wing sailplanes, but Walter believed it was Udet who had brought the Hortens so close together. When he did not receive orders to join Germany's Condor Legion air unit after it began fighting in Spain in 1936, Walter's suspicions grew stronger. Many of his fellow pilots served in Spain, and Walter was trained and ready to go there, but he sensed that in "the background was Udet [saying] let them work further in the field of flying wing designs."[39] Walter was not surprised when Reimar joined him in JG 134 a few months later.

Reimar began supervising the building of the three *H II L* motor gliders in 1936, after Dinort gave orders to supply the Horten brothers with active duty Luftwaffe personnel, aircraft-grade construction material, and a workshop, and promised to provide three new Hirth engines. In essence, the Luftwaffe funded the entire project. Moving to a proper workshop promised to speed the wing construction. To the enlisted craftspersons, Reimar delegated the work of cutting, shaping, and gluing the hundreds of wooden components needed to construct each wing, thus freeing him to spend more time at Dynamit.[40] Dinort wanted the brothers to travel throughout Germany with him, demonstrating the wings in air shows. He was enthusiastic and "air-minded for sailplanes," Walter recalled, and with Udet in the background supporting their activities, they felt very much in charge of their destiny. Said Walter, "We could do what we want. It was very nice, a very nice time . . . [our] best years."[41]

In 1937, the authorities commandeered the Hirth engines that Dinort had wanted for the new *H II L* motor gliders. The brothers finished the wings without motors, and Dinort ordered them to fly the aircraft in the German national glider competition scheduled for August.

Three years had passed since Reimar had soared the *H I* above the Wasserkuppe. He welcomed the chance to demonstrate an improved wing to a national audience, but politics appeared to disrupt the outing before it began. A party official named Seff Kunz complained in a letter to Dinort that the wings were too experimental to compete safely. If his pilots went ahead and flew the aircraft in the meet, they would earn no points. The disagreement could have been a manifestation of the Nazi philosophy rewarding group performance and conformity over individual achievement. Having neither an advanced degree nor affiliation with a major firm, Reimar Horten stood apart from the mainstream aviation community.

Dinort took the Nazi threat seriously and withheld the ground crews needed to help retrieve the gliders after cross-country flights and competition equipment such as barographs to record altitude. Unable to compete, the Horten brothers thus found the meet to be an extended practice session. In truth, the gliders were not prepared to compete. Heinz Scheidhauer, who joined Reimar Horten's group in 1937 and went on to become the chief Horten test pilot, explained years later that heavy noses handicapped the two *H II L* wings. "They could only polish the slopes," he said, skimming low on the air cushion formed between the wing and the ground. What caused the wings to nose down is unclear.[42]

Walter made adjustments to his wing and managed a respectable height gain of more than 2,500 m (8,200 ft.), but the brother were also handicapped by a lack of recent flying experience. Reimar's skills had suffered the most, and he crashed his wing several times while landing on the short, mountain runway. Although not the sort of demonstration to inspire confidence among pilots in their unconventional designs, the brothers' combined performances were enough to earn a second-place award for technical achievement, which included a 1,200 mark cash prize. The prize encouraged Dinort to renew his support, and he ordered Horten to prepare a new wing called the *H III* for the following year.[43]

Workers at Lippstadt could not finish the third *H II L* until 1938. That year, Scheidhauer covered 240 km (149 miles) during a single flight in this aircraft. Others flew the wing, practiced flying in IMC (instrument meteorological conditions), and performed aerobatics during air shows. In March 1939, Kurt Hieckmann, who had piloted an *H II L* at the national meet the year before as a pilot for the Luftwaffe team from Fürth, died practicing aerobatics in an *H II L* when the canopy dislodged and struck him unconscious during a high-speed dive. Hieckmann had completed the following routine several times that day before his fatal flight: tow to 3,000 m (9,840 ft.), disconnect from the tow, half-roll and then dive, begin pullout at 500 m (1,640 ft.), skim across the airfield at approximately 1 m (3.3 ft.) height at 300 km/h (186 mph), loop three times in succession, turn right, turn left, and then land. Pilots flew another *H II L* until nearly the end of the war and often dove the sailplane to 483 km/h (300 mph).[44]

These accounts suggest that the *H II* handled quite well, but several test pilots penned different appraisals. Hanna Reitsch was a civilian with considerable experience piloting many different aircraft. Nazi officials, Hitler and Luftwaffe chief Hermann Göring among them, valued her expertise. During World War II, she served for a time as Hitler's personal pilot, and during a flying career that lasted from 1932 to 1979, Reitsch set more than forty national and international records. After Göring reorganized the national soaring research effort conducted at the Rhön-Rossitten-Gesellschaft along Nazi lines in 1933, renaming it the Deutsche Forschungsinstitut für Segelflug (DFS), he made Reitsch the new organization's official test pilot.[45] Udet asked her to evaluate a Horten wing, and she flew the *H II L*, which bore the RLM registration code D-11-187, in November 1938.

"The takeoff," Reitsch said in her report, "is best carried out with fully-back control column until the aeroplane rises from the ground without change of incidence."[46] She warned that if the pilot did not follow this procedure, then the takeoff run became exceptionally long, in part due to poor undercarriage design. Reimar already understood this and would abandon the tandem wheel arrangement in subsequent designs. Reitsch could not measure dynamic stability because Horten had built the wing without using ball-bearing braces to support the rods connecting the control stick to the elevons. As a result, friction bound the rods and prevented the stick from returning smoothly to a neutral point after the pilot displaced it. Reitsch observed how wind gusts could set the elevons "fluttering," which she deduced, "is probably caused by the lack of static balance of the control surfaces. The [imbalance] of the controls also gives a feeling of lateral instability, which however does not appear in calm air."[47]

When she depressed the left or the right rudder pedal to deploy a set of wing spoilers, Reitsch discovered that the wing turned "immediately about both the vertical and longitudinal [axes]."[48] She continued, "If strong drag rudder movement is applied, maneuverability is good," but cautioned that turns "are impossible with ailerons alone and can only be made using the drag rudder." Reitsch concluded, "The response [to aileron control] is inadequate and unpleasant due to a large negative yawing moment [adverse yaw] which appears when the controls are displaced . . . The true bank cannot be easily obtained [with ailerons alone]."

Reitsch believed that her inability to retract the nose wheel after taking off could have aggravated the resistance to banked turns by pushing aft the center of gravity because the wheel and strut folded forward into the nose.[49] Reitsch did not question in her report whether her small size contributed to the aft center of gravity and exacerbated the wing's sluggish yaw characteristics, but Reimar had calculated the good handling qualities depended on the pilot weighing between 60 and 90 kg (133 to 199 lb.). Reitsch was just over 1.5 m (5 ft.) tall and weighed 45 kg (100 lb.).

Karl Nickel confirmed (and Heinz Scheidhauer later corroborated) the effects of small weight changes on the yaw characteristics of wings similar to the *H II*, during a series of flights made later during the war at the Hornberg glider school, using an *H III f* with a movable weight that Nickel could adjust fore and aft from the cockpit. The picturesque Hornberg airfield was located at the foot of a medieval castle in southern Germany, less than 95 km (60 miles) southwest of Stuttgart near the middle of the Black Forest.[50] British intelligence specialist Kenneth Wilkinson reported after the war that the Hornberg tests began after craftspersons added 35 kg (77 lb.) of ballast "at the back of the center section until the glider was only just flyable . . . A tube 2 m [6 ft. 6 in.] long was then put under the center section with a 10 kg [22 lb.] sliding weight which was kept forward for take-off and landing and moved aft for tests at height."[51]

Reimar had hired eighteen-year-old Karl Nickel in 1942 shortly after he graduated with a degree in mathematics. An experienced sailplane pilot, Nickel agreed with Horten's assessment of the cramped cockpit and poor view inside the prototype *H II*, saying, "to the side and backwards, nothing could be seen [and] I . . . remember quite clearly the unpleasant feeling of being locked up . . . during these flights with one's head inside the wing."[52] However, his account of wrestling the aircraft into a steep turn supports Hanna Reitsch's findings:

> But to my utter annoyance I found that it was extremely difficult to get the sailplane [to turn] "around the corner." . . . [for a right turn, for example, I had to] 1. Push the stick diagonally forward to the left. 2. Wait one second (more or less). 3. Pull the stick diagonally to the right and fully backwards. 4. As soon as the correct angle of bank is reached, [I] move the stick back to the center.[53]

Nickel later explained that these clumsy manipulations were necessary only when he wanted to execute the fastest possible turn.[54] He admitted that a pilot could execute a slower turn by gently moving the stick right to roll right, followed by a touch of drag rudder to overcome the resulting adverse yaw. Still, Nickel's account contradicts Reimar's claim that he needed no drag rudder at all to turn the *H II* in 1935.

Hanna Reitsch had noted at the beginning of her report that the *H II L* "possesses great static longitudinal stability and complete safety in relation to the spin," and she considered this finding important enough to elaborate further. "The aeroplane," during a stall, she said, "cannot by any sort of control movements be made to drop the wing or to spin. With the control column pulled right back the machine pitches slightly forward and sinks without reaching a speed of more than 90 km/h [56 mph]."[55] Nickel, too, had found the resistance to stalls remarkable. He observed how the wing tapered toward the wing tips at a ratio (8.75) higher than any other airplane flying during the mid-1930s. Combined with a leading edge sweep angle of almost 30°, many considered such an aircraft dangerous to fly and "preprogrammed" to wing tip stalls followed immediately by complete loss of control and a potentially fatal spin, "but," he said, "as practical flight tests showed, this bird was highly stable even under the most turbulent conditions. No wing tip stall, no nose-dive or rear-up stall occurred . . . [and] it could not be made to spin, no matter how hard one tried!"[56]

Reitsch considered the wing's stable nature a "great help" to pilots who depended on their instruments when flying sailplanes in IMC, where the moisture could freeze and accumulate on the airframe, rendering the primary flight instruments useless.[57] Pilots had been cloud soaring since the German pilot Max Kegel had discovered strong but turbulent lift after accidentally flying into a thunderstorm in 1926. Pilots won competitions and set altitude and distance records using cloud lift, but they also risked suffering vertigo and frequently encountered violent turbulence, which led officials in the USA to outlaw the practice in 1966, and most European countries to ban it during the 1970s.

To recover from the sudden disorientation of vertigo, the pilot of the *H II L* had simply to pull back the control stick as far as possible and move it fully in either direction, thereby partially stalling the wing and causing it to descend rapidly into clear air while the pilot maintained full control of the aircraft. The wing resumed normal flight as soon as the pilot released the controls. Using a reverse variation of this technique, Karl Nickel successfully *ascended* through more than a kilometer-and-a-half (almost a mile) of continuous cloud, born aloft upon a powerful thermal, in a wing similar to the *H II L* called the *H III f*. Nickel maintained control even when his vital turn-and-bank indicator ceased to function after entering the cloud.[58]

This capability to partially stall the wing while maintaining control was doubly remarkable for the devices Horten did *not* employ to achieve it; for example, the leading-edge slots and boundary layer fences that many designers of both all-wing and conventional aircraft used to smooth airflow and maintain control at slow airspeeds. Slots allow air to flow over a section of wing shaped to continue making lift as other parts of the wing stall. Fences generate a spinning vortex of air whose momentum resists the turbulent flow that leads to stalling.

The stall stability of Horten wings worked when three design elements integral to the bell lift distribution—forward center of gravity, reflex airfoil at the center section, and considerable wing tip washout—combined with the pilot's control stick manipulations—full back and all the way to one side to deflect one elevon full-up. Reimar takes no credit for purposely designing this characteristic into his wings, and it may simply have come about as an unforeseen but happy accident. Both Nickel and Walter Horten claimed that Reimar incorporated these safe-handling qualities into nearly all of the designs that followed the *H II*. Independent research has supported these claims.[59]

Reitsch also noted in her 1938 report that because the *H II L* had no conventional rudder and the drag rudders were relatively weak, compensating for a cross wind during landing by "side slipping cannot be carried out." She nonetheless made executing a no-crosswind landing sound easy. "Landing, even on a small field, is easily made by means of the landing flaps and use of the drag rudders on both sides so that they act as dive brakes. Landing run is normal." Reitsch graded the cockpit on . . .

> comfort—not exceptional . . . view is bad since the edges of the cockpit hood cut off the view at eye level; entry and exit—only possible for athletes; parachute arrangements—satisfactory; arrangement of the instruments—not very satisfactory; arrangement and operation of the retractable undercarriage—only possible for long-armed pilots.[60]

Although the wing's handling and performance did not completely satisfy her, she recognized its potential and ended her report by advising the brothers to continue developing the all-wing concept.

Working with Composites

At the end of 1935, nineteen-year-old Reimar got his first opportunity to earn a salary for his aeronautical engineering expertise. A chemist named Leysieffer, who worked at the Dynamit A. G. firm in Troisdorf, hired Horten to develop new composite materials for aircraft construction. These substances intrigued Horten because he saw their potential to improve all-wing aircraft. They promised to weigh less than traditional wood and aluminum components, and their finish could be much smoother and produce less drag. Leysieffer offered Reimar his own workshop and half-a-dozen laborers, and he agreed to let Reimar design and build a new wing using the materials he would help to develop.[61]

In February 1936, after some preliminary experiments, Horten used two similar composite materials to build much of the internal structure of the wing for a popular training glider called the *Hol's der Teufel* (The Devil Take It!), designed by Alexander Lippisch. The composites that Reimar used were made of paper cellulose filler blended with phenolic resin derived from plants, compressed to about 200 atmospheres (2,940 lb./ft.^{2}), and then heated to approximately 140°C (284°F) for many hours.

Because heat and pressure shaped the compounds, the industry considered them "plastic" according to the terminology of the era, but their composition more closely resembled the composite Bakelite developed in 1909. The new substances proved impervious to both fuel and water and very smooth. Except in durability and strength, they foreshadowed the fiberglass-reinforced plastic that revolutionized the design and construction of sailplanes, and many powered aircraft designed for construction by amateurs, after World War II. Dynamit named these substances Mipolan and Astralon.[62]

Reimar tried to improve Mipolan and Astralon by experimenting with different resin/filler ratios to find a mix that would boost the gluing properties without degrading the resistance to moisture and fuel. Leysieffer's inability to produce uniform sheets of the plastic complicated Reimar's work. There were variances in color, and the varnish did not retain a uniform thickness. Horten conducted "thousands of tests of strengths, of bending, of gluing the material" to measure the compression, shear, and crack propagation properties. The work required considerable effort, which he viewed as wasted time because it did not directly relate to all-wing aircraft.[63]

Despite these exhaustive tests, Dynamit A. G. found it difficult to raise the quality enough to mass-produce finished composite airframes, but the firm eventually succeeded in developing a wing for the *Hol's der Teufel* that weighed 15 percent less than a wooden wing but remained just as strong. Reimar finished building the glider and flew it in May 1936. Technicians then removed the wings and tested one section under increasing loads until it failed. They suspended the other section outdoors for six months to gauge its susceptibility to the weather.[64]

Satisfied that Mipolan and Astralon showed enough promise to continue testing, but impatient to return to developing his wings full-time, early in 1936 Reimar began building an all-wing aircraft using the composite compounds that Walter once likened to "hard paper."[65] As agreed, Leysieffer furnished funds and support, but he encouraged Reimar to develop an advanced, two-engine wing called the *H V a*, rather than a more conservative design. Reimar preferred a simpler design involving less risk, and he hinted at diverging design philosophies between himself and the company management. He had already begun laying out the *H III* all-wing sailplane when the company's eagerness to recoup their investment (forty thousand marks, according to Reimar) in composites research seemed to find its ideal expression in an opportunity to build the *H V a* to fulfill a military requirement.

The Luftwaffe had called for a new bomber escort and multirole heavy fighter equipped with a powerful defensive armament. Military planners believed that the gunners aboard groups of these "fighter destroyers" could wield power-driven turrets to protect a formation of bombers. Reimar reluctantly offered the all-wing *H V a* to fill the role. By placing the turret behind the cockpit and between the two pusher propellers, the gunner would have a clear field of fire aft, unobstructed by a conventional fuselage.[66]

Reimar planned to install a single seat in the *H V a* prototype but added another, probably under pressure from the Dynamit A. G. administration. From a marketing standpoint, two seats were more attractive. Experiencing the wing in flight as a passenger was more likely to impress a procurement official than any spoken or written description of the airplane's performance. Horten made the cockpit deep enough to accommodate the occupants in reclined seats, but he provided no canopy and space was tight. To allow the crew to see outside, he wrapped the leading edge of the center section with clear plastic, following the method used on the *H II*.

Reimar undoubtedly grasped the importance of winning a contract; however, he also remained determined to study new ideas. To better understand and ultimately solve the perplexing Mitteneffekt, Horten fundamentally altered the *H V a* layout compared with previous designs by adopting a double-swept wing plan form, sweeping the leading edge of the center section 25½° and the leading edge of the outer wing panels 40° (Figure 12). He installed landing flaps beneath the center section to measure their effectiveness and prove the devices worked on all-wing aircraft. The Luftwaffe may also have requested flaps to slow the landing speed to meet operational requirements.[67]

Dynamit had manufactured the earliest composites in sheets from which Reimar and his team then cut out and glued together the ribs, stringers, spars, and other components, which they then assembled to build the wing. Technicians apparently formed the components used to construct the *H V a* using molds. Despite an intense development period, the compounds remained brittle and sensitive to cracking in areas where the torsion and bending stresses were high, forcing Horten to use rivets, bolts, and steel wire reinforcements. To build the landing gear and to mount the engines, he had to discard the composites altogether and revert to steel. The two propellers were formed

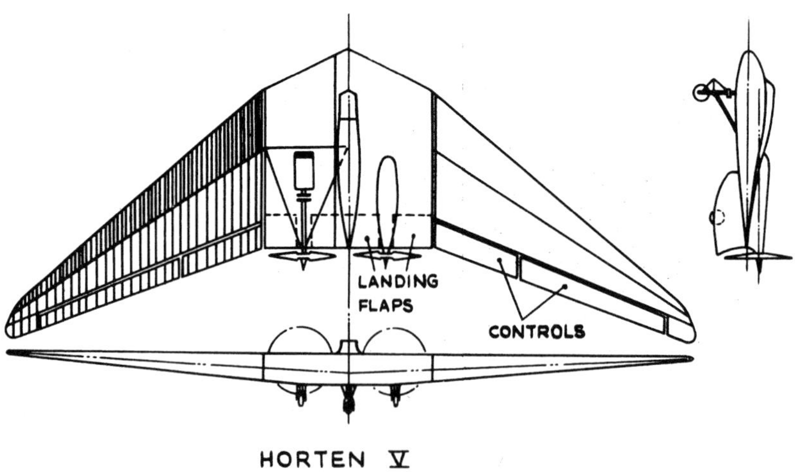

Figure 12. Horten *H V* three-view drawing (Appleyard and Biot, "The Horten Tailless Aircraft," 43).

from Lignofol, a composite composed of "layers of highly compressed beech wood soaked with synthetic resin" and cured under high pressure.[68]

Reimar and his team began building the *H V a* in February 1936. The work progressed well until the end of the year, when Reimar received orders sending him to Berlin for basic training in preparation to join the Luftwaffe. Work almost stopped with his departure, and the Dynamit workers sent postcards to Horten, asking for advice on various aspects of the build. Leysieffer soon arranged to bring him back just before Reimar began a second course of training. He then transferred to the inactive Luftwaffe reserve and returned to the *H V a* overjoyed at the reprieve but dismayed that military planners failed to comprehend the promising nature of his work. "I thought . . . [that by building and flying the *H V a*] I could demonstrate that I am doing new things [in aeronautics] and therefore, I wouldn't have to go into the air force, that I could be free of going into the military in order to concentrate on the experiments."[69] The Luftwaffe discharged Reimar from the reserves in April 1938, but he returned to active duty when the war started in 1939, and rose to the rank of Oberleutnant (1st Lieutenant) before the conflict ended.

Reimar's accounts of the first attempt to fly the *H V a* in May 1937 have varied through the years, but it is clear that the resulting crash, which destroyed the *H V a*, easily could have killed both brothers. Two reliable sources say that poor roll control following the loss of one engine during the takeoff run caused the accident; however, Reimar cited the root cause as his miscalculation of the center of gravity. He had based it on the *H II* but did not grasp the error until the takeoff commenced. Reimar says that he shouted to abort the takeoff, but his brother, who was controlling the aircraft, inexplicably reduced power on only one engine, causing the wing to cartwheel.[70]

A series of minor mistakes and setbacks underlie most aircraft accidents, and this one was no exception. A reliable method for gluing the composite parts still eluded Reimar and the chemists at Dynamit A. G., but the firm had pressured him to fly the *H V a* anyway. The small cockpit, with reclined seating and reduced visibility, may have made the wing harder to control during takeoff. Despite its flaws, the *H V a* represented several milestones: the first aircraft built primarily of composite materials (Figure 13), and Reimar's first multiengine, multiseat design.[71]

The crash climaxed a difficult development period and led Reimar to evaluate the project somewhat harshly. From the *H I*, he had moved immediately to the *H II*, which he should have followed with the *H III*, but pressured by Dynamit, "I had gone to the [*H V a*]. I made a great step . . . and it was a little too much. I needed two years more to be good with this plane."[72] The

Figure 13. Extremely smooth wing skins made of composite materials covering this *H V a* wing are recorded by a camera carefully positioned by the photographer inside this hanger at Troisdorf near Bonn (Reimar Horten via Peter F. Selinger).

experience chastened him to incorporate fewer new ideas into a single airframe and to test them thoroughly before beginning a new project, no matter how restless his advocates might become with the pace.[73]

A postwar study noted that Germany developed several composite substances suitable for making aircraft tooling and structure. The authors of the study judged, "Dynal, a plastic impregnated paper pulp product made by Dynamit A. G., Troisdorf . . ." to be very good.[74] Reimar probably played no direct role in developing Dynal, but his work to develop and use composites to build aircraft helped lay a foundation for more widespread use.

Reimar believed the *H V a* to be aerodynamically sound. When he left Troisdorf during the summer of 1937, Horten traveled to Cologne, capital of the Rhineland province and a major cultural and trading center, and began building the *H V b*, essentially the *H V a* with a more conventional control arrangement and tricycle landing gear.[75] He increased the wingspan slightly and used traditional wood and metal components, rather than composites, and this wing turned out "very well."[76] Reimar and Walter found that it handled nicely in the air and agreed that every landing could be a smooth one by using simple and easily mastered flying techniques. Walter considered the aircraft underpowered, but by this time, he was more accustomed to thc higher power-to-weight ratio found in fighter aircraft. He showed the *H V b* to authorities at the RLM in Berlin late in 1938, but to his consternation, they dismissed it as amateurish.[77]

The wing remained stored in a hanger until the war started, and ground personnel had to push it outside to free space for warplanes. Two years' exposure to the elements took their toll, and technicians employed by the Peschke Company at Minden had to rebuild the aircraft to flyable condition. The RLM granted permission to make the repairs in August 1941. Under Reimar's guidance, the craftspersons converted the aircraft into a single seat configuration, designated the *H V c*. Walter flew the wing to Göttingen for testing, where Reimar also took the controls and ascended in "this wonderful little machine" in 1942.[78]

CHAPTER 3

Competition Wings

FOLLOWING THE *H V a* crash in spring 1937, Reimar Horten remained on inactive reserve and free to work without the interruption of military service. The desire to enter a new design in the national soaring championship in 1938 helped drive him to press on with his fourth all-wing design, the *H III*. Horten's reputation might rebound if the wing did well in the contest and perhaps encourage another sponsor to back his work. Major Dinort again used his authority to provide funds, personnel, materials, and shop space to build the aircraft at Cologne.[1]

No doubt recalling failure of the overly ambitious *H V a*, Reimar laid out the *H III* by simply increasing the span and aspect ratio of the *H II*. He kept the same airfoil arrangement; a reflex profile across the center section that transitioned into a symmetrical profile at the tip. Likewise, he used the same thickness ratio, 20 percent at the root and 10 percent at the wing tip. He widened the span about 20 percent more than the *H II* and narrowed the root chord, thereby increasing the aspect ratio 27 percent. These adjustments increased lift across the wing without paying an exorbitant drag penalty (Figure 14).

Horten divided the elevons into two pairs, and he scheduled the inboard set to move up less and down more when they functioned as ailerons to help reduce adverse yaw. This suggests that he had not entirely solved the adverse yaw problem. Reimar also scheduled the pair to move in unison as an elevator

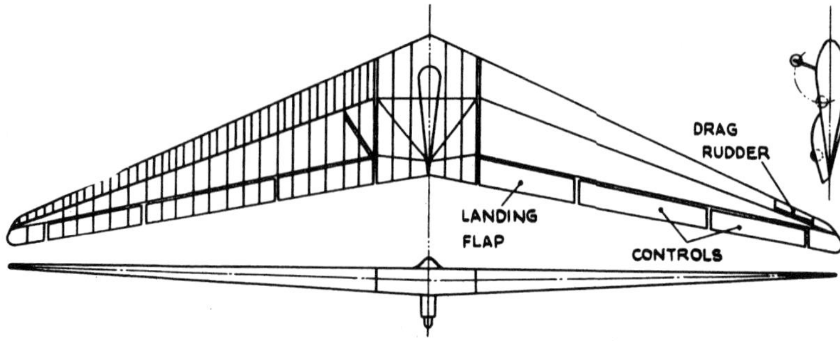

Figure 14. Horten *H III* three-view drawing. Arrangement of landing flap and other controls varied between *H III* subtypes (Appleyard and Biot, "The Horten Tailless Aircraft," 42).

to control pitch, and he gave them more travel than the outboard elevons. This measure reduced the torsional bending loads on the wing tips at high speeds. Hanna Reitsch had been particularly harsh when she described the rough and binding elevon controls on the *H II*, so Horten suspended on ball bearings all of the control rods that connected the pilot's yoke to the surfaces.[2]

The first two *H III* wings were under construction when Reimar decided to modify one of them by mounting a small Vorflügel (forewing), spanning about a meter, on struts above the nose (Figure 15). The pilot could pivot the device up and down to augment the pitch authority vested in the main elevons. If it functioned as Reimar intended, the Vorflügel would delay the onset of turbulent airflow, and loss of lift, over the thick center section and allow the pilot to turn more tightly inside a thermal. Horten also hoped the Vorflügel would help him understand the cause of, and solution to, the Mitteneffekt. He changed the designation to *H III c*, and a pilot first took the aircraft aloft on May 7, 1938. The second *H III* (the first *H III a*) was finished shortly before the contest began in August, with enlarged elevons to reduce adverse yaw during flights with the CG near the aft limit. The *H III* became the most successful of Reimar's designs, in terms of total numbers built (nineteen), and the number of subtypes he developed from the basic design (seven).[3]

Assembling the Team

Supervising construction of the two *H III* wings and other projects left Reimar no extra time, and Walter was away frequently serving in the Luftwaffe. Consequently, the brothers were too busy to practice flying in preparation to

Figure 15. A pilot could adjust the angle of attack, and consequently the lift, provided by the Vorflügel (forewing) mounted above the nose of the *H III c* (National Air and Space Museum, Smithsonian Institution, SI 98-15997).

compete in the 1938 national competition. Walter therefore became the team manager while Reimar assumed no official role other than to observe and advise. Team Horten needed new pilots. One day, a young, noncommissioned military pilot named Heinz Scheidhauer wandered into the Horten workshop at Cologne and began chatting with Reimar. He impressed the designer with his Silver C badge, the mark of a skilled soaring pilot. To earn the badge, Scheidhauer had to fly 50 km (32 miles), gain 1,000 m (3,200 ft.) above his takeoff point, and remain aloft at least five hours. A pilot could fulfill no more than two of the requirements during one flight.[4]

Scheidhauer introduced Reimar to three other pilots who wanted to join the team, too. Dinort added a fourth, but because a sponsor could field only two pilots, Reimar and Walter had to devise a soaring challenge that would eliminate the two least qualified. The men had no experience flying all-wing aircraft, and without some basic training to prepare them, a fair assessment was not possible. With just weeks remaining before the contest, Reimar and Walter were able to tow the men aloft in one of four available wings enough times that each compiled about fifty flying hours, frequently venturing cross-country beyond the home airfield while in IMC flying.[5]

Now the men had sufficient experience to face the challenging elimination test that Reimar had devised. He asked each man to begin from a reasonable height of his choice and fly just 20 km (12 miles) from Bonn north to Cologne. Such a jaunt was no test at all in calm conditions, but it was May, and a steady wind blew from the north. Low wing loading made the *H II* and *H III* good "floaters" that permitted pilots to soar in weak lift conditions and cover respectable distances as long they drifted *with* the prevailing wind. Flying a reverse course was altogether different. Only a skilled pilot could make headway against the wind to achieve cross-country distances.

The pilots would fly three *H II L* wings and the *H III c*, which Reimar thought performed about evenly with the older sailplanes when flown against the wind. Walter towed the four pilots aloft in rapid succession, and without landing, continued on to his airbase in Cologne. After just a half-hour, the pilot of an *H II L* called to report landing near Bonn. Another candidate called three hours later, having doggedly kept his *H II L* in the air over Bonn, trying repeatedly to penetrate into the wind toward Cologne without success. Scheidhauer and Werner Blech stayed airborne, slowly approaching their goal. Both pilots found strong lift emanating from chimneys at the coal-fired power plant near Cologne. By circling within these clouds of warm, sooty air, which were noxious enough to sicken them, they spiraled up to 3,000 m, nearly 10,000 ft.

Blech landed the new *H III c* in Cologne more than six hours after taking off. He had won a place on the team using the experience and knowledge of weather gained while working as a meteorological survey pilot. Blech had also competed in the 1935 national contest, where he placed thirty-second among the field of sixty-three pilots. He improved the following year by finishing fifteenth out of fifty-eight competitors and set an altitude gained record for climbing 3,530 m (11,578 ft.) after takeoff. Long after Blech landed, Scheidhauer finally touched down in an *H II L* on Cologne's east side early that evening after soaring for about nine hours. He had flown the longest distance by remaining aloft until the wind slacked off near dusk.[6]

Reimar had supplied each pilot with the best speeds to fly each type of wing, depending on whether the surrounding air was calm, sinking, or rising. For the *H II L*, he calculated a pilot could fly the greatest distance by maintaining 72 km/h (45 mph), 18 percent faster than the best cruise speed he had calculated for the *H III*, a wing with more area and a lower wing loading. Yet the pilots discarded Reimar's calculations during the competition, relying more on their own skills and experience to plot contest strategy and tactics.[7] Scheidhauer complained that Reimar had based his calculations on optimistic L/D figures derived from Lippisch's method for determining lift distribution, which made no allowances for wing sweep. Karl Nickel and Hermann Strebel later claimed the *H III* had a 24:1 L/D, not the 33:1 that Reimar claimed at that time.[8]

The two new *H III* wings both went to Scheidhauer and Blech. Based on his meteorological expertise, Blech concentrated on altitude events while Scheidhauer worked the distance tasks. Dinort wanted more Horten wings competing and formed another team staffed with personnel from a Luftwaffe unit based at Fürth. The team's pilot, Kurt Hieckmann, drew an *H II L* for the meet.[9]

Compared to the previous year, weather conditions produced strong lift during most of the 1938 Rhön contest. Six pilots put up respectable scores on all but one contest day, and the top two finishers flew every day from July 24 to August 6. Team Horten began with promise when Blech placed in the top five during the first day with a height gain of 2,020 m (6,626 ft.). He ascended to the second highest altitude the next day, 3,650 m (11,972 ft.), and he ended the third day ranked number three for gain of height. Blech did well, in part, because the *H III's* low wing loading allowed him to turn within a tighter radius and climb within smaller thermals than any of the tailed sailplanes. Conditions deteriorated near the contest midpoint, and many pilots, including Blech, chose not to fly for several days. When the weather improved, he immediately set several respectable distance marks and a string of good altitude flights, including a spectacular gain of more than 5,000 m (16,400 ft.) when Blech reached an absolute altitude of 6,380 m (20,926 ft.) the day before the meet ended.[10]

August 6 was the fourteenth and last contest day, the final opportunity for Team Horten to improve its standing. Scheidhauer had remained near the bottom of the leaderboard during the entire meet, not least because he had inexplicably refused to fly on seven of the 14 contest days. Perhaps determined to do better, he and several other pilots misjudged a large cloud as it materialized near the mountain during the afternoon. Competition pilots had routinely followed lift into clouds since the late 1920s, but the technique was risky because there was no way to determine whether the cloud was about to develop into a raging thunderstorm.

Tugs hauled aloft Scheidhauer, Blech, and the others. They quickly found lift and began to disappear one by one up into the vapor when a crack of thunder rolled over the mountain. Meteorologists had not forecast severe weather, but the cloud was fast becoming a strong storm. Turbulent updrafts drove the sailplanes so high that several pilots lost consciousness in the freezing air. Hailstones shattered the canopy over Scheidhauer's head and tore holes in the wing's plywood skin, forcing him to bail out. The storm caught his parachute, and he went missing for hours until a search party discovered the unconscious flyer hanging from a tree near Wüstensachsen, 6 km west of the Wasserkuppe. The flyer's right thumb and index finger were black with frostbite.

Scheidhauer's wing had crashed nearby, and technicians sifting the wreckage recovered the barograph installed in all the gliders before a flight. The barograph log showed the wing rising to 8,000 m (26,240 ft.) and still climbing until the ink nib had dropped off the side of the recording drum. Witnesses saw Blech's wing plummet out of the clouds, his parachute streaming behind the stricken craft, and smash heavily to earth, but they found no pilot amid the wreckage. His body fell about 5 km west of the mountain near Poppenhausen. Officials theorized that he probably died from hypoxia or a broken neck when violent gyrations tossed him from the cockpit.[11] His barograph tracing had also tracked off the recording paper, so there was no way to determine the maximum altitude he had reached. The storm also claimed the life of a pilot flying a conventional sailplane and forced another to take to his parachute.

Both Horten wings bore the marks of intense bombardment by hailstones, which had passed easily through the thin plywood and cotton cloth covering the airframes. The updrafts had propelled the aircraft deep into the bowels of the storm, where their light wing loadings made them vulnerable to the strong winds. The assumption that pilots who got into trouble while flying in clouds had simply to pull back and hold their control wheels to invoke the wing's well-mannered stalling characteristics—a rapid, stable, controlled descent—may have encouraged Blech and Scheidhauer to enter the storm. Too late, they realized that its power far surpassed anything the fragile aircraft could escape or withstand.

Neither Blech nor Scheidhauer earned points for their tremendous ascents because they had bailed out. For losing two fingers from his right hand, Scheidhauer wound up in thirty-ninth position while Blech posthumously finished twenty-eighth, out of a total complement of fifty-four. Hieckmann, piloting the *H II L* for the Luftwaffe team from Fürth, landed 21 km (13 miles) from the Wasserkuppe, and climbed no higher than 500 m (1,640 ft.), during the single flight that he managed to complete on the fifth day of competition. Perhaps to compensate for his weak performance, Reimar recalled that Hieckmann performed a series of dramatic aerobatics while piloting the *H II L* above a crowd that gathered the day after the contest ended. That his rivals in the NSFK captured nearly all of the top ten positions must have stung Dinort.[12]

Scheidhauer had knocked himself far down in the standings when he stayed on the ground instead of flying, but it was not solely his lack of aggressive piloting that had caused Team Horten to perform so poorly. Technological innovation, and Reimar's strange delay in adopting it himself, played a greater role. While Reimar had continued to design for low wing loadings, the designers of tailed gliders pursued a trend that had started ten years earlier when sailplane pilots discovered that higher wing loadings allowed them to fly faster with no loss of lift. By speeding more quickly through the sinking air that prevailed between lift sources, they wound up with more altitude when they located the next source of lift and could resume the process of rebuilding altitude from a higher starting point. Consequently, they spent less time climbing, more time cruising for distance, and could also fly faster into a prevailing wind, like sailors tacking their sailboats into a freshening breeze.[13]

The wing of the winning sailplane in 1938 lifted 17 kg/m^2 (3.5 lb./ft.^{2}), while the tenth place finisher lifted 13.8 kg/m^2 (2.8 lb./ft.^{2}). By contrast, the *H II* lifted 10.3 kg/m^2 (2.1 lb./ft.^{2}), the *H III* even less at 8.3 kg/m^2 (1.7 lb./ft.^{2}), or nearly half the winner's wing loading! It may have been Reimar's strategy all along to concentrate on climbing. The meet organizers specifically encouraged climbing for points, as Reimar recalled in 1951:

> Marks were given for height without distance if the climb exceeded 2,000 meters [6,560 ft]. This means that there was a chance to score in the contest without leaving the site, since it [was] quite possible to repeat these altitude flights several times on the same day.[14]

Reimar may have hesitated to abandon low wing loading design because he felt close to reaching its ultimate potential with a wing called the *Parabel*, under construction in 1937 and completed in 1939. The wing did not fly, but Reimar calculated that if it had, the *Parabel* would have exhibited one of the lowest wing loadings of any cantilever airplane ever built, and its climb rate would have exceeded any conventional type of sailplane.

Emphasizing climb instead of cruise also favored those who handled the wings on the ground by eliminating the time and labor spent recovering the all-wings after a cross-country flight. When flying for altitude, the Horten pilots were likely to remain close enough to simply return and land at their launch point atop the Wasserkuppe, where technicians could have the wings ready for another flight in a few minutes. Disassembling a Horten wing and returning it to the airfield following a cross-country flight was arduous work and could require many hours, if not days, while the ground crew struggled to disassemble the aircraft, move it back to the Wasserkuppe, and then reassemble it for flight. Even when broken down into a center section and two outer wing panels, the size and weight of the components made them nearly impossible to retrieve quickly or easily. The *H II* root wing chord measured 3.5 m (11 ft. 6 in.), and the center section was 2.4 m (7 ft. 10 in.) wide. The *H III* was the same dimension across the center section while the root chord was slightly smaller, and each of these major components weighed hundreds of pounds. Conventional sailplanes could be disassembled into more individual parts that were smaller, lighter, and less bulky. As a result, moving them back to the mountain for another flight proceeded more quickly.

Despite the poor results for the men flying Horten aircraft, the 1938 contest organizers made Reimar and Walter the first recipients of the Otto Lilienthal Prize for technical innovation, which included an educational stipend of five thousand marks. It was enough money to send them to the Berlin Technische Hochschule (technical university) in Charlottenburg, where they studied engineering for two semesters until the beginning of the war in September 1939. Walter again sensed Udet's intervention to place the boys at the prestigious school, but Reimar found the curriculum lacked a focus on aviation.[15]

Surge in Production

Publicity from the 1938 Rhön meet, and perhaps Hanna Reitsch's report about the *H II* (see Chapter 2), which began circulating in December among government and military aeronautics officials, led in 1939 to a burst of *H III* production across Germany. It began when a senior official named Hans Helbig in the Reichsministerium für Wissenschaft, Erziehung und Volksbildung (Science and Education Ministry) invited the brothers to visit the ministry's workshop near the Tempelhof Airfield during fall 1938. Here they soon began working with friends to construct the outer wing panels for at least one, and as many as three, *H III b* sailplanes. Other craftspersons were soon building center sections in southern Germany and additional wing sets at several locations around Bavaria. A group of Luftwaffe personnel in Fürth built one *H III*, according to Reimar Horten, and the DVL acquired two more of these aircraft with funds from the Science and Education Ministry. It seems that Helbig played a role in much of this work. Altogether, at least nine *H III b* wings were finished in 1939.[16]

Between May 1938 and July 1944, craftspersons built more *H IIIs* than any other type of Horten aircraft. This was partly the result of good support from Reimar to the various groups building the wings, but also because the type's reasonable performance and manageable handling characteristics attracted pilots. Reimar claimed for the *H III* a respectable 28:1 L/D at a sinking speed of 0.48 m/sec (1.6 ft./sec.), while others reported a more conservative 24:1 L/D at 0.64 m/sec (2.1 ft./sec.). One knowledgeable expert concluded the light wing loading (8.3 kg/m^2 (1.7 lb./ft.^{2}) enabled a pilot to turn the *H III* more tightly than any sailplane equipped with a tail, a true statement when compared with the wing loads carried by four conventional sport sailplanes similar to the *H III*.

Like the Horten wing, the Göppingen 4 Gövier from Germany, Slingsby Kirby Gull 1 from England, and the Bowlus Baby Albatross and Schweizer SGS 2-8 from the USA were sport sailplanes designed to carry one or two persons and produced in modest quantities. The Baby Albatross had the lowest wing loading of the group, but still nearly double the *H III*. The all-wing aircraft fell about midway among the four conventional sailplanes for L/D and minimum sinking speed.[17] When flying for distance in calm or light winds, or on a downwind course, an *H III* pilot could expect a fairly long flight, and of course, good climbing performance due to the light wing loading.

Maintaining Independence

In autumn 1938, Udet brokered a series of discussions between the Horten brothers and Ernst Heinkel, who had been designing advanced and influential aircraft since World War I. Heinkel's factory now employed thousands, and Udet may have hoped that pairing the young designer with a more experienced manager would make Reimar more productive. Reimar seemed interested, and talks continued for more than a week, but his preference to work as a partner, rather than just an employee, doomed the prospect of collaboration. It seems likely that Walter conspired with Udet to place Reimar with Heinkel, as it was he, and not his brother, who met personally with Heinkel.[18] Following Hitler's seizure of Austria and the crisis in Czechoslovakia in March, Walter may have sensed the inevitability of armed conflict between Germany and her European neighbors, and that when it began, Reimar was a choice candidate for conscription into the army unless already employed in a job related to the war effort.

Reimar attributed Heinkel's failure to offer him an equal partnership to his lack of advanced credentials. "With my PhD," he said, "I could enter in the design studios of a factory and the people there would not say 'he is only a boy designing airplanes.' Instead, this is Dr. Reimar Horten. I would have a [respectable] position."[19] Reimar's sister, Gunilde, also believed that Heinkel took Reimar's lack of credentials into account when evaluating him for employment.[20] A degree meant little in the small workshops and airfields where Horten aircraft were built and tested. It meant a great deal to the professional designers and managers working in the German aircraft industry.

On the other hand, Reimar relished the degree of control and autonomy that came with working alone, but his experiments now required more resources.

> Every man wants to be independent and free . . . [but] if Heinkel would have given me a position [and provided] the men I needed, this then would have been better for me than being by myself and independent. More things could have been accomplished. More experiments could have been conducted and carried out. Development would have gone faster. To build and to test is very expensive.[21]

Reimar recalled vividly years later the lost opportunity, but he also realized that Heinkel could have terminated his research once Reimar had joined his company. The man had designed many exceptional aircraft during his long career, but he never showed any interest in tailless aircraft.[22]

With several important projects underway by 1939, there was no time for Reimar to maintain even basic flying skills, let alone perform the demanding and lengthy process required to test his wings. It was time to take on a dedicated test pilot, and Reimar selected Heinz Scheidhauer.[23] Reimar also assigned Scheidhauer to lead the team competing in the 1939 national contest held July 23 to August 4. He practiced for several months before the meet, but he could not finish higher than twenty-second among forty-one competitors. His one exceptional distance flight to German-occupied Czechoslovakia, which exceeded 340 km (211 miles), actually cost him four crucial contest days waiting for the ground crew to retrieve the wing back to the Wasserkuppe. Scheidhauer claimed that a promising altitude flight ended prematurely because he could not activate his oxygen supply due to the hand injury suffered during the thunderstorm flight the year before. The other three team pilots, flying *H III b* aircraft built earlier that year, also fared poorly.

This was the last time anyone flew Horten wings with light wing loadings in competition; however, Reimar had planned to continue experimenting with the concept, using a remarkable design not fated to fly but still deserving mention. He had started thinking about this aircraft four years earlier in 1935. His work with the two-engine *H V a* had led him to surmise that its double-swept leading edge (25.5° across the center section and 40° along the outer wings) substantially reduced the Mitteneffekt. This apparent breakthrough provided additional incentive to build a new sailplane he called the *Parabel* (Parabola).[24]

Reimar directed his staff to keep airframe weight as low as possible by building the wing using only the minimum amount of glue to fuse very thin and lightweight wooden components. The wooden covering applied to the wing leading edge, for example, varied in thickness from 0.6 to 1 mm (0.02 to 0.04 in.). In another attempt to reduce the Mitteneffekt, Reimar applied a variation of the double-swept, or kinked, wing he had tried on the *H V a* by curving the *Parabel* in one grand and graceful arc from wing tip to wing tip (Figure 16). The architecture was visually striking and difficult to build. Horten estimated that his workers spent 25,000 hours constructing the sailplane from 1937 until they finished in 1939. It is not known who paid for this considerable effort, but the result was a noteworthy expression of lightweight design. Compared to the heavier tailed sailplanes with higher aspect ratio wings, and wing loadings that averaged 18 kg/m^2 (3.7 lb./ft.^{2}), each square meter of the *Parabel* would have lifted just 5.15 kg/m^2 (1 lb./ft.^{2}).[25]

The *Parabel* tipped the scales at a mere 89.1 kg (198 lb.), or about 45 kg (100 lb.) more than most hang gliders that began flying in the early 1970s. Hang gliders became quite popular and many thousands were built, in part because most had nearly the same low wing loading as the *Parabel*. The designers of hang gliders achieved this low value by using materials and techniques not dreamed of in Reimar's time: synthetic fabrics applied over frameworks made of aluminum tubes bolted and riveted together, braced with wires exposed to the slipstream, and only lightweight fabric hammocks, or cocoons, to support the pilot. Working before World War II limited Reimar's choices to natural fabrics, wood, and metal, yet he was able to give the *Parabel* a

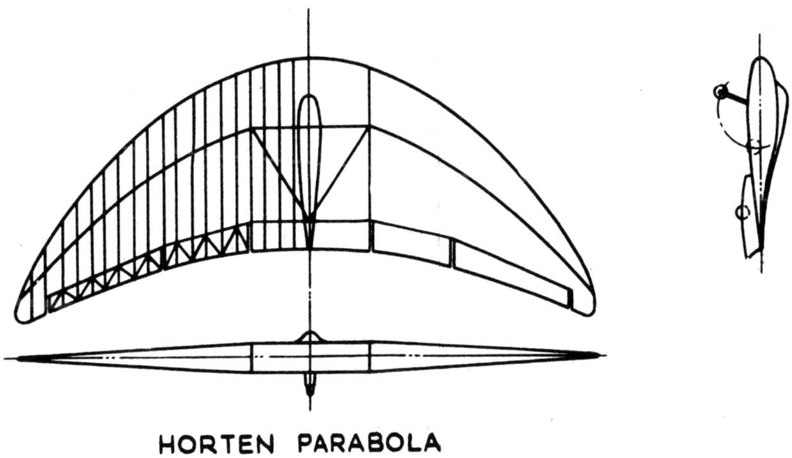

Figure 16. Horten *Parabel* three-view drawing (Appleyard and Biot, "The Horten Tailless Aircraft," 43).

smooth, cantilever structure not burdened by the drag of external supports and equipped with a cockpit that completely enclosed the pilot.

The control layout resembled the system fitted to the *H III*, except that Reimar split the *Parabel* elevons along their trailing edges so that they also doubled as drag rudders. When the pilot deflected the elevons to roll the wing into a turn, the control surface on the wing tip to the inside of the turn deflected upward to drive the wing tip down, but it also split in half, the top half angling into the airflow much more than the bottom half. The elevon outside the turn deflected down without splitting. Both elevons were supposed to move simultaneously to climb, but Reimar used a set of flaps mounted inboard of the elevons to drop the nose and dive.

Using elevons to climb and flaps to dive was another attempt Reimar used to further reduce drag due to control deflections. Raising the nose required more force, and a greater control deflection and hence more drag, than pitching it down. So Horten relied on the wing tip elevons, positioned farther back from the CG due to wing sweep. The greater distance increased the moment arm and hence, the leverage of the elevons. With gravity helping, less force was needed to drop the nose, so Reimar used the inboard flaps for this purpose since they pivoted more closely to the CG.

The *Parabel* spanned only 12.2 m (40 ft.), and a pilot maneuvered it using large elevons for the wing's size. Consequently, Reimar expected the wing to turn about a very small radius. The combination of low weight, light wing loading, and agility would have made the *Parabel* formidable in climbing contests. Theoretically the wing could fly in conditions that grounded conventional sailplanes, but without flight tests to confirm whether the structure could withstand the turbulent air usually found in thermals, there is no way to know how practical the *Parabel* would have been for competition. The start of World War II in 1939 prevented flight tests. The wing was stored improperly, and the thin, wooden structure warped enough to render the glider unsafe for flying, so Reimar burnt it to ashes in 1940. Had it flown successfully and safely, the *Parabel* might have encouraged Horten to experiment further with light wing loading designs.[26]

War

World War II began less than a month after the 1939 Rhön contest ended. Walter Horten left Berlin and returned to Cologne to join JG 26. He eventually became the unit's technical officer early in 1940 and held the post until the end of 1941.[27] Walter was a skilled and trusted pilot, and wing commander Adolph Galland frequently made Horten his wingman during combat missions. Walter was one of the few to survive the assignment, which he said cost twelve other pilots their lives from August to September 1940 at the height of the Battle of Britain.

Reimar's numerous projects underway during the summer of 1939 must have distracted him from noticing the escalating words and deeds that climaxed in war on September 1, 1939, because Hitler's invasion of Poland surprised him, as it did many other Germans. The war abruptly ended his studies at the technical school in Charlottenburg and curtailed the all-wing work for some months. The Luftwaffe ordered Reimar to Halberstadt for an accelerated program of flight training shortly after the war began. By flying whenever the weather allowed and remaining in the air for most of the day, he earned a basic military flight rating by summer. A transfer to Fürth followed, and Reimar briefly trained there at the fighter pilot's school.

In spring, the Luftwaffe began preparing to support the invasion of England. The Horten family suffered its first wartime loss when eldest child Wolfram died during the night of May 20, 1940, after his Heinkel bomber exploded near Dunkirk. Reimar's superiors ordered him to the Luftflotte (Air Fleet) Glider School #1 at Braunschweig in August to modify sport gliders to carry ammunition supplies to the invasion troops. He worked on both conventional sailplanes and his own wings.[28]

Five months later Hitler canceled the invasion of England when the Luftwaffe failed to achieve air supremacy over the country. Reimar then transferred to the glider pilot school at Königsberg-Neuhausen to train about sixty pilots a month to fly transport gliders at night, in tow behind three-engine Junkers Ju 52 transports. The syllabus required that he make about three hundred training flights a month. The air temperature was far below zero at the altitudes where Reimar conducted most of the training, and the gliders lacked heaters. He caught a few hours of sleep during the day but devoted most of his off-duty time to design work and managing the workshops engaged in building his sailplanes. Walter's career also shifted dramatically from the cockpit of a fighter to a desk in Berlin when he left JG 26 at the end of 1940 to join the staff of General Kurt-Bertram von Döring, in charge of the Inspektion der Jagdflieger, the unit within the RLM responsible for the readiness, training, and tactics of the Luftwaffe fighter aircraft forces. These assignments separated the two brothers for six months.[29]

High-Aspect Ratio Wings

Between 1929 and 1932, Reimar had sketched high-aspect ratio wings, and even tested models of them, but when he built the *H I* in 1933, he reverted to a lower-aspect-ratio configuration based on the proven Lippisch Delta I. Horten used this configuration to design the *H II* and *H III* because it provided a sound basis for testing innovations in aerodynamics, cockpits, and landing gear and because it allowed the wings to fly well enough, and often enough, both in competition and in more relaxed settings, to keep Reimar's work visible to the flying community.[30] In August 1939, pilots flying the conventional sailplanes with higher aspect ratios and heavier wing loadings again won the national competition held atop the Wasserkuppe, and for the third straight year they soundly beat a group of pilots flying Horten wings. This loss, plus growing confidence in Scheidhauer's piloting skills, finally motivated Reimar to break boldly from the old configuration and design and build a new wing of much higher aspect ratio and wing loading.[31]

In July 1940, Hitler ordered the German military to prepare to invade England. Reimar was posted to Braunschweig at the time, and between that month and September, he labored to modify two *H II* and five *H III* wings to deliver ammunition to the army troops once they had landed on English soil by making compartments inside the outer wing panels and within the center sections to secure the load. When Scheidhauer tested a modified *H III b* loaded with 400 kg (884 lb.) of ammunition, he found the sailplane behaved well in the air, despite a jump in wing loading from 8.8 to around 20 kg/m^2 (1.8 to 4.1 lb./ft.^{2}). He even soared the overloaded aircraft using thermal lift! Most striking of all was the wing's newfound capacity for faster speeds and better penetration into a prevailing wind; heretofore something no Horten wing had done well.[32]

Scheidhauer's test flight led Reimar to settle on a new wing design with approximately half of the chord and twice the aspect ratio of the *H III*, called the *H IV a*. Halving the chord required thinning the airfoil, which reduced the space available to house the pilot inside the center section, so Reimar again experimented with accommodating the pilot in an unorthodox setting. He was encouraged by students and professors in the Akaflieg Stuttgart (the academic flying group at Stuttgart, Germany), who had designed and built an experimental sailplane called the fs 17 equipped with a prone cockpit accommodating the pilot flat on his stomach (Figure 17). This plane first flew in March 1938. The goal of the project was to understand how flying prone made a pilot more resistant to the high G-forces placed on the body while maneuvering an aircraft at high speeds.[33]

The German gliding pioneer Otto Lilienthal had written presciently about prone flying in the late 1880s, and the Wright brothers flew prone from 1900 until 1908. Lilienthal had theorized that flying prone could increase aircraft performance simply by reducing the drag of a pilot seated upright. And the fantasy

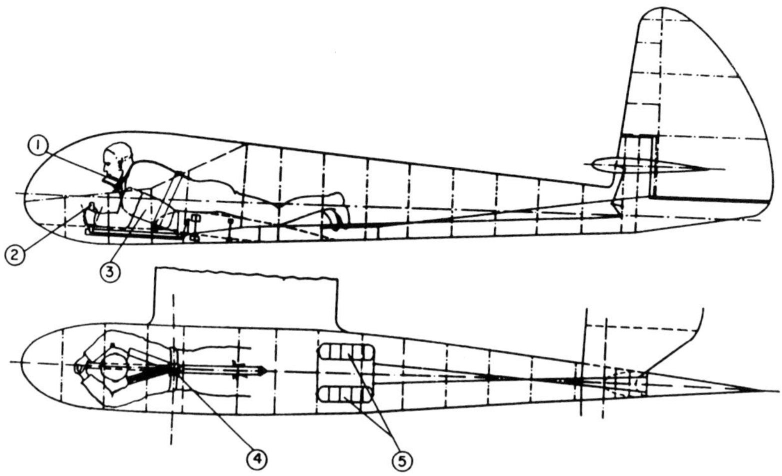

Figure 17. Details of the prone cockpit fitted to the fs 17 aircraft designed by students at the Akaflieg Stuttgart appear in these side- and plan-view drawings of the fuselage: 1. chin cushion, 2. control stick, 3. parachute, 4. safety harness, 5. ladder-type rudder controls with step-in rungs (H. Wiesehoefer, *Luftfahrt Med*, April 1940, 145).

of flying prone like the birds surely helped to inspire more than a few pilots to support the idea.

Examining the fs 17 also exposed to Reimar the handicaps imposed on the prone pilot: great difficulty seeing above the aircraft and to the sides and the physical discomfort and exhaustion of holding up the head for long periods. Flying level in calm air, the pilot's neck axis was oriented approximately level with the horizon, forcing him to only use upper back and neck muscles, among the weakest in the human body, to simultaneously raise and turn his head to scan the surrounding airspace. After only a short time, the prone pilot's head began exerting uncomfortable pressure on the chin, which rested on a small padded cushion. Discomfort to neck muscles and chin grew quickly when maneuvering inside a thermal with other gliders, where the collision risk was significant, and the G load increased during the tight turns required to keep the sailplane within the thermal lift.

Reimar developed an innovative new configuration he called the "kniend-liegende" (kneeling-lying) position that accommodated the pilot in a *semiprone* crouching position.[34] Complete details about the process involved in achieving this breakthrough remain elusive; however, as he had done when working out the major dimensions of the supine cockpit used on the *H II*, Reimar's bed again provided the means to work out important ergonomic details and perhaps facilitated the initial flash of creative insight. He may first have tried out the prone arrangement, but it is easy to visualize the young designer sliding off the bed until his knees rested on the floor, leaving the mattress to support his torso and head at a comfortable angle. Horten used the bed to determine the approximate overall height, span, and length of the center section and where to place the controls and instruments to fall easily to hand while flying.

Scheidhauer probably had a role in working out these details because Reimar sized the semiprone cockpit specifically to fit his test pilot's 1.6 m (5 ft. 3 in.) height and slight build. Horten kept the cockpit small and only widened it by 10 cm (4 in.) when designing the three aircraft that followed the prototype.[35] Sizing the wings for Scheidhauer shows Reimar's deepening confidence in the test pilot's ability and his commitment to designing aircraft to reach the highest levels of performance. Unfortunately, he was also denying a much larger group of taller and heavier pilots access to the aircraft and thereby ensuring that the wing with the best chance of generating widespread interest in Reimar's work would instead for many years remain unknown because only an exclusive few could fly it. Walter saw an important need to investigate the semiprone cockpit for use in German fighters and vigorously supported his brother's work. His interest stemmed from the debilitating physical effects that he had suffered during violent maneuvering in aerial combat while serving in JG 26. Perhaps put off by the size of the semiprone cockpit, the Luftwaffe never looked into it.[36]

Reimar's cockpit solved difficult ergonomic problems and refined the wing's aerodynamic and structural qualities (Figure 18). For example, it concentrated the pilot's mass in a small area and helped to keep the CG well forward.[37] The position comfortably accommodated the pilot by angling torso, neck, and head about 30° to the horizon, which Reimar determined by trial and error. This geometry was critical to ease pressure on the chin and allow the pilot to more easily raise and rotate his head. Horten also experimented with angling the pilot's thighs relative to the horizon, from 90° to less than 30°, and found disadvantages at both positions. Although the high angle reduced the overall body length, it made the ventral fin enclosing the legs and hips too deep. Thirty degrees reduced the height of the fin but also lowered the wing tips very close to the ground, so Horten settled on about 60°. The fin contributed some directional stability, but because it lay close to the center of gravity and center of pressure, the effect was small.

Reimar made the ventral fin serve other purposes as well. It enclosed the pilot's legs, with space remaining at the back to hold several pieces of equipment needed for high altitude flight, including an oxygen tank, regulator, and associated plumbing, plus batteries, wiring, and heating elements to warm hands and

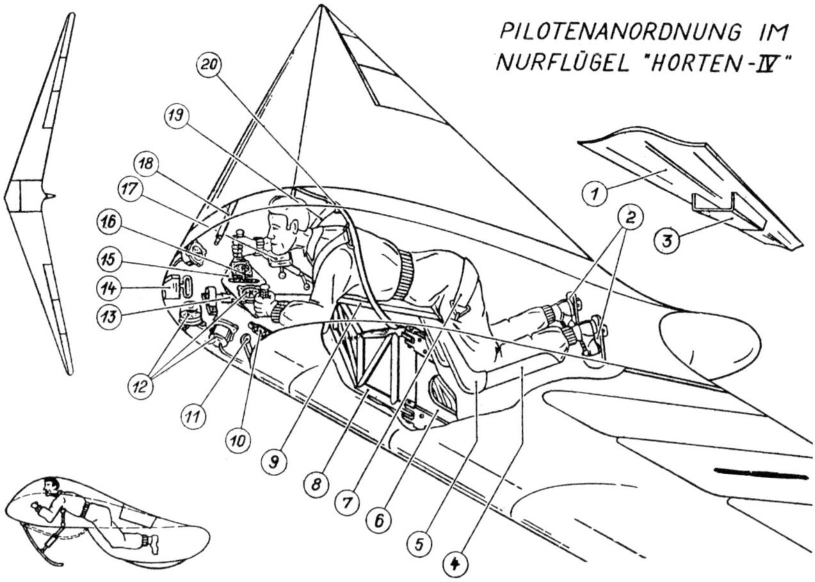

Figure 18. Drawing of semiprone cockpit fitted to the Horten *H IV a* wing. A profile side view of the center section is at lower left: 1. cockpit hatch cover, 2. rudder pedals, 3. parachute tray, 4. ankle support, 5. formed aluminum knee support, 6. wooden spar, 7. harness strap, 8. steel tube framework joining the main spars of each wing panel, 9. padded torso support, 10. dive brake control lever, 11. cockpit hatch and safety harness release lever, 12. instruments including altimeter, airspeed, and variometer, 13. fixed shaft about which the control wheel moved, 14. tow-hook and release handle, 15. control wheel, 16. trim wheel, 17. adjustable chin rest, 18. skid extend and retract lever, 19. shoulder straps, 20. crash pylon (Courtesy of Smithsonian Institution Libraries, Washington, D.C.).

feet. With 5° of wing dihedral, the fin easily kept the wing tips on the long, flexible wing from contacting the ground.[38] The fin also supported one of the two wooden skids that served as landing gear, this one fixed to the fin bottom. Reimar fitted the other skid beneath the wing's nose and designed it to retract rearward into a space just forward of the fin and below the stomach cushion. The pilot folded this skid after jettisoning a set of wheels used only during the takeoff run.[39]

A thick pad near the size of a chair cushion supported the pilot's chest and much of his weight. An aluminum cradle, shaped and padded to eliminate pressure points, cupped the knees, which bore much of the weight of the hips and legs. The feet rested on two pedals inside the ventral fin, which the pilot depressed to control drag rudders mounted at both wing tips. These and other controls are explained later in more detail. Reimar arranged for the pilot to drape his arms over the maze of steel tubes that passed through the center section and connected one wing to the other and to grasp a control wheel shaped like ram horns; there was not sufficient space to install a conventional control stick. Ball bearings supported the wheel, which the pilot could slide with little effort fore and aft on a fixed rod to control pitch and rotate left and right to roll the wing. The designer also provided a small, stuffed-leather chin rest, adjustable for height.[40]

The pilot flew the aircraft with his head right inside the wing leading edge, where visibility down and forward was very good, thanks to the Plexiglas panels with which Reimar had wrapped the leading edge. Reimar put most of the instruments on a small panel a few inches ahead of the pilot's face and fixed a stopwatch and altimeter to the control wheel. He gave the pilot straps to place his shoulders against that restrained forward movement. Back straps belted him securely into the wing and restricted movement aft. The forward end of the access hatch and leading edge Plexiglas bulged enough to allow space for the pilot's upper back and head and gave unrestricted views from one swept wing tip to the other. A tray built inside the cockpit hatch cover above the pilot's waist held a parachute, which the pilot was more likely to use since there was no risk of colliding with a tail during the jump.

Crash safety appeared roughly equal to conventional upright seating, although no pilot wanted to lead with their head during sudden decelerations. The sheet metal keel, stiffened at the front by a single steel tube, offered little protection, despite Reimar's assertions. One postwar British gliding club concluded that flying semiprone in a Horten wing required specialized training for new pilots, but this often applied to pilots transitioning to more advanced, or unusual, conventional aircraft, too.[41]

As remarkable as it was to house the pilot in a restful position and provide a fine view outside, even more impressive was the dramatic reduction in the size of the center section. The small frontal area and reduced chord of the short center section dramatically cut aerodynamic drag. The overall effect of the semiprone cockpit was comfortable, practical, and aerodynamically efficient. Several other German designers considered using semiprone cockpits late in the war; however, no one approached the elegant efficiency of Horten's rendering. After the conflict had ended, the German engineer Hans Amtmann helped the U.S. Air Force experiment with an articulated couch supporting a semiprone pilot, and the U.S. Army identified the semiprone position as worthy of study in a series of 1946 memorandums. Yet neither the German concepts, nor the American experiments, blended cockpit and aircraft as successfully as Reimar Horten did.[42]

A bit of luck gave Reimar the means to begin building the *H IV a* after Hitler postponed the invasion of England in mid-September. Reimar encountered a group of idle men whose commander was delighted to grant Reimar's request to borrow a half-dozen for three months. Several of these men had soared above the Wasserkuppe, so they readily took to the work. Horten began building a mock-up center section before leaving Braunschweig in December 1940.[43] Because he had borrowed state labor and used surplus plywood, the first *H IV a* "cost zero, nothing, nothing" to build.[44] That he had scrounged both labor and materials suggests his excitement to get the project underway. At this early stage, Reimar was naively confident the war would soon end in Germany's favor. When it did, he intended to be ready when soaring competitions resumed with an all-wing sport sailplane that could well be competitive with any conventional type.

Horten followed standard practices to build the outer wing sections, covering the leading edges with thin plywood and the framed spaces behind the main spar with fabric. He enclosed the welded steel tubes of the center section with thin metal, plywood, and Plexiglas. For control, Reimar devised one of the most complex systems ever developed for a small aircraft. Most conventional gliders flew with eight control surfaces: elevator, rudder, two ailerons, two spoilers, and sometimes two flaps. Horten used fourteen, which the pilot manipulated using a control wheel and levers attached to rods, all mounted on ball bearings (Figure 19). This aerodynamic architecture is described below, grouped by type of control.

Elevons

To control pitch and roll, Reimar mounted no less than three elevons along the trailing edge of each outer wing section. He constructed two elevons using wood and covered them with fabric, but he mounted the third and smallest at the wing tip. Here, the components were too thin to be made of wood, so Reimar chose aluminum. The two inboard elevons covered a large area in order to compensate for the high aspect ratio layout, which reduced the distance between the hinge axis of the elevons and the centers of gravity and pressure. The larger elevons boosted the fulcrum effect required to maneuver the wing smartly, but the number of elevons, their size, and the multitude of narrow gaps between them increased drag.[45] Piloting the wing at certain speeds and flight attitudes required physical effort, which Reimar tried to ease by installing bungee cords to adjust, or trim, the control wheel.

The aft end of the wheel joined to control rods that actuated the inboard elevon pair. Rods to the two outboard elevon pairs met at the front of the wheel. The pilot pulled the wheel back to climb, pushed it forward to dive, and rolled it left and right to turn, thus fanning the elevons according to a busy schedule (Table 1).[46]

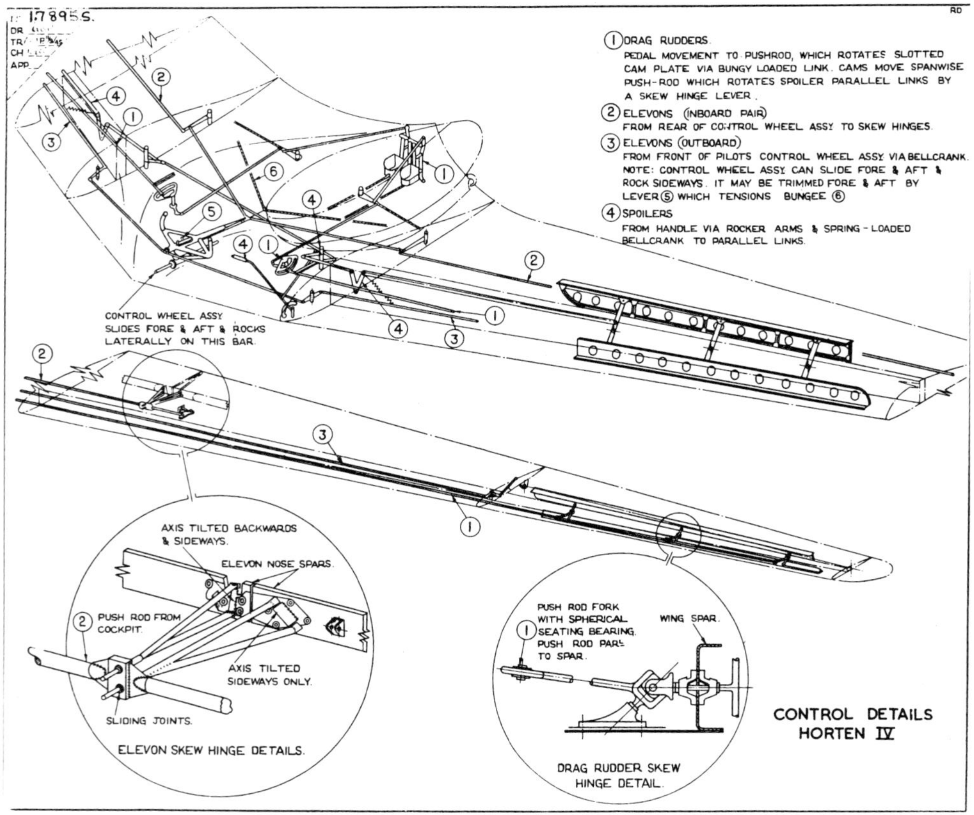

Figure 19. Reimar used a complex arrangement of metal push rods to connect the pilot's control wheel and levers to the fourteen control surfaces fitted to the *H IV* sailplane (Kenneth G. Wilkinson, "The Horten Tailless Aircraft," R. A. E. Report No. F. A. 259/1, Tech. Note Aero. 1703, October 1945, Royal Aircraft Establishment, Farnborough, England, Fig. 11).

Table 1. Elevon deflection angles measured on Horten *H IV a*, Werk Nr. 25.[1]

Control Wheel Action	(+ trailing edge up, – trailing edge down) Left Elevons			Right Elevons		
	Outboard	Center	Inboard	Inboard	Center	Outboard
Wheel full forward	+12	+14	+14	+17	+14	+11
Wheel full back	– 16	– 14	– 3	– 4	– 14	– 11
Wheel full right[2]	+16	+9	+13	– 2	– 7	– 28
Full right and full back	0	– 10	– 2	– 5	– 17	– 33

1. "Aero. 1703," 10.
2. Values for control wheel full left not recorded.

The table shows how Reimar designed the outermost elevons to perform primarily as ailerons working when the pilot rolled the wheel right or left. The middle pair acted as elevators to pitch the wing up and climb when the pilot pulled the wheel back, and the inner pair caused the wing to dive as the pilot pushed the wheel forward. The Royal Aircraft Establishment Farnborough took these readings directly from Werk Nr. 25, the *H IV a* bearing the RLM registration code LA-AC. The asymmetrical deflections noted by the RAE between the left and right sets of elevons indicate poor rigging, or an adjustment to compensate for asymmetry caused by a weight discrepancy, or aerodynamic misalignment between the two wing halves.

Taking another approach to reducing adverse yaw, Reimar used a type of elevon design named after the British engineer, Leslie George Frise (pronounced "freeze"). The Frise elevon weakened adverse yaw because it combined a particular leading-edge shape with placement of the hinge points on both the control surface and the wing to which it was attached (Figure 20). Reimar magnified this effect by gearing the outer elevons to swing differentially; that is, the up elevon moved less than the elevon deflected down.[47]

Drag Rudders

The drag rudders consisted of two small aluminum spoilers that Reimar set into the top and bottom of each wing tip. They functioned on the *H IV a* much as they had on earlier Horten wings, augmenting the elevons to provide good control during turns and countering the adverse yaw produced by the elevons. By using his feet to manipulate stirrups mounted inside the ventral fin, the pilot deployed the rudders one at a time, or simultaneously to function like spoilers, and adjust rate of descent and glide path.[48]

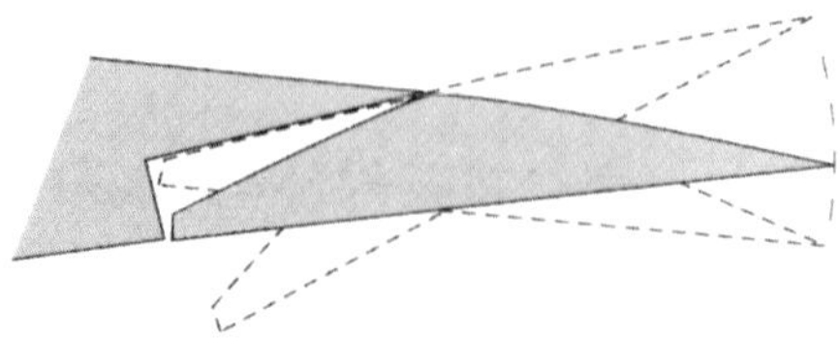

Figure 20. This drawing shows a Frise elevon deflected at various angles to illustrate how the shape of the nose, and the point through which the elevon is hinged, help to counteract adverse yaw by promoting drag when the elevon deflects up and by reducing drag when it deflects down (Martin Simons).

Spoilers

Reimar fitted two main spoilers on each wing section to protrude through the top and bottom wing skins. The inboard end of each spoiler began about a meter-and-a-half (almost 5 feet) from each wing root where it joined the center section and about 1/3 of the distance from the wing leading edge to the trailing edge. The pilot could pull a handle near his right shoulder to deploy them simultaneously, slowing the aircraft or adjusting its rate and angle of descent. Each spoiler spanned just over 2 m (6 ½ ft.), considerably larger than those fitted to the *H III*, and supplied considerably more braking power than needed to land the *H IV a*. Reimar believed that Blech died at the end of the 1938 Rhön contest because the *H III* spoilers were too small to break the thundercloud's hold on his sailplane. German soaring pioneer Heini Dittmar disliked the Horten wings for this reason and because they could not be put into a full spin to quickly escape from bad weather, as could be done with many conventional sailplanes.[49]

***H IV a* in Flight**

When craftspersons finished building the first *H IV a* (Figure 21) at Königsberg-Neuhausen in May 1941, Scheidhauer began flying it using tows from a winch, but this launch method proved unsatisfactory. Following a tow behind a Heinkel 46 powered aircraft that hoisted him to 3,000 m (9,840 ft.), Scheidhauer flew for an hour. Upon landing, he exclaimed to Reimar, "this is a very good ship."[50] Less than two months after completing the prototype, the Luftwaffe began moving Reimar's flying school from Königsberg-Neuhausen to Frankfurt-am-Main. At about the same time, and 240 km (150 miles) north at the Peschke works in Minden, efforts began to restore the neglected *H V b*. Reimar decided to relocate here and bring with him the prototype *H IV a*, six key members of his team including Scheidhauer, and several engineers and draftspersons. He and Walter drafted a telegram for the RLM personnel department that ordered Reimar to move to Minden along with his team and their equipment.[51]

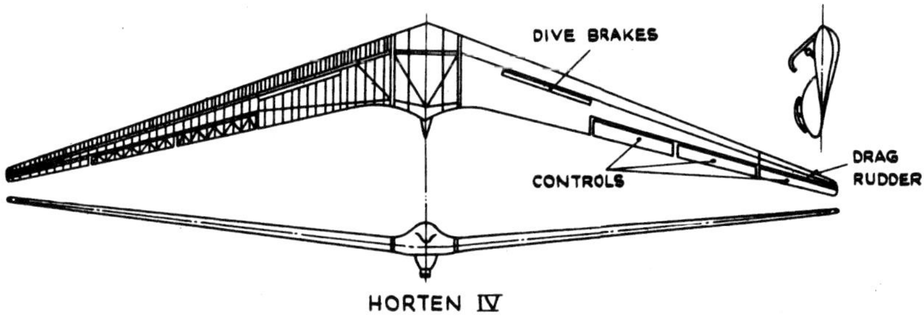

Figure 21. Horten *H IV* three-view drawing (D. C. Appleyard and M. A. Biot, "The Horten Tailless Aircraft," 42).

The Horten group was unpacked and ready to work in September 1941. Thereafter, test pilot Scheidhauer carried out an intensive six-month evaluation of the *H IV a*. Reimar frequently towed him aloft behind several Focke-Wulf Fw 56 Stösser utility airplanes sent to the base for repair. The operation had no official sanction; still, Scheidhauer was able to fly the new wing as much as fifty hours a month.[52] He often soared above Lake Steinhuder about 36 km (22 miles) northeast of Minden. Reimar was interested in Scheidhauer's findings, but because he immersed himself in studying mathematics, spending about half of each week at Bonn University, he was not present for many of the flights.[53]

One of the first challenges a test pilot must confront when evaluating a new aircraft is determining safe maximum airspeeds. Scheidhauer came dangerously close to exceeding the "speed of wood" when he dove the *H IV a*

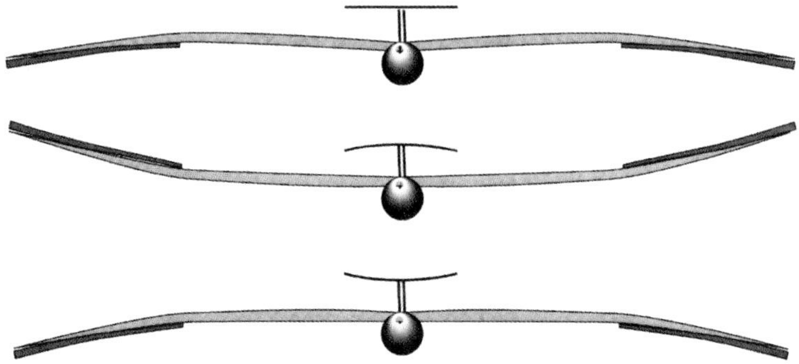

Figure 22. This head-on view of a conventional sailplane illustrates the destructive properties of wing flutter. The momentum generated by beating up and down as often as several cycles per second could cause the ailerons (or elevons on a Horten wing) to flap and twist the wing about its lateral axis (Martin Simons).

to such a velocity that the wing tips began to oscillate rapidly in a motion known as *flutter*, which combined up and down flapping with twisting about the wing's lateral axis (Figure 22). The flutter occurred at between 140 and 150 km/h (87 to 93 mph); however, Scheidhauer gingerly pushed the wing to 180 km/h (112 mph) before pegging this as the maximum velocity, or never-exceed speed, although Horten had calculated that a higher speed was possible.

The fluttering frequency varied from 60 cycles per minute on the prototype to 110 on the three *H IV a* wings that followed. Simple mass-balances installed only in the two outermost elevons may have rendered the four aircraft more susceptible to flutter; however, the primary cause was aero-elasticity, or lack of enough stiffness in the wing structure to resist aerodynamic forces. The wing's sweep probably exacerbated this flaw. Reimar had worked hard to increase performance while minimizing induced drag by thinning the *H IV a* wing to a 21.8 aspect ratio. Unfortunately, the cantilever wooden structure lacked the stiffness to support such a long and thin structure.[54] Reimar wrote after the war that stopping the flutter during flight required the pilot to tap simultaneously both drag rudder pedals, but it is not clear how widely this remedy was known.[55]

Except for the flutter, further testing at Minden corroborated Scheidhauer's initial glowing appraisal of the *H IV a* and confirmed that it outperformed nearly every other sailplane in the world, but Reimar must have questioned whether he paid too high a price in handling qualities to achieve such performance.

A pilot named Lehmann conducted a series of tests in May 1943 that helps answer the question. Technicians loaded an *H IV a* with ballast until the CG fell at a point near the aft limit. Lehmann took the sailplane up and found the wing oscillated badly in pitch and that gusty conditions made flying unsafe. Yet he was able to recover from stalls and could not force the wing into a spin. Lehmann needed only to tap the drag rudders at the beginning of turns to counter adverse yaw, and he found just a slight delay when turning solely with ailerons. Lehmann concluded that the Frise treatment Horten had applied to the leading edges of the outermost elevon pair contributed to this good behavior. Lehmann estimated an L/D of 37:1 at a horizontal velocity of 85 km/h (53 mph) while sinking at .64 m (2 ft.) per second and determined that the slowest rate of descent was .55 m (1.8 ft.) per second while flying at 60 to 65 km/h (37 to 40 mph).[56]

These were exceptional numbers, the lift-to-drag ratio in particular was very good; however, Reimar wisely sought to supplement Lehmann's test data by comparing the wing to several conventional sailplanes. He arranged for Scheidhauer to fly the *H IV a* alongside a Kranich, Weihe, Condor III, and Reiher. Each aircraft was progressively better in performance, but the Horten wing bested them all in both L/D and minimum sinking speed.[57]

Late in May 1943, in what Walter described as a "private competition between the Hortens and Akaflieg Darmstadt," Scheidhauer flew the fourth and last *H IV a* to be built (Werk Nr. 26 bearing German registration LA-AD) alongside the sailplane that had repeatedly demonstrated the highest L/D of any in the world, the Darmstadt D 30 (Figure 23).[58] Facing him and his wing was a formidable combination of man and machine. Hans Zacher was an experienced and respected engineer and test pilot who had helped develop the Akaflieg system into a successful educational and entrepreneurial enterprise.[59] Students and staff of the Akaflieg Darmstadt had begun developing the D 30 in 1933. The design matured over the years into an aeronautical tour de force by the time it flew in 1938. Aspect ratio measured an astonishing 33.6, the longest and narrowest wing fitted to a sailplane until well after World War II.

By contrast, the wing of the 1952 Boeing B-52 jet bomber had an aspect ratio measuring 8.5, the 1967 Lockheed U-2R reconnaissance jet measured 10.6, and the 1994 Nimbus 4D unlimited competition sailplane measured 39.1. To support the D 30 wing structure, Akaflieg Darmstadt designed a cantilever wing constructed mostly of wood but supported internally by an aluminum box spar. Space remained inside the spar to install a mechanism allowing the pilot to alter the dihedral angle of the outer wing sections to vary the glider's stability during flight. Soon after trials began, Bernhard Flinsch set a world record by flying the D 30 to a declared goal and returning to his launch point, a round-trip distance of 306 km (190 miles).[60]

Despite poor weather, which the pilots may have chosen to fly in because German aircraft of all types, including sailplanes, were vulnerable to Allied fighter aircraft that often roamed the skies searching for targets of opportunity on the ground and in the air, Scheidhauer and Zacher flew their sailplanes together eight times from May 23 to 26, 1943. Scheidhauer reported that during one flight, the pilots circled each other inside a thermal while climbing to 1,100 meters (3,608 ft.) without either aircraft gaining significant height over the other, except when rain degraded the airfoil's lifting properties, causing the *H IV a* to descend. Once the rain ended, Scheidhauer claimed he climbed the wing to a height of 50 meters (164 ft.) above the D 30 after a half hour of flying.

Zacher agreed with the Horten pilot's observations, adding that during other flights, both sailplanes sank at the same rate but at different airspeeds, the result of their different wing loadings. This figure for the D 30 stood at 24 kg/m^2 (4.9 lb./ft.^{2}), for the *H IV a* it was 17.1 kg/m^2 (3.5 lb./ft.^{2}). The greater load on the D 30 wing allowed Zacher to fly faster while losing the same amount of height as the all-wing aircraft. Reimar reported that the *H IV a*'s maximum L/D was determined to be 32 when the wing flew at 73 km/h (45.3 mph). The wing's

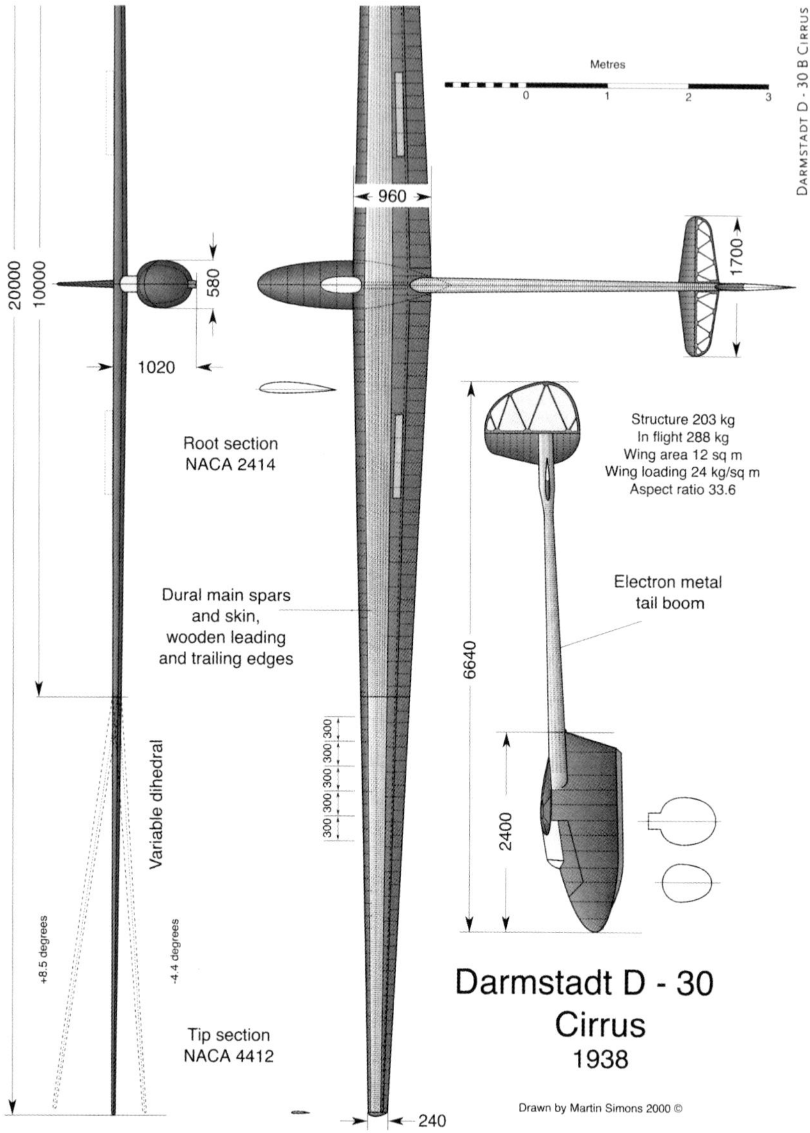

Figure 23. The strong dural (aluminum alloy) wing spar allowed the Akaflieg Darmstadt to design the wing of the D 30 without external struts that produced unwanted drag (Martin Simons).

minimum sinking rate, .55 m/sec (1.8 ft./sec.), occurred at 55 km/h (34 mph), whereas the D 30 reached its minimum sinking speed 10 km/h (6 mph) faster.[61]

Evaluated strictly in terms of performance, the two sailplanes each possessed unique advantages. Whereas the *H IV a* climbed better at slower speeds and in weak thermals, in competition, the cruise speed advantage of the D 30 gave it the edge, allowing the pilot to cover more sky while searching for thermals and losing less altitude. The key to this superiority was not that the D 30 weighed more—the *H IV a* was heavier by 38 kg (84 lb.). The Darmstadt sailplane excelled because its higher aspect ratio wing gave excellent lift with lower induced drag. The D 30 also graded higher for docile handling qualities. There is no record that pilots flying the D 30 encountered the type of potentially dangerous, and unnerving, flutter that plagued the *H IV a*.

There was no single cause for the Horten wing's elasticity; rather, it stemmed from a combination of structural and aerodynamic design characteristics. An aluminum box spar made the D 30's long, thin wing rigid enough to withstand flutter; however, Reimar used laminated wood spars in the *H IV a*. While easier and less costly to fabricate, wood was considerably less stiff than aluminum.[62] Horten used three different construction materials—steel tube, wood, and aluminum—each with different bending properties that could combine in unusual ways. Moreover, each joint connecting the main components, center section to the main wing panels and the wing panels to the wing tips (four joints in all), was susceptible to flexing under flight loads. Finally, the considerable washout twisted into the wing may have set off the flutter, or worsened it.

At Walter's urging, Reimar signed an affidavit attesting with "reservations" to the results of the fly-off with Zacher, but he anticipated a rematch between the D 30 and a new wing he was already designing called the *H VI*.[63] This never took place, and as the years passed, Reimar realized that the aeronautical community would judge the quality of the *H IV a* based largely on this single evaluation conducted during wartime in mediocre weather conditions. He began making excuses for the outcome, saying the *H IV a* was damaged shortly before the fly-off, which limited its performance. He also questioned the accuracy of Zacher's measurements pegging at 37, and both Horten brothers suggested that during several flights, Zacher had flown with an unfair advantage by riding the wing tip vortex shed by the *H IV a*, in the same way that large birds often fly in a vee formation while migrating.[64]

Reimar clearly had expected his wing to do better. To understand why, we must look back to June 10, 1942, and examine a flight between two landmarks made by Heinz Scheidhauer in an *H IV a*. In the still air just before dawn, Scheidhauer deployed his spoilers and descended from above the Hoher Hagen Mountain until he was "exactly" level with the top of the Gaußturm, a lookout tower outside the city of Göttingen. He then closed his spoilers and did not use them again for the reminder of the flight. Setting course for the Göttingen airfield, the test pilot skimmed so close over the forest that he had to maneuver around the taller trees. On he flew for "10 uncomfortable minutes." Scheidhauer nearly turned back to avoid power lines that flitted past beneath him "like big hemp ropes." He flew a few feet above smooth terrain over the final 200 m (656 ft.), hopped the hedge that marked the airfield perimeter, and lowered the skid before touching down on the grass runway.

Later that day, Reimar and Scheidhauer studied aeronautical charts of the area made before the war. They concluded from the height differential between the top of the tower and the airfield, and the distance flown, that the *H IV a* had demonstrated an L/D of 37:1. This flight by Reimar's most trusted test pilot, together with Lehmann's findings in 1943, led Reimar to believe that the *H IV a* performed considerably better than was demonstrated during the fly-off with Zacher. The story took an ironic twist forty years later when Scheidhauer and Karl Nickel revisited the 1942 flight. The men studied a new set of aeronautical charts, based on recent surveys, and determined that the top of the lookout tower had been much higher. They made new performance calculations using the corrected height figure and deduced an L/D of less than 32:1![65]

It is worth noting that Scheidhauer flew the last 200 m (650 ft.) to a landing in ground effect, meaning he was low enough to ride the cushion of air that formed between the *H IV a* and the ground, and possibly skewing the L/D falsely higher. More than six decades later in 1982, Jochen Ewald explained that to measure sailplane performance accurately, hundreds of flights were required to eliminate the statistical errors caused by even light turbulence or inconsistent piloting. Only in 2005 did this tedious work ease somewhat with the introduction of electronic global positioning navigation systems.[66] Little wonder, then, that a few comparison flights carried out in 1943 during poor weather conditions could generate controversial results.

Horten completed three more *H IV a* airframes in 1943, and pilots flew the four wings regularly until the war ended. Although the flying community was interested in experiencing firsthand the aircraft's numerous innovations, there are no complete flight records to corroborate Reimar's excessive claim that as many as a dozen pilots combined to fly the four sailplanes about 1,200 hours.[67]

Prone and Semiprone Evaluated

In spring 1944 at Göttingen, before Karl Nickel's first flight in one of three *H III f* wings fitted with a prone couch to train pilots for the semiprone *H IV a*, skeptical friends questioned whether he could maintain a feel for the aircraft while lying on his stomach. It would be impossible, they warned, to fly instinctively by sensing slight deviations in heading or attitude. Nickel would need to consider beforehand each control movement, and centering thermals, a basic and vital skill, would be "out of the question from the start."[68]

Nickel's friends also feared that during cross-country glider flights, which often lasted for hours, the effort to keep one's head elevated would exhaust even a thick-necked flyer, particularly one subjected to multiple G-loads from turning tightly and climbing inside thermals. Blood would pool after several hours and the limbs would fall asleep, they said. Thus immobilized, the pilot would be unable to hoist himself out of the aircraft after landing.

Nickel's flight answered every question and concern. "I climb from behind [onto] the center-section of the flying wing to step inside and lie down," he later recounted. A chest-type parachute pillowed his torso. The "lying-trough," as he called it, "is well-upholstered with foam rubber and artificial leather . . . there is the chin-rest which is easily adjustable. The designer has thought of everything and wants to accommodate the pilot in comfort." Seat belts were fitted but functioned in reverse. "They are fastened over the back and are released automatically as soon as the cockpit [hatch cover] is opened. How Practical!" said Nickel.[69]

Pilots controlled the conventional German sailplanes of the day with a stick fixed below the seat, pivoting left and right to control roll and fore and aft to control pitch. To control all the Horten wings equipped with prone and semiprone cockpits, Horten designed a control wheel shaped like a ram's horn. "Will it be possible," Nickel wondered, "to get quickly accustomed to this?" Once airborne, he had the answer.

> All of a sudden I am completely baffled: there is nothing unusual, it's exactly as flying while sitting in a seat! I feel the stick force, the sailplane reacts to the smallest control movements. I completely forget that I am lying horizontally in space, that the control [wheel] looks so strange, that the H III is no normal aircraft. [It is as though] I had been flying in prone position for years . . . The first gusts are felt and are counteracted automatically, without thinking. I see my hands moving to act in the correct way, but there is no conscious command from the brain. The bird feels good . . . [and this] reaction comes so strong and unexpected that I wish to sing at once . . . I am so delighted . . . there is nothing to learn about prone flying and everything is so simple.[70]

After gusty winds forced a landing, Nickel returned to the air a half hour later:

> I am floating again in the air . . . flying over the houses and streets of Göttingen. Wonderful, this marvelous view down through the acrylic glass pane. Exactly as on a street map I can track the roads and alleys with my finger. Seemingly just in front of my face there is that hive of activity. Magnificent to soar and glide high over the rooftops, horizontal in space like a bird. This sort of flying really is the only natural way, how could anybody doubt it ever? The view is unobstructed on all sides through the large canopy, but the most astonishing aspect is the excellent view downward . . . Slowly we are losing altitude. It's time for a thermal to appear. Oops, here it is. [I apply drag] rudder and aileron, slowly pull up, it's just the same as with any other sailplane. Only the banking at first seems to be excessively large.[71]

Nickel at first tended to make shallow and flat turns, but he soon regained his "feel" for banking into turns during a flight of two hours. His mind began to accept the new set of visual cues of a standard turn he was seeing through the clear leading edge of the *H III f.* Banked turns of 60° became easy and "remained the only difficulty I encountered and it didn't occur anymore during later flights."

Nickel also flew the *H IV a* and judged it "quite comfortable," calling the pilot's view good, and "important in particular during aero-tow. Especially with tailless sailplanes a good view of the towing aircraft can be decisive against flying too low and, consequently, being dragged down by the downdraft behind the tow aircraft."[72] During a tow, a pilot had to hold position above the turbulent wake from the tow airplane, but the noses of many conventional gliders obscured the critical visual cues a pilot needed to maintain the position. Nickel's overall impression of the *H IV a* was "very pleasant, the control action and control forces including the air brakes were finely tuned. The [semi] prone position for the pilot was extremely comfortable . . . The flight performance, finally, was superior to all other sailplanes of that time, except the Darmstadt D 30."[73] The *H IV a* was easier to fly than the *H III f* training wing, gushed Nickel. "Any sailplane pilot with about 20 to 30 hours of flight experience (and the permission [from an instructor] to [launch using airplane] tow flights) could have flown," the *H IV a*, he said.[74] Scheidhauer considered this wing "as good as any fuselage machine in steering, better than all in blind flying due to great stability . . . I can go into every cloud, full of trust in my bird."[75]

To the question of comfort and fatigue on long flights came a convincing answer on August 7, 1943, after Hermann Strebel flew an *H IV a* for ten hours over the Wasserkuppe. Nickel took off several hours later in an Olympia Meise sailplane of conventional layout, equipped with standard upright seating, and remained airborne for seven-and-a-half hours. After he landed, Nickel "went to [Strebel] limping with aching backside. But he approached me laughing and completely fresh, and could only shake his head to my envious questions: 'No, no bruises, no limbs which went to sleep, no stiffness of the neck, nothing!'"[76] After the war, British test pilots found that "elbow and shoulder movement was restricted, which constrained one to stay in the same position all the time" while flying an *H IV a*; however, this comment stands apart from the near unanimous praise for the semiprone cockpit heard from other pilots. Perhaps the British pilots were larger and heavier than Scheidhauer, for whom Horten had specifically tailored the cockpit.[77]

Horten *H VI*

Horten's penultimate sailplane design was the *H VI*. This wing could be flown farther for each foot of altitude lost than any other sailplane in the world, exceeding even the D 30, but the aircraft was not practical for routine use. Developing the aircraft during the last two years of the war with Allied fighters frequently overhead hunting for targets was both difficult and dangerous. Always multitasking, Reimar begun to design the *H VI* even as construction commenced on the *H IV a* in 1940. Two years after building began in 1942, the first *H VI* flew bearing prototype designation *V1* (experimental prototype 1) and the RLM design number 253.[78] The second and final example soon followed, the *H VI V2*.

The RLM usually assigned the Versuch, or prototype designations such as V1 and V2, to new military designs, but in this case, Walter had somehow manipulated the system to create an official cover that obscured the true, nonmilitary nature of the *H VI* project. Reimar recalled having "hid in the hills" to build the wing, and quarreling with his brother who was very reluctant to "take the responsibility, and the risk" to hide the work.

Groping after the war to justify the project, Reimar claimed he was testing another cure for the Mitteneffekt. In fact, he increased slightly the wing area of the *H VI* at the aft end of the center section, a subtle modification that appears as a graceful curving taper to a point when seen in plan view. All models of the turbojet-propelled *H IX*, including the glider prototype, display the shape, too; but given the structural, aerodynamic, and performance differences between the sailplane and the jet, the claim seems weak.[79] Walter added to the subterfuge by submitting a bogus proposal to study a "long-range, high-flying, reconnaissance aircraft," knowing after May 1943 that his brother's real purpose was building the "best sailplane [in] the world" to surpass the D 30.[80]

Reimar wanted to give the *H VI* the same aspect ratio and wing loading as the D 30, and then let Scheidhauer prove during another fly-off that the all-wing was superior *because* it had no fuselage and tail. Reimar believed that without the nonlifting components (tail and fuselage) to encumber it, the *H VI* would perform about 15 percent better. The designer reduced the *H VI* chord at the wing root by 14 percent compared to the *H IV a*, while growing the wingspan to 24.25 m (79 ft. 7 in.), which was a 16 percent increase above the *H IV a*. In the end, the total wing area of the *H VI* dropped about 5 percent below the smaller wing (Figure 24).[81] In terms of both wing loading and aspect ratio, Reimar came within 5 percent of the D 30.

Horten again used wood in the wing spars. To make them more rigid and resistant to flutter, he deepened the wing root and combined a laminated and compressed wood called Lignofol with laminations of Scotch fir to make it strong and doubled the number of massive steel fittings machined to receive the hardened bolts connecting the wing panel to the center section.[82] Horten's team built the slender wing tips from elektron, a lightweight and strong alloy developed in Germany and comprised principally of aluminum and magnesium. Each tip spanned 2.9 m (9 ft. 6 in.) but the designer crammed into their miniscule width (barely 38 cm, 15 in. at the widest point) not only thin-chord elevons but also spoilers set into both the top and bottom surfaces and drag rudders that popped out of the top surfaces when the pilot needed them.

The high aspect ratio made the wing tips too thin to be made of wood and still contain elevons, drag rudders, and spoilers. The tips had to withstand the forces encountered during normal operations, such as a tip touching the ground during takeoff and landing, and the manipulations of the ground crew who had little else but the wing tips to grab when positioning the aircraft. Reimar designed sturdy tabs to plug into sockets in the main wings, making the tips easier to install and dismantle. Unfortunately, the resulting joint provided another source of flutter. Photographs of the *H VI V2* made after the war show an alarming degree of bending along the entire wing, but particularly in this area.

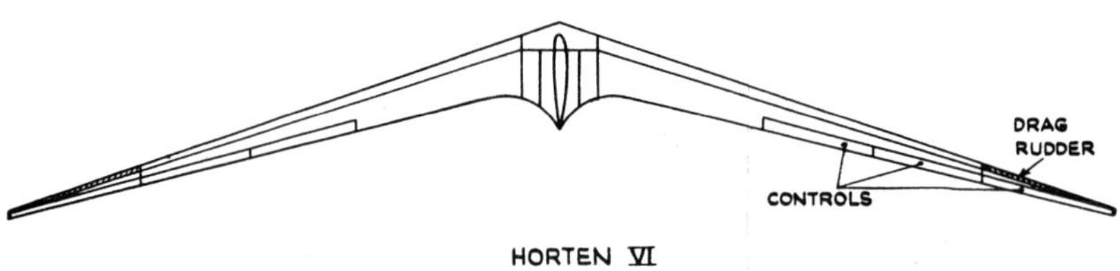

Figure 24. Horten *H VI* three-view drawing (Appleyard and Biot, "The Horten Tailless Aircraft," 42).

Craftspersons worked about eight thousand hours to complete an *H VI*.[83] To assemble each main wing panel, which spanned 11.3 m (37 ft.), workers had to glue together hundreds of handmade components, including the 147 ribs that formed the airfoil shape at the leading edge between the main and auxiliary spars and aft to the trailing edge. The two inboard elevons required another fifty. Thus, the entire wing required 394 ribs! Every rib differed slightly from its neighbor in length and curvature. All were handmade from three thin layers of plywood stiffened with six additional pieces of wood; the whole assembled using jigs to ensure that the shape conformed to Reimar's construction drawings. To finish a wing panel, workers covered the ribs between the main spar and the control surfaces with fabric and applied chemicals that tautened and waterproofed the material, saving considerable weight and easing access to the inside of the wing for repairs and control adjustments.

Steel tubes welded to form a truss work framed the *H VI* cockpit. Reimar again accommodated the pilot semiprone, but he acknowledged that the cluster of instruments inside the *H IV a* had partially blocked the pilot's view; therefore, he mounted all but the compass and the turn-and-bank indicator inside the leading edge of the *H VI* and facing forward. So the pilot could read them, craftspersons reversed the characters on the face of each gauge and mounted small mirrors inside the leading edge. Reimar installed equipment to help pilots brave the elements by adding a complete oxygen system for high-altitude flying and providing the hardware to install a battery and wiring to power heated socks and gloves.[84]

Like all his sailplanes, Reimar designed the *H VI* to operate from grass airfields. Although skid landing gear was standard practice within the soaring community, Horten employed both skids supported on shock-absorbing steel springs and wheels, as he had done with the *H IV a*. Ground crew attached temporarily a small two-wheel nose dolly to a retractable hardwood nose skid to ease ground handling and for taking off. Just after breaking ground, the pilot dropped the dolly and raised the skid into a recess in the ventral fin as springs closed two sculpted aluminum doors over the recess to cut drag. This tidy arrangement pared the weight of a permanent wheel and strut and saved space. Along the bottom of the ventral fin, Horten hung another hardwood skid. A tube-type tire and wheel cushioned by rubber biscuits protruded through the middle of the skid. Craftspersons tacked and screwed sections of canvas over the gap between skid and fin to streamline the assembly.[85]

Constructing the *H VI V1* began at Göttingen in spring 1943, but only weeks later, the brothers moved the project to Aegidienberg, about 20 km (12 miles) southeast of Bonn. Reimar loaded a trailer with the *H VI V1* main spars and parts of ribs and other materials, and with about ten men accompanying him, he drove to the new workshop and met a group from the Lower Rhine Flying Club. The combined team resumed work and finished the aircraft near the end of May 1944.[86]

Now an old hand with more than 1,190 hours of sailplane flying time in his log book, Scheidhauer nonetheless cautiously approached his first flight in the *H VI V1* on May 24, 1944.[87] Unlike the *H IV a*, he would need more than two fingers to fly this larger wing. No other Horten sailplane had more inertia in the roll axis, requiring considerable muscle on the control wheel to overcome. In contrast, pitching the wing up and down required a light touch on the wheel because the wing had so little mass ahead of and behind the centers of gravity and pressure. This quick-pitch response allowed Scheidhauer easier control of the wing's position inside a thermal, but adjusting to the disharmony between the pitch and roll control forces would take time.

Reimar was away at school and did not witness the first flight, but Scheidhauer recorded his impressions succinctly.[88] "An extraordinarily beautiful bird," he wrote in his pilot's logbook. "Hopefully the small bugs are smoothed out soon, to start with the performance measurements. This year we <u>must</u> beat the D 30."[89] Commenting later and off the record, he called the handling qualities "very poor" compared to the *H IV a*.[90] Scheidhauer carried out more flights before the war ended, but the Horten brothers disagree on how many. Reimar estimated the test pilot made five to ten flights, always taking off well before sunrise to avoid Allied fighter aircraft, remaining airborne from 1 to 1.5 hours and clocking a total of about ten hours in the air. Walter recalled that Scheidhauer flew the *H VI V1* six times, but regardless of the actual number, he gathered enough data for Reimar to calculate an initial L/D of 42:1.

The number grew to forty-five over the years; however, bearing in mind that Reimar's calculations did not account for wing sweep, a more accurate L/D is probably in the high thirties. Reimar organized a flight during March 1945 to compare an *H IV a* flown by Hermann Strebel with the *H VI V1* piloted by Scheidhauer. Both men were experienced sailplane pilots who understood the quirks of flying the Horten wings, so Reimar expected a thoughtful appraisal of the new wing.

It was early in the morning and quite dark when the gliders left the ground behind their towing aircraft. Takeoff had been timed so that the wings broke into sunlight at 3,000 m (9,840 ft.). Flying later in the day risked a dangerous encounter with Allied fighter aircraft, and the thermals and other turbulent air that often developed could skew the results. The pilots dropped off tow, and about a half-hour later, Strebel landed first, followed after several minutes by Scheidhauer. The air raid siren sounded as the ground crews frantically rolled the aircraft into nearby hangers to end flying for the day, but not before the test pilots had observed that "at all speeds the *H VI* was very much better than the *H IV*."[91]

There was, however, a cost to be paid for the wing's exceptional aerodynamic performance. The *H VI V1* suffered from flutter in conditions similar to the *H IV a*, and for the same reasons. "We had great difficulty with the spar bending moments on the H VI," Reimar admitted after the war.[92] The composite material Lignofol was a stopgap measure; "however, [the wing] bent very much and we needed, for this aspect ratio, better materials."[93] The D 30 had shown that a wing strengthened with aluminum could resist flutter, but Reimar for some reason did not follow this practice. He also failed to use an aerodynamic solution in vogue on many aircraft of the era. Horten equipped none of the three control surfaces on each wing with weights to balance each surface about its hinge axis. As a result, the momentum of each wing beat tended to drive the control surfaces up and down and amplify the flutter.[94] Thus flawed, the wing was "of interest only as a high performance sailplane for record breaking purposes. It is too costly to build and difficult to handle for general use."[95] Yet in the time remaining before the war ended, Reimar had no chance to organize attempts to set records with the *H VI V1*—building the military wing propelled by jet turbine engines consumed nearly all his time.

CHAPTER 4

Jet Wings

THROUGHOUT THE 1930s and early war years, Reimar focused primarily on designing motorless aircraft. In 1943, his career took a radical turn when he received official permission to develop the first all-wing military aircraft propelled with jet turbine engines. Only a handful of other men from around the world designed jet aircraft that flew during World War II.

Reimar had known that the Luftwaffe needed a new and more advanced fighter since late 1940 when Walter revealed to him the true cost of the Battle of Britain. Six weeks of action had claimed all but nine pilots from a group of twenty-six replacements in Walter's unit alone. Many units suffered similar losses. Those pilots not dead were prisoners for the remainder of the war, and all of their aircraft were beyond reach of salvage teams because they had fallen on British soil or into the English Channel. The mount flown by these pilots, the Messerschmitt Bf 109, had several weaknesses. The aircraft was not designed to escort bombers, and it lacked the fuel capacity to fly long missions into England or engage in more than a few minutes of combat. It was also vulnerable to a single bullet hole in any part of its liquid cooling system, which was often enough to stop the engine cold.[1]

Walter's battle experience allowed him to frame a realistic set of goals for a new fighter and gave him incentive to press hard for it. Since his earliest days in the Luftwaffe, he had thought about "a tailless fighter aircraft."[2] Now he told Reimar that the overall performance of a worthy replacement for the Bf 109 must exceed the newest version of the British Spitfire. The new fighter wing had to possess enough range to fly a considerable distance over water and deep into English airspace. More range also provided extra time to loiter; for example, to extend fighter protection to the bombers during escort missions or to pursue more distant air and ground targets of opportunity.

"The only [solution to these requirements]," responded Reimar, "is to have a twin-engine plane of long range and long endurance . . . it should be all-wing."[3] Unfortunately, such an advanced, high-performance design lay beyond the designer's experience. He still had much to learn about aerodynamics, structures, and propulsion, yet Reimar had no access to wind tunnels, few personal connections to more experienced designers within the established aeronautical community, and none of the technical and logistical support available to a large firm such as Heinkel or Messerschmitt.

On his own, Reimar would not have ventured away from the comfortable milieu of sailplane development to work on another powered wing if his brother Walter had not appealed to Udet. Following Walter's permanent transfer to Berlin in May 1941, he convinced the Nazi leader in August to approve restoring the *H V b* (carelessly parked outdoors and exposed to the weather for several years) and then thoroughly test the aircraft to give Reimar more experience with powered aircraft, and the confidence to begin a jet project.[4] Walter got the Luftwaffe to defer Reimar's training duties at Frankfurt-am-Main, freeing him to concentrate on rebuilding the wing. The *H V b*, described by the RLM as the work of an amateur four years earlier, had now become an important step toward an all-wing jet fighter.[5]

To "obtain materials, components, subassemblies, parts," and other items needed to return the *H V b* to flying condition, it was necessary, Reimar said,

> to hold a priority letter before we could purchase [the items] anywhere. Walter always handled this stuff and he would obtain an official request for the materials and then I would go out and purchase whatever we needed. We never had any difficulty [getting the items] because Walter, acting in his official position as an officer in the Inspection of Fighters [Inspektion der Jagdflieger within the RLM] could always arrange things in our favor.[6]

Walter's future bride also helped. At the beginning of summer 1941, Walter had resumed a friendship with Udet's secretary, Sabina von der Gröben, which had begun when the two met in 1938. Udet often arrived at his office late for appointments with Walter, or missed them altogether,[7] whereupon Sabina and Walter would chat for a while, and then she would "type out a telegram stating what Walter required and then this telegram would be sent out. Sometimes Udet signed this telegram after he returned later and other times the secretary signed it for him and sent it out."[8]

In September 1941, Reimar collected three draftspersons, plus Scheidhauer and the crew helping him test-fly the *H IV a*, and moved to Minden to begin restoring the *H V b*. As the work progressed, Horten converted the cockpit to a single-seat configuration and made other changes until he finished the job during the winter of 1942. The resurrected and revised aircraft was designated the *H V c*.

Late in 1941, Walter heard that German propulsion engineers were making remarkable progress developing jet turbine engines. He recognized the benefits of joining these power plants with an all-wing airframe and passed the startling news to his brother.[9] The biggest disappointment thus far in his career had been the *H V a*—too much unproven technology forced into one airframe. Reimar was now more cautious, and when Walter urged him to procure the jet turbines and proceed directly to design the fighter, he chose to build a new interim wing called the *H VII* to study aerodynamic behavior at 300 to 400 km/h (185 to 250 mph), before attempting to design for jet speeds in the range of 800 to 900 km/h (500 to 560 mph) and higher. The maximum force of air on the airframe, or dynamic pressure, quadruples with each doubling of speed. Reimar needed to proceed carefully and observe how these increasing loads affected the wing's wood and steel tube airframe. Without access to a wind tunnel, the only reliable approach he could take was to study a wing in flight.[10]

Reimar rounded up an additional fifty or so civilians and soldiers and began to build the interim wing early in 1942. Like the *H V c*, the *H VII* was a two-engine pusher configuration. Reimar continued to favor this arrangement because, he said, a tractor setup with the propellers at the front "makes the plane unstable" by forcing the wing to fly through the airflow stirred by the propellers.[11] He toyed with mounting two Schmitt-Argus pulsejet engines on pylons above the wing or a pair of Junkers Jumo 004 jet turbines slung on pylons beneath the wing. Horten also considered installing Daimler-Benz reciprocating engines in a back-to-back, push-pull arrangement, as the Dornier firm had done with their Do 26 seaplane. In the end, Reimar abandoned all of these ideas as too heavy and chose instead to use a pair of Argus As 10 reciprocating engines each producing 176 kW (236 hp).

Reimar wanted to operate the *H VII* from grass airstrips, and a lighter aircraft was less apt to mire in soft earth. Even with strict weight control, flaps were required to slow the wing down. Flight trials showed the devices reduced landing speed to 80 km/h (50 mph), yet the clean design and surplus of power allowed the wing to cruise at 300 km/h (186 mph). The speed range was "very good," according to Reimar.[12] The *H VII* equaled the *H V c* in span, covered less than 14 percent more wing area, and weighed about 7 percent greater, yet three times more horsepower propelled the aircraft.[13]

The designer was now ready to transform wood, metal, Plexiglas, and other advanced materials into the most powerful and complex Horten aircraft yet built, but a key question remained: who would organize the support needed for the project? Besides designing and guiding construction of the *H VII*, Reimar was supervising several other sailplane projects. He was also busy studying mathematics at Bonn University during the fall of 1941 and at the University of Göttingen the following winter.[14]

The job of acquiring much of the monetary, material, and human resources to build the *H VII* fell to his brother. Reimar claimed that Walter's organizational skills had begun to blossom years earlier when he supplied Reimar with the resources needed to finish three *H II* wings, the first *H III*, and the mockup of an experimental control layout, all while serving full-time duties in the Luftwaffe between 1936 and 1938. Although Reimar credited his brother with organizing these projects, he could not explain his methods, saying, "only he knows."[15]

Walter obscured authorization to begin building the *H VII* behind an old work order number and directed the release of construction materials early in 1942 from his office in the Inspektion der Jagdflieger (also called Luftwaffen-Inspektion 3, Air Force unit Inspection 3, and abbreviated L In 3 on the RLM organization chart). Sabina no doubt continued to speed the paperwork through the bureaucracy. Her role had become even more important after Udet committed suicide in November 1941. As the effort gathered momentum that spring, Walter created a Sonderkommando, or Special Command, attached to L In 3, with help from a member of the Luftwaffe Quartermaster staff named Colonel Arthur Eschenauer.[16] To hide existence of Sonderkommando L In 3 from the RLM, Walter stamped telegrams to the unit with the word *Geheim* (secret) whenever it seemed safe to do so. According to Reimar, top RLM officials refused to support developing all-wing aircraft, making the subterfuge necessary to build such resource-intensive projects as the *H VII* during wartime. They were "not technical men and knew absolutely nothing of development nor did they want to learn" about new technology, he remarked.[17]

According to Reimar, the guise was effective because the RLM unknowingly paid the Pesckhe firm to labor about one thousand hours manufacturing the aluminum center section and assorted metal fittings for the *H VII* and parts for other wings. Reimar recalled that as the construction moved further along,

> Walter then began locating engines, wood, metal, instruments, fuel tanks, landing gear from whatever sources within the Luftwaffe and Air Ministry [RLM] where he could find them . . . Walter would classify his activities as secret so when people asked [we would refuse to answer] . . . because all this information is classified![18]

Reimar requested space and personnel from the Aerodynamische Versuchsanstalt (Aerodynamic Experimental Center or AVA) at Göttingen and received unskilled laborers, drafters, metalworkers, and plenty of carpenters. The test facility was working mostly with metal aircraft at that time, so there was a surplus of wood workers. In total, about fifty men initially joined Horten's group at Minden, and more soon followed.

Walter's organizational skills extended to day-to-day management, thus freeing Reimar to focus on design. Reimar recalled specifically telling his brother that he "wanted to work in the technical matters and did not want to be occupied with the personal problems of the workers. Walter, therefore, looked after all the men including their personal problems."[19] Indeed Reimar reveled in the freedom from military duties and petty distractions. "Yes, yes," he recalled, "[intellectual] freedom is one thing I enjoy . . . when I had to work seven and nine hours a day to look for the aerodynamics of a plane and get it to perform, that was my best time . . . I was able to go to my room and carry out my calculations which I would then give [to the drafters] the next day."[20]

Small Shop Network

Reimar's preference to work alone on several projects simultaneously during the war led the designer to supervise a network of small workshops scattered around Germany. More than one project rarely occupied a single shop so that each group of craftspersons could focus with fewer distractions. In this way, Reimar explained, "projects were completed on time and one project was not worked at the expense of others . . . Separation, for us, was the best form of organization." Separation also allowed Horten to focus on design. As he put it, "if I had all my work centralized, people would come and want this project or that project finished first, and so I'd be getting down with matters of production and not with matters of development, which I needed to pursue in order to develop the all-wing plane. This trap I could not allow myself to get caught in."[21]

A decentralized approach also dovetailed with Reimar's preference for smaller and less complicated wings with shorter design/build/fly cycles, which were less likely to tax shops staffed by small groups working with limited resources. "If I have a small plane," he said, "I could change the design year for year, the development is quicker, is better. If I have a great plane like Northrop with eight turbines, I cannot make year for year a new design [but I would require] five years for a new design."[22] Reimar was referring to American designer Jack Northrop, who needed seven years to develop the YB-49 all-wing bomber powered by eight jet turbine engines and first flown in 1947.

A small shop also drew less electrical power (Reimar estimated about 2,000 kilowatts), making them easier to set up, less costly to support, and less likely to attract official attention. The network helped to reduce the fiscal signature of Horten's operations. "Yes, yes, that is one of the reasons! To put the workshops in another province, which then would be in another administration. This way there would be no interference" from the RLM.[23] "We had to camouflage our work . . .," noted Reimar, so "that [they] would not disturb [us]."[24]

Horten's operation was flexible enough that he could move several gliders, along with test pilots and support personnel, to the Wasserkuppe from April 1943 to the end of August and continue to test-fly them because for a time the fuel and oil required to operate the tow airplanes was too expensive at Göttingen, where they had been based. In 1943, the government ordered the twenty-seven largest aircraft manufacturers to disperse their production facilities and hide them in forests and underground hangers as protection against the intensifying Allied bombing offensive. The order caused Reimar much less disruption than it did some of the larger factories—he had been running a dispersed operation for years. What made decentralization difficult was the time and effort required to manage such a scattered effort. Reimar faced problems similar to those confronting the established aviation firms: efficient and timely transportation of components to central points for major assembly and effective communications between the satellite firms.[25]

To communicate, Reimar took to the roads. To make a full circuit of his shops in 1943, he had to drive about 1,000 km (620 miles) and make the drive usually twice a month, and at night to reduce the chance of attracting Allied fighters. He managed the construction of several wings at a workshop in the Hessian Forest at Bad Hersfeld, about 320 km (198 miles) southwest from Berlin, and 40 km (25 miles) from the Wasserkuppe. The Bad Hersfeld shop was 110 km (68 miles) from his quarters at Göttingen, but pulling a trailer loaded with equipment that included woodworking tools and a lathe slowed the car to 70 km/h (43 mph) and the trip took an hour-and-a-half. The depletion of petroleum supplies compelled Horten to convert the car to burn gas from a wood-fueled generator.

Reimar sometimes traveled with a man to take over for him at the wheel so the designer could steal some rest. He drove himself to Bonn University to attend classes, sometimes twice a week during a semester. He claimed to have worn out as many as twelve automobiles during the war. Reimar believed that his visits, although demanding physically and mentally, reduced the engineering overhead. He cited the *H VII*, which had required five hundred drawings to build, and the *H IV a*, which required about two hundred drawings when built by teams working without his direct assistance. Reimar halved these numbers with his presence onsite and reduced both the number of drafters needed and the time required to draw the components so that craftspersons could build them.[26]

Mismanagement Leads to Opportunity

By the end of 1942, more than a dozen different models of German two-engine airplanes were operational or in the final stages of development.[27] Each was designed to fulfill roles that often overlapped, but even worse, the aircraft carried modest payloads at sedate speeds over relatively short distances. During a tirade delivered on March 18, 1943, Luftwaffe Chief Hermann Göring berated representatives from the aircraft industry and the Luftwaffe for recklessly expending vital resources on redundant designs with obsolete performance, adding that few of the major components and subsystems required to build these complex machines were interchangeable. By contrast, the Allies fielded fewer different aircraft types, and most exhibited superior performance.

In his speech, Göring singled out the De Havilland Mosquito, a British two-engine, propeller-driven aircraft that frequently made him "go beserk" and turn "green and yellow with envy."[28] Mosquitoes could carry up to 1,800 kg (4,000 lb.) of bombs at speeds beyond 644 km/h (400 mph) and at ranges as great as 2,206 km (1,370 miles). The aircraft had been operational since September 1941, and on January 31, 1943, the first Mosquitoes bombed Berlin and interrupted a parade presided over by Göring. The Nazi leader's diatribe climaxed with a decree: henceforth he would approve no new two-engine airplane unless it possessed a range of 3,000 km (1,860 miles) and could carry 1,000 kg (2,210 lb.) of bombs a distance of 1,000 km (620 miles), at a speed of 1,000 km/h (620 mph).[29] Göring's hoped to goad the aviation industry into fielding a German version of the superlative Mosquito.

Walter considered it impossible to meet every one of Göring's requirements, and Reimar agreed that reciprocating engines could provide neither the speed nor the range, but the jet turbine engine had forever altered the performance landscape. Just months earlier, the Horten brothers had visited the Junkers firm and obtained performance data for the company's 004 jet turbine engine. The numbers convinced Reimar that an all-wing airframe propelled by these motors could come close to meeting Göring's demands. The jet engine consumed too much fuel to meet the range requirement, but Reimar thought that an all-wing jet with low-wing loading would carry double the bomb load and compensate for missing the range goal. Upon receipt of Reimar's initial proposal early in 1943, Göring's technical officer, Major Ulrich Diesing, circulated the paper within the RLM and set in motion a series of evaluations of the Horten all-wing concept.[30]

Early in 1943, Professor Josef Stüper, who worked for the AVA at Göttingen, flew the *H V c* as many as twenty times, as directed by the DVL, which wanted a firsthand account of how the wing handled in the air with an engine out compared to conventional aircraft. The *H V c* was the only Horten aircraft tested by AVA staff. Both Walter and Reimar had flown "this wonderful little machine" in the winter of 1942, and it seemed ideal for introducing the all-wing configuration to others.[31] A postwar report paraphrased Stüper's evaluation, saying, "it was hardly fair to compare [the *H V c*] with conventional aircraft with many years more development behind them but it was, nevertheless, a good example of tailless design and a perfectly practical aeroplane—if anyone wanted tailless aeroplanes. His main suggestion for improvement was in the rudder control."[32]

Stüper gave high marks, the report continued, for longitudinal stability, and he rated lateral stability satisfactory with no tendency to Dutch roll. He found good damping about the yaw axis, an area where he had expected trouble. Stüper could easily move the controls and found them effective except for the rudder, which he said was stiff. It is interesting that Stüper did not mention adverse yaw.[33]

This favorable assessment never reached the mainstream aeronautical community, which disputed what many flights in Horten aircraft had demonstrated: that Reimar Horten's designs were safe to fly and merited further study. Results from wind tunnel tests conducted on non-Horten flying wings continued to sway many in the RLM who seemed to believe that all tailless aircraft flew poorly. The official opinion on the all-wing configuration remained negative.

In April 1943, experts gathered in Berlin to assess the current state of all-wing aircraft design during the Sitzung Nurflügelflugzeuge (Meeting on

Only-the-Wing Aircraft) sponsored by the Lilienthal-Gesellschaft für Luftfahrtforschung (Lilienthal Society for Aviation Research). Alexander Lippisch attended along with scientists, engineers, and mathematicians. The Horten brothers delivered two papers, an historical summary of their work over the previous decade leading up to the design of the all-wing jet fighter and the only known technical paper authored by Reimar Horten. In it, he refers specifically to a jet-powered all-wing fighter designed to meet Göring's extreme requirement for a new two-engine bomber.[34]

Reimar's technical paper contrasted sharply with the conclusions presented at the meeting by Professor A. W. Quick. A wind tunnel specialist at the DVL facility in Berlin, Quick warned that a swept-wing tailless aircraft was unstable at high angles of attack because of increased spanwise airflow due to wing sweep, which was likely to cause premature airflow separation, or loss of lift, along the outer wing. He also noted with particular concern "that once the wing reached and exceeded the point of maximum lift [with increased angle of attack], the wing sweep caused pitch instability that could only be cured with less sweep or by adding a tail unit."[35]

Quick warned that these aerodynamic characteristics would cause loss of control during a stall, likely followed by an unrecoverable spin. However, he was ignoring the considerable body of real-world experience that proved otherwise, gathered from pilots who for years had safely flown Horten wings such as the *H III*, *H IV a*, and *H V c*, under the very conditions that Quick and Dr. Rudolf Göthert, another DVL aerodynamicist, called dangerous. Quick based his conclusions on tests conducted in a wind tunnel using a model that differed from the Horten designs in aspect ratio, wing taper, degree of washout, and control surface layout. Göthert also published findings that drew the same false conclusions. Walter recalled how the AVA warned the industry against swept all-wing designs, prompted, in part, by Quick's and Göthert's skewed analyses.[36]

Perhaps Reimar believed that pilot experiences would carry more weight if he could duplicate them in wind tunnel tests using an accurate model of a Horten all-wing aircraft. Still denied access to state-owned tunnels, Reimar decided in 1943 to design and build his own flying wind tunnel called the *H VIII*. He envisioned an all-wing aircraft with a large compartment (the tunnel) open at both ends, fitted beneath the wing's center section. Reimar wanted to mount inside the tunnel various airfoils and aircraft models and then fly the wing to generate the airflow necessary for testing. He also needed the tunnel to calculate the correct location and shape of the turbojet air intakes to be fitted to the *H IX* jet. Reimar wanted to use the aircraft to "develop further the concept of laminar flow . . . to investigate the profiles of other planes . . . And I wanted to do this in natural air without any turbulence from a propeller."[37] He believed that his "wing tunnel" offered the opportunity to "fly with the speed of about 300 km/h [186 mph], but the [air velocity inside the tunnel test section could be] two-and-a-half to three times [higher]. We get a high-speed tunnel with low energy [input, and we can] fly at different heights, different Mach number, not supersonic but . . . high subsonic Mach number."[38]

Reimar claimed that the AVA facility at Göttingen had nothing comparable, and only a few other tunnels in Germany could perform as well. The *H VIII* would have been large, doubling an *H III* wing in both span and area, and require six 236-horsepower reciprocating engines driving pusher propellers to fly. Although construction started, Reimar could not obtain enough high-quality wood in long, straight lengths to finish the wing, and a craftsperson had assembled only the welded steel tubes of the center section when American army units overran the building site at Göttingen in May 1945.[39]

Prandtl Impressed

On February 27, 1943, Walter arranged for Scheidhauer to demonstrate a wing at minimum altitude and airspeed for a group from the AVA that included Ludwig Prandtl. Although the flight occurred before the Sitzung Nurflügelflugzeuge at Berlin in April, Reimar wrote nothing about it in either of his papers, perhaps because he had already submitted them to the conference. While flying an *H III d* motor glider under power low over the ground, Scheidhauer suddenly pulled back on the stick, literally whipping the wing almost straight up into a stall. If the aircraft lurched out of control, as Quick and Göthert said it would, Scheidhauer had little chance to escape the ensuing crash by bailing out at this altitude.[40] Walter remembered how Prandtl visibly recoiled during each maneuver, but three times the test pilot reared the wing up to a near vertical attitude, stalled, and then recovered immediately and easily to normal flight. Karl Nickel remembered there "was no tail-slide or roll-off, but rather the [*H III d*] fell down into the normal flying position without loss of altitude and continued her horizontal flight."[41]

According to Walter, Prandtl was "flabbergasted" and moved to ask why the demonstration contradicted wind tunnel tests.[42] Reimar answered that the wing had remained under Scheidhauer's control because the twist (washout) he had imparted to the wing tips allowed the flight controls to remain effective throughout a stall. Both the degree of twist and its effect on stall control were characteristic of the bell distribution. Reimar had also placed the CG well forward, as Walter recalled his brother's explanation, to neutralize the pitching moments about the center of gravity and center of pressure. Walter and Reimar concluded that Prandtl had not considered this critical factor because he had no experience designing aircraft. Instead of grasping the complexities of an aircraft maneuvering under actual flight conditions, Prandtl's ability to visualize was more theoretical and limited to the wing "flying fixed in a wind tunnel. There is quite a difference, quite a difference," said Walter.[43]

Walter claimed that following Scheidhauer's demonstration, the AVA withdrew its recommendations against swept all-wing aircraft to the aviation industry, while Reimar recalled that the flight moved Prandtl to write a memorandum to officials in Berlin who controlled the wind tunnels, recommending tests of the Horten bell distribution. I found no documentary evidence supporting either of these actions. Reimar became convinced that RLM officials continued to question his work because he lacked education and title.[44]

> We had performed all the measurements, we had the numbers, the theory and everything worked. You know what we didn't have? We did not have credentials . . . [and as a result], people viewed us, viewed me as an amateur, merely playing with sailplanes.[45]

Proposal, Acceptance, Deadline, and Delay

Six months after the head of the Luftwaffe issued his ultimatum on the minimum performance of new twin-engine aircraft designs, Reimar and Walter met with Göring to discuss the *H IX* proposal on September 28, 1943.[46] Lippisch joined the group, too. To have heard their pitch despite the recent controversy about Reimar's ideas was a sure sign of the outright desperation beginning to grip the Nazi leadership. By the end of summer 1943, the increasing tempo and intensity of the day and night strategic bombing campaign conducted by the western Allies was affecting key areas of German military production and devastating civilian population centers. Attacks on the factories building fighter aircraft caused a drop in production of two hundred per month in July and August, while the bombing of Hamburg from July 24 to August 3 killed more than forty thousand people and displaced most of the population of more than one million. The Luftwaffe had transferred increasing numbers of fighter units from the Eastern and Mediterranean fronts to combat the bombers, only to incur greater losses of aircraft and trained crews without appreciably lessening the bombing.[47]

The Luftwaffe chief listened for several hours as the brothers explained their work and Walter demonstrated the semiprone pilot position that Reimar had designed for the *H IV a*. Reimar projected 960 km/h (595 mph) as the wing's top speed.[48] When Göring questioned their background and experience, Reimar answered, "we had learned and developed all our techniques from sailplane construction and flying beginning with the Rhön competition in the mid to late 1930s."[49] Göring remained relaxed and informal, Reimar remembered, "and he was open [to our ideas] . . . and I could [speak] to him like with my father," but he remained critical of the aircraft industry.[50]

Göring asked Reimar for a glider prototype of the *H IX* ready to fly in three months. Reimar called that goal impossible to meet and asked for six months, which Göring accepted, and promised to provide 500,000 marks to continue the project if the flight occurred before March 1, 1944. The terms delighted Reimar, who would soon have "a basis for starting my work in an official and highly organized manner."[51] The brothers soon formed the company Horten G.m.b.H. with ambitions to develop advanced aerodynamics, structures, and propulsion. Based on the *H IV a* experience, Reimar must also have seen possibilities to divert some of the resources from the *H IX* to other projects.[52]

By starting with a glider prototype, Reimar would be able to test the basic characteristics, and refine them if necessary, before confronting the work of installing the engines, fuel tanks, and other equipment. He also needed time to work through other difficulties, too, such as determining whether wood was a suitable material for the jet's outer wing panels. For a time, his calculations showed that the aerodynamic loads at jet speeds would require wing skins too thick to be practical, but the material was appealing because Reimar and his team at Göttingen had neither skills nor equipment for working aluminum into such large components as wing panels. Horten was comfortable with the materials and processes he had used since finishing the *H II* in 1935: steel tubes for the center section and wood for the wings.

One morning, Walter picked up his brother at Minden and flew him to the test center at Peenemünde on the southwest Baltic Sea coast. They gathered with a crowd to witness the former champion sailplane pilot, Heini Dittmar, demonstrate the rocket-propelled Messerschmitt Me 163A prototype. Messerschmitt had built the airplane's wings from wood, and Dittmar demonstrated that morning how the material could withstand speeds as high as 1,000 km/h (620 mph). Reimar returned to Minden with his faith in wood renewed. Although using wood in such high-speed applications was unusual, Eric Schatzberg has pointed out that Germany accelerated its use of the material for aircraft construction toward the end of the war, whereas the USA stopped using wood in aircraft in 1943.[53]

Construction of the *H IX V1* glider moved apace. Göring promoted the Horten brothers for meeting their six-month deadline after Scheidhauer flew the glider on March 1, 1944. The flight was not without drama because the wing possessed less L/D than any other Horten sport sailplane. As a result, there was no extra glide in this glider and approaching to land had to be planned with care. Compounding the problem was the airfield at Göttingen, which consisted of a grass circle 1,100 meters (3,608 ft.) in diameter, ringed with hangers. Overcoming the drag of the soft grass and accelerating to sufficient speed to clear the relatively tight confines of the airfield perimeter with the Horten wing in tow required a powerful airplane, so the Luftwaffe lent the team a two-engine Heinkel He 111 bomber stripped of all unnecessary weight. Reimar decided to lighten the big Heinkel by draining half its normal fuel capacity, yet even in this lightened configuration, technicians had to attach booster rocket motors to the bomber, which ignited during takeoff to further shorten the ground roll. The glider was relocated to Oranienburg Airfield near Berlin, 260 km (160 miles) northwest of Göttingen, which had longer, safer concrete runways, and which became the permanent test site for the *H IX* project.[54]

Reimar recalled that several curious Luftwaffe and RLM pilots flew the *H IX V1*, including his brother and Erwin Ziller, whom Walter selected to test-fly the jet version. These flights revealed several problems. Reimar judged excessive vibration in the nose wheel as a small concern, easily remedied, but the aerodynamic issue was more serious. With no control inputs from the pilot, the glider exhibited the oscillation that Reimar described as Dutch roll. Walter recalled that the glider "was stable, it [flew well], but in pumping air [turbulence], a little [unstable] for gunnery."[55]

After a team from the DVL fitted the *H IX V1* with instruments and measured the frequency and duration of these oscillations, their tests showed that a pilot could steady the wing by pushing both rudders simultaneously. Based on these findings, Horten believed that the oscillations could be "controlled in enough time to hit a target, if the target was known before hand."[56] He failed to mention that tapping the drag rudders would also have slowed the jet, perhaps enough to negate its speed advantage over Allied aircraft.

The yaw oscillation was a problem that could jeopardize the jet wing's primary mission; nonetheless, Reimar hurried to finish and fly the first jet wing, designated the *H IX V2*. An RLM official had recommended the BMW 003 jet turbine engine, but the designer never seriously considered it for several reasons. The 003 had been delayed in development while the Junkers 004 was further along. Horten also saw an opportunity to demonstrate the superiority of the all-wing layout over the conventional by using the engine that also powered the Messerschmitt Me 262. Designers had mounted two 004 engines in streamlined pods slung beneath the wings of this tailed fighter-bomber, and pilots had begun to fly the airplane during operational sorties in April 1944. Reimar correctly regarded this aircraft as the yardstick by which the German professional aeronautical community would measure the success of the *H IX* program.[57]

On paper, the *H IX V2* outperformed the Me 262. Compared to the primary fighter variant called the Me 262A-1a, the Horten wing carried nearly 22 percent more fuel, but because Horten used a larger wing area to carry the load, he reduced the wing loading by nearly 43 percent compared to the Messerschmitt jet, which led to several operational advantages. The Me 262 required a 1,050 m (3,445 ft.) concrete runway to take off, but Reimar designed the *H IX V2* to operate from shorter grass runways. Walter claimed that nearly any Luftwaffe airfield could have accommodated the all-wing aircraft, whereas only three Luftwaffe airfields had runways long enough to operate the Me 262. The Allies soon discovered these bases and bombed them regularly, he said. When the mission required it, Göring could probably have accepted a longer takeoff distance in return for carrying a heavier bomb load.[58]

Walter may have overstated the case somewhat, but grass airfields were certainly plentiful throughout Germany compared to the concrete type, which was expensive to build and maintain and more difficult to repair if damaged by bombing. The *H IX V2* appeared to give the Luftwaffe more tactical flexibility than was available using the Messerschmitt jet. Although foreign-object damage to the fragile jet turbines that powered the Horten wing would have been a concern when flying from grass airfields, the *H IX V2* squatted on its landing gear at an angle that raised the jet intakes a couple of meters above the ground and clear of most debris.

Lighter wing loading would have given the *H IX V2* a faster rate of climb and a higher operating ceiling, increasing both tactical effectiveness and range over the Me 262. But Walter felt that flying at altitudes above 12,000 m (39,360 ft.) without a pressurized cockpit could cause pilot discomfort and lessen the wing's effectiveness. This concern may have led to trials of unknown scope and duration, using the *H IX V1* glider to test an experimental pressure suit. Photographs of this suit appeared without explanation in the book *Nurflügel*.[59]

Progress to integrate the 004 into the wing in the latter half of 1944 slowed to a stop after poor communications with Junkers about the size and configuration of the engine forced Horten's team to rebuild major portions of the aircraft and delayed the first flight by three to four months to February 1945. Although Reimar claimed friendship with the man in charge of developing the 004, Anselm Frantz, the sizing error may have occurred because the

Junkers firm regarded Horten's jet program as less than credible. Reimar could also have used faulty data supplied by Walter to calculate the volume needed. Walter had provided his brother the engine thrust curves, for example, after tracing an original graphic held against a window.[60]

Reimar had wanted to carefully revise the aircraft to accommodate the larger motors, but there was no time. Because he had buried them inside the wing, Reimar faced a significant challenge, not just to enlarge the mountings but to reengineer the location of many major components, leading to changes in the exterior dimensions and shape of the aircraft that required new aerodynamic and performance calculations. By comparison, changes to the wing-mounted engine nacelle of a conventional multiengine aircraft would have been confined largely to the engine mount and exterior cowling covering the engine, leaving the wing, fuselage, and empennage unaffected.[61]

To save construction time and control costs, Reimar scrounged parts from wrecked aircraft. He took the complete tail wheel, support strut, and hydraulic actuating mechanism from a Heinkel He 177 heavy bomber and mounted this assembly into the nose of the *H IX V2*. Heinkel had designed the bomber to weigh more than thirty-five tons fully loaded with fuel and ordnance; hence, these components could handle the *H IX V2's* projected weight—more than 8,800 kg (19,000 lb.) with a full load.

Lack of a tail at the end of a conventional fuselage promised to make landing the jet wing an unorthodox affair. Many tailless and semitailless aircraft share this characteristic, tending to alight first on their nose gear, and at higher speeds than similar conventional aircraft. Reimar anticipated a touchdown speed of 120 to 130 km/h (74 to 81 mph), a velocity high enough to stress the *H IX V2* nose wheel assembly as the uneven surface of a grass airfield sent shocks and vibrations into the airframe.[62] The reduced pitch authority characteristic of tailless aircraft could lengthen the takeoff run, and Reimar compensated by lengthening the nose wheel strut, which increased the angle of incidence and boosted lift. This was particularly important since Horten balanced his wings with the CG well forward in order to prevent stalling.[63] Yet he had to keep the nose strut short enough to fit within the confines of the center section when retracted, while the two main undercarriage units had to be long enough to prevent overhanging components, such the tips of the swept wing, from snagging the ground.[64]

Reimar broke from tradition and tucked not one but two sets of spoilers into the top and bottom of each *H IX V2* wing tip to act as drag rudders and designed them to open sequentially depending on the wing's velocity. If the pilot pushed on the pedals lightly, only the smaller pair opened, but pushing farther, as he might do when turning at a slower speed, popped the second, larger pair out of the wing.[65] Horten installed an elevon and two flaps into each side of the wing's trailing edge to complete the control group on the jet. He felt that he should ease the pilot's workload by reducing the force needed to activate the controls at high speeds, so he designed a mechanism to vary the control stick leverage. When the pilot moved his stick at the high-speed setting, the control surfaces moved 1/3 less than their travel at the low speed setting. A control stick with variable leverage was also tried in the Me 262.[66]

Jet Wing Aerodynamics

Reimar wrote that he had choreographed a bell lift distribution for the *H IX V2*, but after studying drawings and wing configuration data calculated during the war by Karl Nickel, Reinhold Stadler concluded otherwise.[67] The Horten distribution was characterized by a symmetrical airfoil profile at the wing tip and a reflex profile at the center section. To accommodate the unexpected increase in the diameter of the Junkers 004 engines, Reimar had to shift the reflex profile toward the wing tips a distance of 400 mm (16 in.) and insert a new profile at the center section. As a result, the *H IX V2* lift distribution differed substantially from the bell, as Stadler explained:

> The geometrical twist was linear. The combination of [linear] geometrical twist and wing section [profile] distribution does not result in the bell shaped lift distribution. The Horten formula does not describe which part of the total twist is covered by zero lift angle of the wing section [profile] and which by the twist. But Horten normally used this reflexed center section [profile] and the symmetrical [profile] at the tip. In my understanding Horten was not fixed to the bell shaped lift distribution, but played with other distributions too in order to get the optimum layout.[68]

Twisting the wing to a linear geometry, or washing out the wing at the same angle from rib to rib, was considerably easier to calculate, and Reimar had no time to calculate a nonlinear wing twist in his haste to redesign the *H IX V2* center section. He still worried over the Mitteneffekt and attempted to restore lift by making the maximum airfoil thickness fall across the aft center section, giving the wing its characteristically pointed tail. Supposedly to find out more about the effect of these changes on the jet's stability, Reimar configured the center section of the *H VI* sailplane in similar fashion.[69]

Wind Tunnel Substitutes

Walter would not accept Reimar's claim that his sailplanes contributed to the German war effort and became "the backbone of the *H IX*" test program when he could not get access to a wind tunnel.[70] Nonetheless, several gliders proved useful to the work. Reimar used an *H II* to refine the shape of the intake openings planned for the third prototype, called the *H IX V3*, and to study different center section shapes for the *H IX V6*, a trainer and night fighter version. He also modified an *H III f* to observe and adjust the layout of the drag rudders he designed for the *H IX V2*.[71]

Tufting to visualize airflow was a favored research tool. Reimar directed his team to glue up to several hundred tufts of wool at regular intervals onto a wing surface. The designer could designate a section of particular interest, say the air flowing behind the drag rudders, and leave the remaining wing untouched. Horten then observed how the tufts moved about in flight, sometimes observing them from a two-seat Kranich sailplane or a Fieseler Storch powered airplane as it followed the wing through various maneuvers.

Reimar frequently photographed the tufts with what Walter described as a "grosse Kamera," a large-format camera equipped with a powerful lens and developed for military reconnaissance work. Reimar had obtained it from the Bildschule Hildesheim, a photography school. Technicians could set up the camera so that Reimar could trigger the shutter remotely from inside the cockpit. For some tests, Horten fitted to the wing an airspeed indicator and a wind vane indicating the direction of air flowing at that spot. The camera could record the wing, tufts, airspeed, and vane in a single image. Reimar sometimes painted the tufted areas in contrasting colors to make them stand out on film. He could later refer to this photographic record to determine how well various aerodynamic configurations functioned; however, the technique had limits. Reimar's team once tried to study an entire *H II f* covered with tufts from wing tip to wing tip, but the focal length of the camera lens was too short to image all the tufts at one time.[72]

First Flight

Early in October 1944, trucks towed the *H IX V2* center section on its landing gear, accompanied by the outer wing panels riding a trailer, to Oranienburg Airfield. Technicians joined the major components and carried out other final preparations. Lt. Erwin Ziller recorded in his logbook the first takeoff and landing in the jet five months later on February 2, 1945. The day was overcast, making it harder for patrolling Allied fighters to ambush the wing as the test pilot concentrated on his flying. Ziller knew that the fortunes of Horten G.m.b.H.

rode on the outcome to the flight. The wing shared the airfield with two other jets already in production, the Arado 234 jet bomber and reconnaissance airplane, and the Me 262. Complicating the situation was the nearby location of quarters frequented by Göring and his staff. Everyone on the airfield would watch closely to see how the *H IX V2* compared with the other jets, and even a minor incident during testing would spread rapidly by word of mouth, possibly eliminating the wing from consideration for a military contract. Fortunately, Ziller did not have to bear the anxiety alone. From the control tower, Reimar talked him through the entire half-hour flight as the pilot ascended to about 1,000 m (3,280 ft.), retracted the landing gear, executed climbing and descending turns, and then landed uneventfully (Figure 25).[73]

Immediately after the flight, Reimar pressed Ziller to describe the effects of aileron and inboard flap deflections on adverse yaw and whether Ziller had to apply the drag rudders to counter it. Horten also wanted to determine whether he had designed the correct differential gearing between the ailerons to minimize adverse yaw, 3° up-aileron for each 1° of down-aileron.

Although Ziller's answers to these questions are not known, it seems that both men were satisfied with the jet's basic aerodynamic qualities. Measuring the heat radiated by the engines and its effects on the wooden skin behind the

Figure 25. Technicians prepare the Horten *H IX V2* jet wing for flight (National Air and Space Museum, Smithsonian Institution, SI 77-14586).

exhausts was also a high priority, and Walter planned to inspect the wing for damage in this area after the next several flights.[74]

Both Reimar and Walter were away from Göttingen during Ziller's next flight, but they gave the test pilot specific instructions: limit his time aloft to thirty minutes, climb no higher than 4,000 m (13,120 ft.), use full power to climb but reduce the throttles in level flight to remain below 500 kp/h (310 mph). Several days later, Ziller called Reimar with the results from his sortie. The test pilot was familiar with the Me 262, and he reported that the Horten jet could climb faster, and land more slowly, than the Messerschmitt fighter-bomber. The *H IX V2* seemed to have adequate basic handling qualities in the air, and the brothers must have been happy with the test program's initial progress. Unfortunately, Ziller also reported the troubling news that he landed the wing hard enough to bend one or more landing gear struts. Fourteen days were lost while technicians repaired the damage.[75]

Ziller's third flight on February 18 ended in a crash that killed the pilot and destroyed the jet. He had made three passes at an altitude of about 2,000 m (6,560 ft.) to allow a team from the Rechlin Flight Test Center to observe the wing. Ziller then lowered his landing gear and began a spiraling descent before crashing near a railroad embankment. Inspecting the wreckage confirmed that one engine had failed in the air, something the early jet turbine power plants did often and unexpectedly. Ziller's lack of experience with all-wing aircraft, undetected damage from his hard landing two weeks earlier, and the poor weather conditions all seemed to contribute to the accident. Reimar acknowledged that one engine thrusting at full power would have caused more yaw than a pilot could compensate for, using the drag rudders and elevons; however, the brothers maintained adamantly that Ziller could have reduced power on the functioning engine and flown the jet to a survivable crash landing.[76]

There had been official interest in whether a pilot could control the *H IX V2* if an engine failed, and Walter informed the ministry of Ziller's crash without revealing the details. During a visit to Oranienburg in November 1944, Göring had surprised the Horten team by asking Scheidhauer to demonstrate stopping one engine while flying the Horten *H VII*. The pilot obliged and then turned and climbed with apparent ease, but the flight ended awkwardly. Scheidhauer could not restart the engine and crash-landed, albeit unhurt and with little damage to the wing. Göring was impressed and dismissed the crash as something expected of new aircraft types, according to Reimar.[77]

The years of laboring to build the jet from scrounged materials and components had come to naught. Reimar Horten's most ambitious design now joined the growing list of wonder weapon concepts that failed in practice during the final desperate months of the Third Reich. At the time that Ziller crashed, the Nazis had designated a firm to prepare to manufacture the *H IX*, in parallel with additional developmental test flights, but Walter realized that "all our work was over at this moment."[78] The jet wing project had indeed ended, but not before it had mesmerized Göring with grandiose possibilities. At a meeting on March 12, he proclaimed that all types of conventional aircraft were destined to metamorphose into the all-wing configuration. Then Göring announced that "the Chief of the Air Equipment Branch should consider immediate inclusion of Horten development and production work into the Führer's emergency [aircraft] program," and he ordered pilots to begin training to fly the Horten wings.[79]

Göring also urged "the best experts in the [aircraft design] field . . . [to join] into one working association for only-wing aircraft," organized to operate under the auspices of the Reichs-forschungsrat (National Research Council) and in cooperation with the Horten brothers.[80] Indeed Chris Wills, whose father, Philip, had helped to establish the sport of soaring in Britain, and who accompanied several teams of intelligence experts touring Germany at the end of the war, calculated that more than sixty Horten powered wings and all-wing sailplanes were planned or under construction when hostilities ended in May 1945. Yet Göring's support for their work in no way altered the reality that both Reimar and Walter now faced. Hitler's regime was in its death throes, and although they continued to work, the brothers accomplished nothing significant in the last few months of war. For them and many other Germans, the primary goal had become survival.

Engineer Felix Kracht worked at the branch of the DFS in Ainring, Germany. He had codesigned an experimental, high-altitude reconnaissance aircraft called the DFS 228, equipped with a prone cockpit and propelled by a rocket engine. As Kracht put it,

> At this time—around early 1945—no one was thinking about an aircraft for the Front anymore. You could see the war effort being supported by America and what was happening to our homeland due to the USAAF's bombing. Despite this hardship you had a task. You did it, to make it go no matter what was happening around you. If you didn't, you'd likely end up on the Russian Front. So we did what we were told to do.[81]

By 1945, Göring and many of his sycophants also sensed their own impending doom. With this awareness came desperation to develop weapons that could delay or deflect defeat. The weapons development community met this demand with a flood of proposals. Some were prescient concepts that other nations would develop into successful aerospace technology after the war. Many other proposals constituted little more than scribbled notions and hurried sketches of bad designs that ranged in quality from the experimentally feasible, but operationally impractical, through the morally reprehensible, to the utterly ridiculous. Yet nearly all these concepts appeared capable of altering the balance of power on the battlefield, and therein lay the source of their appeal to the Nazi officials, who lacked the qualifications to judge their merit, but were nonetheless desperate to seize on them for salvation. Members of the German technical and scientific community well understood the dynamics of this desperation and manipulated it to enhance their own survival and that of their brethren. Alexander Lippisch described how

> the difficult military situation demanded the drafting of many of the remaining civilian employees for service in the "Volkssturm" (people's army), which was regarded by the powers-that-be as the last line of defense. This meant that the aeronautical engineering students also had to join. Only work of highest priority rating could protect these students from being drafted. Aeronautical students of the Institutes of Technology of Darmstadt and Munich approached me in the hope of obtaining work in connection with the P 13 [aircraft project], since the end of the war obviously was imminent anyway. I created the project of a wooden flying glider version of the P 13, which the students were to build under the direction of my assistant, Heinemann, in a hanger [at] the small airfield in Prien on the Chiemsee . . . We succeeded in obtaining deferment [from combat] for the students.[82]

Drawing large numbers of aviation workers into the army had begun years earlier in December 1941 when Hitler tried and failed to capture Moscow, and occurred again in October 1942 as fighting raged at Stalingrad. A year later, Hitler ordered another sixty thousand workers from the aircraft industry into the German army. As Russian forces were fighting in Rumania and Bulgaria, Walter recalled that a Nazi official visited Göttingen around August or September 1944 to announce the Horten staff would be drafted to fight with the German army on the Eastern Front.[83] The draft never happened, but Reimar's status as student and independent maverick designer made him vulnerable to conscription; therefore, he wanted to stay as busy as possible until the war ended.

Busy to the End

Hitler announced the program to mobilize all fit German males into the Volkssturm (people army) to defend Germany, on September 25, 1944. In addition to a land component that consisted of a hastily organized and trained army of "volunteers," there was also an aerial component planned to be organized around the Volksjäger, or People's Fighter aircraft. The Volksjäger had to be easy to construct by unskilled workers using nonstrategic materials and simple to fly for the Flieger Hitlerjugend (Flying Hitler Youth) aged sixteen to eighteen, whom the Nazi officials were quite willing to throw into aerial combat with almost no training.

Despite a statement made by historian Ulrich Albrecht, there is no evidence that Reimar actually submitted an official proposal to develop a Volksjäger, although he apparently made drawings and a preliminary study, perhaps to buttress his value to the Third Reich as an aircraft designer rather than a soldier for the Eastern Front. A single, three-view drawing identified as the *H X Volksjäger* was published in *Nurflügel*, and Reimar wrote a seven-page

Figure 26. The swept wing of the Horten *H XIII a* hides the cockpit gondola suspended beneath the apex of the trailing edge in this photograph taken from above (National Air and Space Museum, Smithsonian Institution, SI 90-9456).

proposal, *Nurflügel-Jägerprojekt "Horten X,"* dated March 6, 1945, describing a semidelta wing fighter aircraft propelled by a Heinkel-Hirth S 109 011 jet turbine engine and equipped with a semiprone cockpit. The design was similar in some ways to the *H X* depicted in *Nurflügel*, but it also differed from that aircraft in important details.

Reimar began working on a project during 1944 that suggests he had visualized a time after the war when aircraft were being designed for supersonic flight. As usual, he began with a model tested at Göttingen that was 1.5 m (4 ft. 11 in.) long with slightly less span and no vertical stabilizing surfaces but with as much as 10 or 20 degrees of dihedral. Both the leading and the trailing edges were highly swept, far beyond any other Horten design of the period. Following attempts to fly the model, Karl Nickel constructed a smaller example that could be hand-tossed into the air, and he demonstrated it to Reimar at Göttingen and the Hornberg gliding school. The model research soon led to a piloted version designated the *H XIII a*.[84]

Using his secluded shop at Bad Hersfeld, Reimar selected a small group of men from his staff at Göttingen, picked up a surplus pair of outer wing panels for an *H III*, and set to work. Reimar insisted that he could not have finished the *H IX* without building the *H XIII a*, but its influence on the jet wing is difficult to perceive. There is little doubt that the *H XIII a* represented a step toward supersonic flight, which had edged closer to reality during the war after it first appeared on the intellectual horizon during the 1930s. The *H XIII a* provided basic data about the difficulties of operating potentially transonic aircraft with highly swept wings at subsonic speeds. Maneuvering for a landing was of particular interest because Nickel's flying model had demonstrated dangerous stall characteristics. A relatively slow glider that could simulate the reduced speed necessary to land supersonic aircraft suited the research objective.[85]

Reimar built the *H XIII a* by omitting a center section and joining the two *H III* wing panels at their roots. After considering several different angles, he finally set the angle of the leading edges to an acute 60° (Figure 26). The aircraft had the lowest aspect ratio, 4, of any piloted Horten aircraft, and a correspondingly weak L/D between 12 and 16.[86] The pilot sat inside a cockpit suspended beneath the base of the vee formed by the wing's trailing edges. A rear hatch provided access to the cockpit as well as a handy means of escape if the experimental aircraft became uncontrollable. Visibility from the dangling cockpit was so poor that Scheidhauer refused to fly it, and Hermann Strebel got the job. He made the first flight at Göttingen on November 27, 1944, and then filed this report:[87]

> Weight and c/g [center of gravity] was 393 kg [869 lb], and 4,725 mm [189 in] respectively, wind W, 2 m/sec [5 mph], take off to the NW . . . The take-off was normal, tow speed 130–140 km/h [81–87 mph], good aileron and elevator control, weak rudder, adverse yaw light. Poor forward visibility, but adequate for the tow. Released at 2,700 m [8,856 ft], and flew at speeds between 70 and 150 km/h [43 and 93 mph], straight ahead. Heavy elevator! Ailerons normal when slow, but at 150 km/h, roll control is almost non-existent. With the stick all the way back, I get 70 km/h with no stall tendencies. The rudder is still weak, but coordinated turns can be made. A light roll oscillation of about 5° occurs with 45° bank; aileron application will dampen these somewhat. The approach is difficult, because the extended [landing] skid and the overhanging center section blocks the view of the runway, making landings in unfamiliar areas very difficult. The aircraft touched down fast and bounced a few times, the third bounce resulting in a [long] flat glide with only 50 km/h [31 mph] indicated. The ineffective spoiler could not prevent the long float over the ground.[88]

The *H XIII a* was challenging to fly, and it is remarkable that pilots took it aloft for as many as twenty flights, according to Reimar's recollection. He calculated that the landing angle of attack could vary from 25° to 35° degrees and that the wing would only reach its maximum lift coefficient when pitched up to an angle of attack of 40° to 45°. He found that at 30° degrees, coefficient of lift ranged between .8 and 1. The *H XIII a* showed that very low aspect ratio wings could be designed to fly at speeds slow enough to safely land. Horten used the data mined from this project to begin to design and build a powered version, the *H XIII b*.[89]

In the second half of 1944, technicians began to secretly build this supersonic fighter at Bad Hersfeld. Reimar easily paid for the effort with funds from the *H IX* program. Models, drawings, and Reimar's recollections portray a powered aircraft that picked up where the *H XIII a* glider had left off. The arrowhead-shaped craft was not completed before the war ended; nonetheless, it was an interesting concept bristling with unusual features that Horten imagined were required to operate at speeds faster than sound.[90]

Horten swept the leading edges at a sharp 70° (10° beyond the *H XIII a*), and this rakish angle held the span to just 7 m (23 ft.). The trailing edge also was sharply swept, thus the wing was not a true delta form. The aspect ratio was a very low 1.52, but the designer expressed doubts that he had selected the correct value. He knew that designers in Germany were working to develop aircraft with unusually low aspect ratios, but no one shared their research. Reimar made the airfoil symmetrical and 45 percent thick at the root, tapering to 7 percent at the tip. Horten planned to use the powerful Heinkel He S 011 jet engine that engineers promised would produce 1,300 daN (2,868 lb.) thrust.

Reimar calculated that the *H XIII b* could not exceed the speed of sound unless the pilot dove the aircraft, but once beyond the sound barrier, he believed the pilot could maintain supersonic speed in level flight. Prandtl suggested that a speed as high as Mach 1.5 was possible. At such velocities, drag rudders were unworkable, and Reimar envisioned using a vertical stabilizer and a conventional rudder to control yaw, in Walter's words, for "better handling in aerobatics," a more stable gun platform "to hold the aim," and good control during flight in turbulence.[91]

Reimar worked with the notorious Dr. Siegfried Ruff to develop a prone cockpit capsule design for the *H XIII b* that would be filled with water to prevent injuries to head and limbs from accelerations of up to 10 G. A pilot inhabiting such a space would have required underwater life support and special equipment for controlling and navigating the aircraft. Both Horten brothers had known Ruff since at least 1930 when he had taught Reimar flying lessons.[92] The physician became a notorious figure after the war when it came to light that he collaborated with other Nazi doctors in 1942 to conduct grotesque experiments on the prisoners from concentration camps to determine the effects of exposure to extreme altitudes and cold. Many of the test subjects suffered agonizing deaths.

Reimar and Ruff drew up a cockpit arrangement with a narrow fuselage cross section. After pondering how the pilot might safely eject at supersonic speeds, Horten decided the capsule must drop down to avoid impact with the vertical fin and the hot jet exhaust of the *H XIII b* and to eliminate the challenging requirement to boost the pilot, and the tremendous weight of his liquid-filled cockpit (as much as 500 kg, 1,105 lb.), up and away from the aircraft. One source reports plans to arm this aircraft with both bombs and heavy automatic cannon.[93]

Yet another late war project that occupied Reimar was the flawed attempt to adapt the laminar airfoil used on the North American P-51 Mustang fighter to an all-wing sailplane. The idea presaged more successful work undertaken by Richard H. Johnson, Harland Ross, and Dick Lyon in the USA during the late 1940s. Probably with Walter's help, Reimar learned that designers had shaped the Mustang's wing to maintain smooth laminar airflow over more of its surface than contemporary airfoils. This characteristic reduced drag, and therefore increased speed. To keep his staff at the Bad Hersfeld workshop busy,

Horten had them build a wing similar to the *H IV a* but shaped to the airfoil copied from a crashed Mustang. Analysis of drawings has shown that a portion of each wing section matched the Mustang's airfoil. Reimar was uncertain about the effects of flying an airfoil designed for high-speed flight at slow sailplane speeds. Hoping that flight tests could answer the question, a group at Aegidienburg built an *H IV b* and Scheidhauer flew it. The wing tended to spin easily, and the wing tips fluttered alarmingly at speeds above 110 km/h (68 mph). In February 1945, Hermann Strebel was making another test flight when he became disoriented after flying into clouds, lost control, and died in the ensuing crash.[94]

Although not an aerodynamic success, Reimar used the *H IV b* to practice novel construction techniques. The laminar profile made the wing too thin to build with wood, so Horten recalled his experiences helping the Dynamit A. G. firm to develop composites during the mid-1930s. He built the wing spar, the aircraft's structural heart, using laminations of aluminum, and he fashioned the wing's leading edges from corrugated cardboard stiffened with resin and sandwiched between thin plywood sheets. His workers cast this composite in 2 m (6 ft. 6 in.) sections, using female and male molds to shape the delicate airfoil curves.

Near the town of Rottweil, 380 km (240 miles) south-southwest from Göttingen, men at a workshop in the village of Gut Tierstein had started building ten of the *H IV b* wings when the war ended.[95] The plan was not to produce flyable aircraft but rather to save the men from military service. In discussing projects undertaken at the end of the conflict, Walter remarked that he and Reimar had to "make sure [we] hold busy our people . . . we could not say 'now we have nothing for you to do.' [If we ran out of work] they have to go to the Russian Front, oh, they didn't want it."[96]

Reimar supervised another series of laminar flow tests late in the war, and roughly parallel to the *H IV b* work. The objective was measuring the quality of laminar airflow over an *H III g* sailplane by modifying the cockpit to hold an observer who wore headphones to monitor variations in the sound produced by "the different laminar [air] flow over the wing itself. The noise from the laminar flow from different parts of the aircraft could be learned this way! And we could test the aircraft now under different [lift coefficients] and speeds." It was relatively easy to detect a distinct difference between smooth laminar airflow and turbulent airflow. Karl Nickel piloted the wing during these tests.[97]

Horten G.m.b.H. received its last contract on March 12, 1945, to design a long-range bomber propelled by six jet engines, which never proceeded beyond preliminary calculations about size, speed, range, etc. In April 1945, Reimar and Walter learned they had won a five-thousand-mark cash prize for successfully flying the first all-wing airplane propelled with jet turbine engines. A man from Frankfurt named Freudenberg had offered the award, but the brothers had no chance to collect it. Reimar's most prolific period of design activity was ending.[98]

War Years Retrospective

More than half of the sixty-five wings that Reimar Horten designed and built in his life were constructed during the war. He estimated that 400 to 450 men, mostly soldiers, had worked on Horten projects during the final years. To complete so many aircraft, Horten had worked twelve-hour days for much of the duration of the conflict, and in the final months, sixteen-hour days. The wartime aircraft encompassed a broad range—from the graceful, competition sailplanes that spanned almost 21 m (70 ft.), weighed less than 337 kg (750 lb.), and circled inside thermals at a sedate 60 km/h (37 mph), to the fire-breathing jet wing that weighed more than seven-and-one-half tons (6,750 kg/15,000 lb.) ready for takeoff and was possibly capable of speeds between 800 to 900 km/h (496 to 558 mph). However, statistics alone reveal little about practical, real-world performance beyond the basic fact that nearly all of Reimar's wings actually flew. It is important to determine whether pilots could fly the wings as often as necessary to fulfill the various roles for which they were designed.

This is difficult to answer for several reasons. As the war ended, Allied soldiers and civilians displaced from countries invaded by the Nazis and brought to Germany as slave labor destroyed the wings whenever they found them. Test pilots and engineers studied several of the sailplanes that survived the conflict, but their findings are confusing, as explained in the next chapter. No intact, powered Horten airplane survived; thus, independent analysis was impossible. The *H IX V2* was the technological peak of Reimar's wartime career and the only powered wing designed to satisfy an operational military requirement. Ziller's two successful test flights demonstrated that a competent pilot could take off in the jet, perform basic maneuvers, and then land, but to make valid assumptions about the jet's potential for long-term practical success, other data are required.

Göring had argued that speed would be the most important attribute. The *H IX* design had high-speed potential, and Ziller reported that it could climb faster than the Me 262, but Göring's desperation to field another "wonder weapon" did not entirely displace Reimar's common sense. He prohibited Ziller from pushing the experimental wing to its maximum speed, for several reasons, not least because Horten had optimized the shape and size of the turbine air intakes for starting the engine and climbing the wing after takeoff, but for flight at high speeds, they were twice as big as necessary. Then there were the oversize 004 engines that made the center section considerably thicker than Horten originally planned and probably robbed some high-speed performance. Although Reimar had described the high-speed potential of the *H IX* in his original proposal, this was not the only measure of practical success.[99]

Aerodynamic stability was also a key requirement. The pilot had to accurately fire the guns and drop the bombs, yet the *H IX V1* glider prototype had already exhibited Dutch roll instability about the yaw axis. The Horten brothers planned to arm the *H IX* with a pair of large 30 mm cannons with low firing rate and sharp recoil, which would have exacerbated the weak yaw control and rendered the jet ineffective as an interceptor. Walter actually insisted that his brother fit a large, vertical fin to increase the wing's stability. "For shooting, exact shooting, you need [the rudder],"[100] remarked Walter, but Reimar refused and the war ended before further gunnery drills could prove Walter's argument.[101]

DFS engineer Felix Kracht assessed harshly the potential of the *H IX* to fulfill the fighter-bomber role, pointing out that in the design of aircraft, the ends must justify the means. "You can make almost anything fly but you must consider . . . what are your needs and requirements in terms of [operational] service." Kracht understood how important flight stability was to successful fighter aircraft, "for accurate shooting, especially when using big ammunition . . . Without a stable base [from which to shoot] one should not even start a design."[102]

Reimar may not have realized that jet propulsion actually increased the *H IX's* instability. Al Bowers, the Chief of Aerodynamics at the NASA Dryden Flight Research Center in California, explained the effect of jet turbine engines on the yaw stability of tailless and all-wing aircraft by using as examples the pusher propeller-driven XB-35 and jet-propelled YB-49 wings built by Northrop. He described how the lip of the air intake bent the airflow streaming into the engines, exerting force at a point in front of the wing's CG, which is "destabilizing," asserted Bowers. He continued:

> Note that the YB-49 had to use [vertical stabilizers], and the XB-35 did not. It was NECESSARY to put verticals on the YB-49 to try and get some

> directional stability back. The YB-49 was very weak directionally; hence, it was a poor bombing platform. This [instability] gets worse at higher power settings, higher [speeds] and lower angles of attack.[103]

Given these flaws, how could the *H IX* have progressed to the point that more than sixty were planned or under construction when the war ended? Kracht thought the RLM's plan to manufacture the *H IX* for operational service an ill-conceived decision based on political influence. The brothers were, he said, "very clever in maintaining personal relations with some important people in the RLM. So many doors were opened to the Hortens which normally would not have been accessible to normal people."[104] Kracht had the wrong agency but he was right about their special access. Only Walter's credibility as a veteran fighter pilot and his personal connections within the Luftwaffe and the Nazi hierarchy could overcome general resistance to all-wing aircraft within the RLM and win Göring's approval to build and fly the *H IX*. Had the war continued, further testing would have exposed Horten's jet wing as the innovative, but seriously flawed, concept that it was—lacking solid directional stability, it was destined to fail as a cannon-toting fighter interceptor, and its suitability for the bomber role might have suffered as well.

CHAPTER 5

End and Beginning

GERMAN FORCES continued to fight until the official surrender on May 8; however, the war ended for Reimar Horten on April 1, 1945, when American or British troops picked him up near Göttingen.[1] For more than a decade, his grasp of all-wing aerodynamics had steadily evolved, driven by a succession of piloted wings. Now he needed to set aside his work and concentrate on survival. Walter joined him within days, and for several weeks they were absorbed into the incoherent mass of several million German prisoners of war. The Allies were aware of their aeronautical expertise, and soon they grouped the brothers with other members of the aviation industry. They were flown to London and joined Kurt Tank, head of the Focke-Wulf firm, twenty to thirty members of his staff, Willy Messerschmitt and his staff, and men from Daimler-Benz, the V-2 missile program, and other projects. Reimar remembered the men occupied an expansive, multistory house for more than a month where "about all we did initially was to play bridge."[2]

After about six weeks, the Horten brothers returned to Germany in early June, accompanied by British aeronautical and intelligence experts. They visited several workshops, and Reimar went to see his mother in the small village of Dietingen, a few kilometers from Rottweil, where the brothers had relocated their parents from Bonn to escape Allied bombing. He discovered that his father had suffered a heart attack when so-called displaced persons, whom the Nazis had relocated from the countries they occupied, pursued and shot at both Max and Elizabeth. Max was still in the hospital, and time ran out to visit him before the brothers had to return to London. Reimar resumed waiting, and later claimed that instead of interrogating him, Allied intelligence personnel discussed Reimar's work at length with Franz J. Berger, who "apparently had told them anything they wanted to know," whether his limited expertise qualified him to comment or not.[3]

Berger had worked alongside Reimar and Walter since at least 1928, helping to build the *H I* through *H III* sailplanes, and the powered *H V a*, and he later joined Sonderkommando L In 3 to help build the *H IX*. He was skilled at drafting construction drawings and had worked at several satellite workshops; nonetheless, to Reimar, Berger was neither "colleague [nor] partner," and the designer had shared with him none of the intricacies of all-wing design. Karl and Gunilde Nickel insist that Berger wanted only to convince the Allies that Reimar's work was important and should continue. Reimar desperately wanted to keep working, but he believed that Berger irreparably harmed his prospects for employment.[4]

In March, members of the British Ministry of Aircraft Production, U.S. Army Air Technical Intelligence, and the U.S. Navy spent four days talking to Berger and collecting notes, drawings, photographs, and other materials at the family residence in Bonn. According to a British source, Walter and Reimar were interrogated in May, but this seems to have been cursory compared to Berger's exhaustive debriefing. Reimar recalled only one visit during this time by an intelligence specialist, the noted aerodynamicist and cofounder of the Jet Propulsion Laboratory in California, Theodore von Kármán, who only asked about their treatment and personal needs.

The British reported another interview in September by their own intelligence specialists from the Tailless Aircraft Advisory Committee. A member of this group, Kenneth G. Wilkinson, published a seventy-seven-page report in October 1945 that described Reimar's work and the aircraft he designed and built up to 1945. According to Reimar, the errors and flawed conclusions that riddled the report were the result of the Berger interrogation.[5] Wilkinson praised Reimar for "a remarkable record of progress made in spite of obstacles" and for grappling "with aerodynamic and engineering problems on a bewildering variety of aircraft."[6] Reimar, said Wilkinson, "remained independent and original in his thought throughout and got little help from outside." Nevertheless, Wilkinson was struck by the "utter irrelevance of much [of his work] to the German war effort." The sailplanes, in particular, "were a dead loss to Germany." Wilkinson missed or ignored Reimar's claim that the lack of access to wind tunnels had driven him to rely heavily on the sailplanes to evaluate his ideas, writing instead that the brother's "general attitude seems to be that tunnel tests take too long and give too little information."[7]

Reimar responded years later by saying, "I built sailplanes to test my ideas and theories and to improve on my previous design[s] . . . not . . . for my ego or personal satisfaction or gain. I investigated things . . . [and] once that investigation had been accomplished, why not use the aircraft for other purposes? There was absolutely no reason not to fly these aircraft and [allow other pilots to fly them] . . . [because their experiences] would also bring me important knowledge which I sought."[8] Wilkinson was correct about the irrelevance of Reimar's sailplane work to the German military, but he missed both the resources and the concealment the war had provided to develop many projects without official interference, which Horten used to advance the stability and control of all-wing aircraft to a very high level by 1945.

The British and Americans might have understood Reimar's accomplishments if they had flown and evaluated the various Horten aircraft immediately after the war. For a time, the Allies seemed eager to do so, but during the final months, marauding Allied troops and displaced persons had destroyed nearly all of them.[9] This happened despite Gunilde's efforts to lead Allied authorities to the sites of a few hidden Horten aircraft by referring to a list of the aircraft given to her by Reimar.[10] The few sailplanes still intact took their place in the queue of thousands of pieces of German military equipment, vehicles, and aircraft gathered by intelligence teams for shipment to Britain and the USA. By this process, several Horten sailplanes eventually made their way into the hands of groups that could study them.

Over the course of one month, however, interest in doing so peaked at the highest levels before fading to nothing. At the beginning of June, the general of the U.S. Army Air Forces, Henry H. Arnold, "urgently required" plans and documents related to the "Horten Flying Wing."[11] On June 15, 1945, a list of Horten sailplanes that included an *H III f*, *H III h*, *H IV a*, and an *H VI* were described by the U.S. 9th Air Force Headquarters, Air Technical Intelligence branch, as "high priority" targets that must be obtained for study.[12] Less than two weeks later, the same headquarters unit abruptly "decided that the Horten Tailless Gliders are of no value to us."[13] Efforts to speed the Northrop all-wing bomber program into service in the USA probably stimulated Arnold's initial interest in Reimar's work. Perhaps the discovery of German research data showing the potential of conventional tailed bombers equipped with jet engines and swept wings to provide a faster and more practical alternative convinced Arnold to start backing away from an all-wing configuration.[14]

Several Horten aircraft eventually made their way to England and the USA, but test pilots flew only one *H IV a*. Allied military personnel eventually located the unfinished prototype *H IX V3* jet wing fighter and shipped it to the U.S. Army Air Forces test and evaluation center at Wright Field, Dayton, Ohio. A plan to finish and fly the jet came to nothing, and the air force put the aircraft into storage. When the Korean War started in 1950, the air force transferred its small collection of Horten wings, four sailplanes and the jet, to the Smithsonian Institution in Washington, D.C.[15]

Work, School, Unemployment

Late in 1945, the Royal Aircraft Establishment set Reimar up to work at Göttingen under their supervision on the *H XVIII* long-range bomber and to begin crafting several proposals for all-wing airliners, but the British lost interest and released him at the end of 1946. In December, he earned his PhD in mathematics from the University of Göttingen, having resumed his studies after the war ended with help from Prandtl.[16] During autumn 1947, while working at Bonn University, Reimar was approached by the head of Fairey Aviation Co. Ltd., Sir Richard Fairey, whose son had flown an *H IV a* at Cologne shortly after the war and reported favorably to his father on the wing's design and flying characteristics. The British government's interest in developing supersonic aircraft helped fuel interest in the German designer. Fairey had probably read in Wilkinson's report some information about Reimar's model work and flight experiments with the *H X* and the *H XIII* to investigate transonic flight. Reimar was eager to meet Fairey. "My only baggage would have been my slide rule and the formulas for [calculating] lift distribution," he said.[17]

Reimar was even more eager to work for Fairey after visiting England, but enough Fairey employees refused to work with a German to torpedo the arrangement and Horten returned home. Years later, he still felt disappointment at not finding meaningful employment in Europe.[18] German society had devolved into a primitive state of subsistence and jobs were scarce, yet Reimar was quite fit physically and intellectually and his vision of potential for the all-wing concept burned bright as ever. The designer soon determined to distance himself from Europe and begin anew.

Argentina

Compared to Germany, Argentina is a land of contrast. Some of the world's highest mountains tower over its western border, while in the east the Atlantic

Ocean washes almost 5,000 km (3,100 miles) of coastline. Often the northern regions are humid and tropical while the south is dry and routinely experiences near-arctic conditions.[19] These climates and terrains might seem uninviting to a German, but on April 10, 1948, after following a serpentine trajectory through Italy, Spain, Morocco, Senegal, and Brazil, Reimar Horten landed in Buenos Aires, Argentina.

Reimar arrived during what historian Uki Goñi identified as the most active year of a carefully organized program of cooperation between Argentina and Adolf Hitler's National Socialist Party that had started well before World War II ended. As armies of the USA, Britain, and Russia advanced on Germany from the east and west in early 1945, Nazi party members working in politics, the military, and industry began to panic at the looming certainty of prosecution as war criminals.[20] The Allies managed to arrest, prosecute, and convict more than 140, but from three hundred to as many as several thousand managed to escape to Argentina using elaborate networks established and maintained on the orders of army general Juan D. Perón, who became president of Argentina in 1946. Perón wanted to improve Argentina's capacity to develop advanced science and technology by welcoming German technical specialists, but he also supported the Nazi political ideology. Perón rescued many of the most notorious Nazi war criminals, including Josef Mengele, the doctor who experimented on children at the Auschwitz extermination camp. He also saved Adolf Eichmann, an officer in the infamous SS (Schutzstaffe/protection detachment), the massive domestic and foreign political and security organization under Heinrich Himmler. In 1941, Eichmann took charge of the SS Rasse und Siedlungsamt (Race and Resettlement Office) to organize and oversee the "industrial extermination" of millions of Jews and political prisoners.[21]

Perón had made the Argentine air force independent of army control in 1945, and he took a special interest in developing the technology for military jet aircraft. In 1947, Perón set up a special "Nordic Route" to funnel jet aircraft specialists from Germany into Argentina via Sweden and Denmark. Reimar would probably have taken this route had the two nations not closed it in November. Perón's agents set up a new network through Italy the following year. Reimar used this path after contacting Kurt Tank, former head of the Focke-Wulf aircraft company in Germany during the war, who appealed directly to Perón for Horten's immigration. Perón had brought in Tank the previous year to replace the French aircraft designer and war criminal, Emile Dewoitine, who had been developing a jet fighter. Reimar used another name when he boarded an FAMA (Flota Aérea Mercante Argentina, the national airline of Argentina) transport for the flight to Buenos Aires.[22]

Hours after landing, and accompanied by interpreter Gisela Hilger, Horten met Perón, followed shortly by friends of Perón's wife, Evita, who pressed a thousand pesos into his hand before he flew on to his final destination, Córdoba.[23] The city was nestled close to the eastern slopes of the Sierras de Córdoba, a range of mountains more than 300 km (186 miles) long. Much of the Argentine aviation industry clustered around Córdoba, and Reimar was soon working at the Fábrica Militar de Aviones (Military Airplane Factory) or FMA, under Kurt Tank, who wasted no time putting the designer to work.

Tank's chief of design at Focke-Wulf, Hans Multhopp, had laid out a new fighter in Germany shortly before the war ended, and Tank had already resumed the project with Perón's full support when Horten arrived in 1948. Reimar convinced Tank to build an interim glider prototype, designated the I. Ae. 33B, and members of Tank's staff helped Horten ready the airplane to fly by August 1948. Staff at the Instituto Aerotécnico Córdoba (Aeronautical Engineering Institute of Córdoba, abbreviated I. Ae.), the research arm of FMA, joined with Tank's people to build the glider. The first flight exposed control problems, and after a series of modifications, flight tests resumed to measure the results.[24]

Poor control characteristics continued to dog Tank's jet project, called the Pulqui II, and the development phase lengthened. Reimar claimed that Tank had deviated from Multhopp's original design and changed the airfoil. The landing gear arrangement was flawed as well. Two prototypes crashed, killing the test pilots, and the jet was not presented officially to Perón until early 1951. Two years later, the fighter was still not ready for production when Perón halted the project, largely because of a worldwide drop in agricultural prices following the end of the Korean War in July 1953 that also slowed armament development programs throughout Argentina.[25]

The fall of the Perón government in 1955 ensured that work on the Pulqui II never resumed. Meanwhile, the USA had offered the North American F-86 Sabrejet "for the price [the Argentines] would have had to pay for British engines!" to power the Pulqui II.[26] Reimar concluded that the Pulqui II failed because Tank was not a "complete" designer—he lacked competence in certain critical areas of design, such as aerodynamics.[27] In 1949, Horten had naively proposed to Tank a new jet fighter "a little bit like the *H IX*" that he claimed could become "friendly competition" for the Pulqui II.[28] Tank showed no interest, and Reimar dropped the idea, and by 1951, the two men had ended their professional relationship.[29]

Despite the failure of the Pulqui II, Reimar's social life was improving. He had developed a relationship with Gisela Hilger, the translator who had accompanied him to meet Perón. Her father was also a German émigré and both were glider pilots. Several years earlier, Hilger had set the national women's soaring record for altitude, about 3,000 m (9,840 ft.), and for duration when she remained aloft for more than five hours. A mutual attraction developed, and the couple married in June 1949. Gisela continued to fly until her husband grounded her when she became pregnant. She bore him a son named Dieter in June 1950 and a daughter, Karen, in April 1953.[30]

Reimar had never stopped thinking about wings for sport soaring. In late 1948, he requested permission to work with the I. Ae., which functioned as the research arm of the state-run FMA, to develop an all-wing sport sailplane that bore some likeness to the high-aspect ratio *H IV a*. The new wing had the same area but less span, a lower aspect ratio, and a completely revised cockpit designed to accommodate two persons sitting in tandem. He positioned the rear seat 0.3 m (1 ft.) above the front seat to allow clear sight of the instrument panel mounted ahead of the front seat. An engine driving a pusher propeller could be fitted in the space taken by the rear seat. Horten increased sweep by less than 2° over the *H IV a*, but wing loading jumped 33 percent as the gross weight climbed 100 kg (221 lb.) to 430 kg (950 lb.).

Of course, the wing lacked any vertical control surfaces, although a blunt fin projected behind the cockpit and probably contributed a small degree of stability. Reimar dispensed with the center section traditionally assembled by welding steel tubes together and used wood throughout the aircraft. The wing was designated the *I. Ae. 34 a* and named *Clen Antú* (an Argentine Indian term meaning "Sunbeam"). The designer envisioned six design goals for this new aircraft (Figure 27).[31]

First, performance must be good enough to soar for extended periods to give student pilots as much time in the air as possible and to reduce the costs of air towing. Predictable and sedate handling qualities with no nasty surprises would satisfy the second requirement. Horten wanted even the newly fledged pilot to feel at ease, particularly when flying solo. The third requirement centered on cockpit ergonomics. Soaring in clouds was an advanced skill to which many pilots aspired. To meet this need without pushing up building costs or slowing construction, Reimar left space in the cockpit for only a single instrument panel in front of the student. Flight trials would be required to confirm whether an instructor, seated behind the student, could see the instrument panel well enough to pilot the aircraft.

Fourth, the wing must tolerate the stress of aerobatic flight, and fifth, the process of converting the sailplane into a motor glider had to be relatively easy. The engine would displace the rear seat, and Reimar tried to design the center section to accommodate various brands and types of power plants, such as a

Figure 27. The multipurpose *I. Ae. 34 a Clen Antú* in flight. Reimar designed the center section to allow the instructor pilot to sit above and behind the student in the forward cockpit (Reimar Horten via Peter F. Selinger).

car or motorcycle engine weighing less than 80 kg (177 lb.) and producing between 15 and 30 kW (20 to 40 hp). Horten calculated that a 22 kW (29 hp) engine, for example, could allow the wing to lift off after a ground roll of 130 m (426 ft.), if flown at the maximum flying weight of 500 kg (1,105 lb.). This same small motor would give a climb rate of 2.5 m/s (8.2 ft./sec.), and a speed of 160 km/h (99 mph) in horizontal flight. The sixth goal that Reimar tried to design the *I. Ae. 34 a* to meet was good performance during soaring competitions, both with and without a motor. The wing had to be capable of satisfactory thermaling performance at slower speeds, and cruising between sources of lift at a reasonably fast clip. He also provided space for parachutes and supplemental oxygen for both occupants. To control the wing, Horten included the now familiar array of elevons, flaps, drag rudders, and dive brakes.

Simplicity and versatility sum up the design philosophy that Reimar applied to the *I. Ae. 34 a*. The designer tried to appeal to the average soaring club whose members were usually weekend budget flyers with only basic carpentry skills. Construction required no exotic woods or complex and expensive welding equipment, and Reimar eased the tolerances required to assemble noncritical components. He replaced the traditional landing skid with two rubber-tired wheels mounted in tandem to reduce the number of people required to handle the wing on the ground without sacrificing off-field landing capability. It was said that four people could take the wing apart, or assemble it, in a "few minutes."[32]

Dual or solo trainer or competition mount, aerobatic ship, sailplane, or motor glider—Reimar wanted to fill all of these roles, but he realized that he could not waste time developing a series of prototypes. He was still something of an outsider trying to work in a strange land without benefit of Walter's organizational and political skills, yet dependent on the whims of both a corrupt and tenuous government and a small community of sport sailplane pilots to support his work. Reimar needed to win local advocates for the wing who could fly it and then tell the larger soaring community what a fine aircraft it was.

No doubt with anticipation, Reimar finished laying out the aircraft, and craftspersons began to build it early in 1949. Reimar now confronted the acutely primitive state of the Argentine aeronautics industry, decades behind in the quality of materials he had known in Germany. Spruce wood and thin sheets of birch plywood were vital to building strong, lightweight wooden aircraft, yet spruce was simply unavailable and the birch plywood was weak and brittle. The critical glue required to join the components could be produced using readily available substances and should have been Reimar's least worry, but the air force general in charge of the I. Ae. ordered the chemistry department to manufacture it. By the time a batch was carried in the heat of day to Reimar's workshop, it had begun to set and some batches were useless.

Construction was delayed while waiting for glue to arrive in usable condition, and the designer recalled that other wooden airplanes, including some two-engine models, had to be scrapped after inspections revealed failed glue joints. Reimar may have thought this difficult period would pass after test pilot Edmundo Osvald Weiss took off on June 20, 1949, from Merlo, 70 km (43 miles) southwest from Córdoba, at the controls of the prototype *Clen Antú* for the wing's first flight. Weiss had made history two years earlier when he flew the first jet aircraft developed in the southern hemisphere, the I. Ae. 27 Pulqui I. The willingness of a well-known pilot such as Weiss to test the Horten sailplane illustrates both the quality of Reimar's work and the state's high regard for it, which the designer would shortly come to loathe.[33]

Not long after Weiss's flight, government officials snatched up the sailplane, moved it to a remote village, and set it on public display outdoors where it remained for months exposed to the elements. Perón considered such seemingly senseless acts necessary to maintain political support among the general population at a grassroots level. He needed to focus their attention on his accomplishments; unfortunately, such demonstrations wasted resources and effort and demoralized the design teams and craftspersons building the nation's aircraft. Reimar witnessed this happen all too frequently. Sometimes, he recalled, "Relaciones Publicas," public relations, consumed the second aircraft built in a series as well as the first![34] He went on to explain how the "most important part of the [aircraft] construction [business] here is called public relations . . . it is very much like propaganda. To the people [in charge of] the factory, this was more important than flying . . . they then could tell the public what was being done in the factory . . . To show them how their taxes were being used."[35]

Technicians repaired the *I. Ae. 34 a* in time for soaring enthusiasts to fly the aircraft when they gathered at Merlo in 1950 to compete in the national soaring competition. The wing seemed to perform well, and several pilots soared the aircraft for at least three hours. The institute built two additional glider versions of the *I. Ae. 34 a*, and Reimar reported that the aircraft satisfied all the design goals that he had set. Pilots continued flying the wings well into the 1950s, but no one tried to fit a *Clen Antú* with a power plant. Reimar claims that he measured a lift coefficient of 1.7. He studied the effects of a Junkers-Doppelflügel, a double, or auxiliary, wing mounted beneath the wing root, to determine if the device could increase lift. He also studied the effects of numerous modifications made to the drag rudders and further refined the glider's flight characteristics. Reimar used balsawood to construct some of these alterations because it was easy to work and he could make the changes more quickly.[36]

Although the I. Ae. supported these experiments, Horten believed that they could have done more. The directors of the institute, said Reimar, took the position that "military work was more important than soaring" research.[37] He also deplored the small scope and capabilities of the Argentine aviation industry and the sometimes primitive working conditions in which it worked. Longing for an organization more capable of developing all-wing aircraft compelled him to approach Northrop Aircraft Inc., "but they declined. They sent my letter [of inquiry] back . . . [but not] for lack of [the correct] address or postage. Written on the back side of the letter was 'not for Northrop.'"[38] Nickel recalled that Jack Northrop responded personally, paraphrasing his reply this way: "You [Reimar] and I, we both know how a true aircraft should look . . . Unfortunately it is beyond my power that we two could work together."[39]

Reimar remembered the Northrop rejection arrived shortly after he left Tank's group, sometime around 1950. By then, Jack Northrop's dream of producing all-wing bombers was close to termination. Northrop, the president, chief engineer, and founder of Northrop Aircraft Inc., had contracted to finish and fly the first all-wing bomber in 1943, but Northrop engineers spent the next three years attempting to ready the wing for service. In groping for answers to the mounting delays, the U.S. Army Air Material Command had approached the Waco Aircraft Company on May 5, 1946, to ask if the firm could "repair and test . . . Horten Models 3 and 229," but the company refused.[40] The following year, Northrop asked the AAF to supply them with Horten gliders "for inspection and detailed analysis."[41]

After Northrop and the Southern California Soaring Association filed a joint petition, a Horten *H III h*, *H III f*, and *H VI V2* arrived at the Northrop complex in Hawthorne, California, in 1948, on loan from the air force. The Horten wings remained there for several months while engineers, aerodynamicists, scientists, pilots, and enthusiasts from the following organizations inspected them: Society of Automotive Engineers, Institute of the Aeronautical Sciences, Southern California Soaring Association, Douglas Aircraft Company, North American Aviation Corp., Lockheed Aircraft Corp., Consolidated Vultee, and Northrop.[42]

The three sailplanes eventually passed to the Smithsonian National Air and Space Museum, along with a Horten *H II L*, but a different fate befell the Northrop bombers. Based on the knowledge that designs with conventional fuselages and tails were faster, could carry the atomic bomb, and would be available sooner and at less cost, an influential group of air force officers recommended canceling the Northrop program in November 1948.[43]

Government Support for Soaring

Early in 1951, Reimar became aware that members of Argentina's sport soaring community wanted more government support. From March 21 to 25, about one hundred pilots and other representatives from thirty soaring clubs scattered around Argentina gathered near Córdoba. Reimar joined them, along with two meteorologists, Walter Georgii from Germany and Italian meteorologist Plinio Rovesti, and a test pilot who was also from Italy, Adriano Mantelli. The charismatic Secretary for Aviation in Argentina, Cesar Raul Ojeda, led the group and enthusiastically accepted their petition for a new institute dedicated to soaring. Ojeda held the rank of brigadier major in the Argentine air force, and his influence was considerable. The group ceremoniously laid a foundation stone for the institute near the Embalse Rio Tercero airfield in the mountains west of Córdoba. The Argentine historian and engineer Fernando Siarez has suggested that the men sought to duplicate the old German DFS.

His assertion has merit since Georgii had managed the DFS in prewar Germany and developed the science of forecasting the meteorological conditions necessary for soaring flight to a greater degree than anyone else before World War II.[44]

Soon the institute had both a name, the Instituto Argentino de Vuelo a Vela (Argentine Institute of Soaring, abbreviated INAV), and the buildings and equipment required to build prototypes. To breathe administrative life into the organization, the attendees nominated Georgii to the director's post and chose José Ortner as his deputy. Georgii had visited Argentina with a soaring expedition from Germany in 1934, and Ortner was an experienced soaring pilot and president of the Club Argentino de Planeadores Albatros (Albatross Soaring Club) in Buenos Aires, the most influencial group of soaring enthusiasts in Argentina, according to Reimar. Rovesti became the head meteorologist, Mantelli assumed the duties of chief test pilot and manager of the motor-glider group, and the Argentine pilot, Claudio Dori, took the job of chief flight instructor.

They named Reimar to manage the department of experimental design and construction. Karl Nickel had married Gunilde Horten in August 1948, and both he and Heinz Scheidhauer joined Reimar at Córdoba in 1951 to lend their expertise and experience. The institute's initial goals included developing new sailplanes and conducting meteorological studies above one of the world's great soaring "engines," the mighty Andes Mountain range. The staff also wanted to study better methods for training sailplane pilots and then transitioning them to fly powered aircraft, and to develop techniques and technology for improving a sailplane pilot's chances of landing safely away from airports, or landing-out, which happened often during cross-country flights.[45]

Two British publications, *Aero* and *Sailplane and Glider*, greeted the new organization with fanfare. Founding this institute "is of the greatest significance not only for the Argentine, but also for the world gliding movement, as with this step the Argentine seizes an important position in the international gliding world."[46] The comment bespoke not only the sport's international resurgence since the end of the war but also the Argentine soaring community's desire to keep pace as Europe and the USA continued to advance soaring technology. To demonstrate its enthusiasm and commitment, Argentina offered to host the second postwar international soaring contest scheduled for summer 1952 and to pay for each competing team's expenses.

Despite the generous offer, soaring officials chose Spain over Argentina, but this decision hardly dampened Reimar's excitement. For the first time, he was positioned to parade the capabilities of his all-wing sailplanes before an international audience of pilots, designers, enthusiasts, and the soaring press. Aviation minister Ojeda enthusiastically sponsored construction of two Horten wings to compete in Spain. The *I. Ae. 34 m* (*m* for *monoplaza*, Spanish for "single-seat") wings were largely identical to the two-seat *I. Ae. 34 a*. Reimar removed the wooden landing skid and one of the fixed wheels and replaced them with a wheel mounted on a retractable strut hinged at the nose of the center section. When the pilot folded the nose gear into the belly and a streamlined door closed over it, drag was reduced. Horten also removed the landing flaps and adjusted the drag rudders to allow the pilot to turn more sharply. Time was short to test the sailplanes. They flew on October 19, 1951, and February 9, 1952, and a team of veteran pilots, José Cuadrado, José Ortner, and Eduardo Bazet, were selected to represent the nation and to fly with the full support of the government.[47]

Nineteen countries entered the contest—Argentina, Australia, Belgium, Brazil, Canada, Denmark, Egypt, England, Finland, France, Germany, Holland, Italy, Norway, South Africa, Spain, Sweden, Switzerland, the USA—and more than fifty pilots gathered at Cuatros Vientos airfield outside Madrid to compete from July 3 to 12, 1952, and determine the best soaring pilots and most capable aircraft in the world.[48]

From the start, the contest challenged everyone involved. Meet organizers struggled to efficiently communicate with pilots speaking so many different languages, and Spain's poor rural road network and primitive telephone system made retrieving both contestants and sailplanes at the end of a task difficult.[49] The Argentine team endured additional hardships due to lack of both competition experience and familiarity with the Horten wings. Ortner landed his *I. Ae. 34 m* in terrain littered with enough rocks to destroy it. He flew a backup sailplane of conventional configuration and finished fourteenth. Bazet landed late in the evening on the fourth contest day, rolled into a ditch, and tumbled into a complete somersault. The accident broke his wing and barred him from further flying.

Most of the teams incurred an "astounding record of damage" during landings in the Spanish backcountry, but the Horten wings seemed particularly vulnerable. Even the presence of Reimar's former test pilot, Heinz Scheidhauer, could not prevent a genuine debacle. He had accompanied the team to consult, not fly, since he was not an Argentine citizen.[50] An Australian pilot who participated in the competition offered a pithy and sarcastic summation. "When is a Flying Wing a non-flying wing?," he asked. "Practically all the time in a country where the fields are tiny and full of rocks, and a poor little Ala Volante [the *I. Ae. 34 m* wing] has a short tummy and long legs and keeps tripping over."[51]

Two other commentators agreed that the tall and spindly nose landing gear that Horten had fitted to the short center section was too delicate. Historian Ricardo Burzaco noted another design flaw: "The main problem that the all-wing had," said Burzaco, "was a bad look down. This was one of the motives that caused the [Argentine] pilots [not to want to] fly them," and when they did, to have difficulty landing.[52] Reimar had placed the *Clen Antú's* cockpit so far behind the wing leading edge that it blocked the view ahead and below that was so crucial during landing, offsetting any benefit from upright seating.

In hindsight, it seems obvious that the semiprone cockpit would have afforded the pilots better vision, but after World War II, Reimar inexplicably abandoned it. Perhaps the Argentines were not comfortable crouching on their knees to fly, or Reimar's desire to broaden the appeal of his wings to a larger community that included novice pilots overrode a return to semiprone seating.[53] Casting further doubt on the Horten wings, José Cuadrado of Argentina finished a quite respectable fourth place overall while flying the Slingsby Sky, a British sailplane with conventional fuselage and tail layout.[54]

Exposing these weaknesses in the all-wing sailplanes under the intense media scrutiny that accompanied the world-class competition marked an important turning point in Reimar's career. The Argentine soaring community turned away from the wings designed for competition, and government support for his all-wing projects dwindled as well. During the fifth national contest held in Argentina in 1953, seventeen teams flew thirty-six gliders during the ten-day competition. Only Scheidhauer flew a Horten wing (*I. Ae. 34 m*) but managed only one good flight, and he did not place high in the final contest results.[55] Reimar turned to developing noncompetitive designs aimed at amateur builders, along with a series of conventional designs sponsored by the INAV for use by flying clubs.[56]

H IV a in the USA

Even as interest in all-wing sailplanes for competition was dying out in Argentina, it was growing fast half a hemisphere away in the USA. In August 1952, less than two months after the contest in Madrid, a German test pilot named Rudolf (Rudi) Opitz flew a Horten *H IV a* against the best American pilots at the USA National Contest in Grand Prairie, Texas. *Sailplane and Glider* had begun to stir interest in this match-up months earlier after it called the *H IV a* the "most efficient sailplane in the world."[57] Opitz lent credibility to this claim when he won against forty other pilots at the Wright Memorial Glider

Meet near Dayton in May 1952, despite missing the first day of competition. Three impressive flights earned him the victory: an endurance flight lasing 7 hours 24 minutes, a cross-country flight that covered 55 km (88 miles), and a climbing flight to 1,666 m (5,480 ft.).[58]

Opitz had found his way into the semiprone cockpit of a Horten wing because of good fortune and his skill and experience handling unusual aircraft. Opitz had worked for Alexander Lippisch during the late 1930s. He tested the semitailless and rocket-propelled Me 163 fighter during World War II, an experience that a number of other German pilots did not survive. Opitz came to know Reimar when both men served at the air base outside Königsberg from 1940 to 1941. Opitz later made several flights at the controls of an *H III*. His familiarity with rocket-propelled aircraft earned him work for the U.S. Air Force at Wright Field, Ohio. There, he met Captain Hollis Button in 1951. Button owned a damaged *H IV a* and needed someone to repair it.[59]

Late in 1945, the Austrian soaring pioneer, Robert Kronfeld, had imported this *H IV a* to England from Germany. He died in the crash of another tailless aircraft before he could fly the Horten wing, and it passed through several owners until Button purchased it in 1950 for $2,500 and shipped it to his farm in Valley City, North Dakota. Button crashed attempting his first flight. After searching more than a year for someone qualified to repair it, he met Opitz, who agreed to undertake the work if the owner allowed him to fly the *H IV a* during the summer contest season. Opitz, who had completed his cabinetmaker's apprenticeship in 1928, repaired damage to one wing—shattered ribs and a broken drag spar—by using the good wing for patterns. Button wanted to insure the aircraft, but even Lloyds of London refused and Opitz flew without it.[60] The pilot fortunately had the skill, experience, and small size necessary to push the wing to its highest performance potential.

Early in June 1952, after the Wright Memorial meet, Opitz soared the *H IV a* ahead of eighteen other competitors to win the Fifth Mid-West Regional Soaring Contest held at Toledo, Ohio.[61] Dayton and Toledo were the best performances yet for a pilot flying a Horten wing in a soaring competition. Perhaps the news of these victories encouraged both Alexander Lippisch and August Raspet of the Aerophysics Laboratory at Mississippi State College in Starkville to attend the nineteenth annual USA National Soaring Competition held in Grand Prairie, Texas, the following month.[62]

This prestigious meet usually allowed most of the pilots to join Team USA and to compete internationally. Opitz, his wife, and crew chief, Willi Elias, had to rely on fellow sailplane enthusiasts from Dayton, Ohio, to raise the cash needed to pay the team's way to Texas. In early August 1952,

Opitz and his crew loaded the wing onto a trailer and drove south to a place where both climate and the competition were hot. Temperatures averaged 10° above normal and exceeded 38°C (100°F) on eight contest days before reaching 42°C (107°F), a Grand Prairie record, on August 23. Laden with moisture from the Gulf of Mexico, the air was ideal for producing thermals when cooked by the central Texas sun.[63] The weather helped pilots set ten records in 1952, one every day of the meet, including a world record out-and-return flight on August 23. More records were set than at any previous USA national. Thirteen flights exceeded 483 km (300 miles), and the thirty-four men and women who competed flew a combined 41,966 km (26,066 miles).[64]

Just getting the *H IV a* airborne safely and consistently demanded exceptional skill. The German's first launch of the contest demonstrated the wing's finicky nature, for as it left the ground, a clump of grass lodged hard between the takeoff dolly and the nose skid.[65] Opitz tried to dislodge the clump and jettison the dolly, but he could not. Continuing the flight was impossible, and he returned immediately to the airport where a long delay ensued while he waited in line for the next tow. To make extra points, Opitz declared a goal and return flight to Grand Prairie, but he started too late in the day. The lift died, he landed out, and a long, hot, and tiring retrieve followed. Opitz was so exhausted by the ninth day of the contest that he failed to spot the turn-point landmark during a cross-country flight and again landed out.[66]

USA law prohibited Opitz from obtaining the radio operator's license, forcing him to announce to his crew a likely objective before taking off. Flying mute forced his crew to drive much further if he veered from his declared course to chase stronger lift. In fact, all the contest crews bore the physical strain of exposure to the heat. The daily distances flown by the top five competitors during the meet often exceeded 483 km (300 miles), about 6.5 hours of flying time, which could translate into as much as 644 km (400 miles) of driving. Nearly every team had to rely on two drivers, but heavy sailplane trailers and secondary roads paved with gravel slowed them down. All but three of the top ten finishers missed at least one contest day due to exhaustion or because their crews got lost.[67]

Opitz sometimes struggled to control the wing in flight. "All of a sudden, it would have an oscillation; I called it a flight-path oscillation . . . triggered by something—by a gust that hit the aircraft, or by a control movement."[68] The oscillation usually persisted for ten to fifteen seconds. This unusual form of aerodynamic flutter, low in frequency but high in amplitude, apparently induced little or no vibration at the wing tips but quite a lot at the center section. Opitz encountered the phenomenon when entering strong thermals at high speed, a common tactic for flying during any contest.

He spent considerable time studying the available literature about Horten aircraft to discover a cause and cure before finding a flight test report written by pilot Lehmann and published in *Flugsport* two years before the war ended. Lehmann, too, had encountered the oscillation while testing an *H IV a* rigged with the CG well aft. He had found that moving the CG forward by adding weight to the nose reduced the effect. It was contrary to most flutter modes and probably resulted from the combined effects of the glider's concentration of mass, the aerodynamic elasticity of the airframe, wing sweep, control surface placement, rigging condition, and other factors.[69]

Opitz had no desire to degrade performance by adding otherwise useless weight to the sailplane's nose, so he experimented to work out his own fix. He limited his maximum speed to 129 km/h (80 mph) and approached every landing with plenty of altitude, using both drag rudders to steepen his descent, for he had observed that the wing never oscillated with the controls configured in this manner. Where other pilots might have refused to fly an aircraft exhibiting such a potentially dangerous oscillation, experience had given Opitz confidence to handle the problem.

Opitz recalled that since the 1930s, many gliders designed for competition had possessed a few poor or unusual handling qualities. Experienced pilots usually accepted such behavior as the cost of high performance and adjusted their flying styles to accommodate it. Having faced, and mastered, a similar vibration while flying a Lippisch Storch tailless glider, Opitz adjusted to the quirks of the Horten wing, and his reward was good maneuverability. "Adverse yaw was very, very acceptable; it was day and night to the Horten III I flew. So it was otherwise a very pleasant aircraft to fly," Opitz said.[70]

During eighty or so hours of competition flying, Opitz frequently landed out in the *H IV a*, but thanks to the predictable handling, Opitz "never scratched it."[71] Visibility was the key. Compared to the Argentine pilot's view from their *Clén Antú* wings in Spain, Opitz could see his touchdown point until the moment of ground contact. Of the semiprone cockpit, Opitz said that despite the chin rest, he needed "to grow a beard, as stubble becomes irritable in flights of 6 hours or more."[72]

For Opitz, the most harrowing moments of the contest came on day four, August 22, 1952, and had nothing to do with the Horten wing. The day's task was to declare a goal and fly to it. Longer distances earned extra points, as long as pilots made their goals. Opitz declared Odessa his goal and launched. The flight went well until somewhere near Abilene when the reckless pilot of a propeller-driven military trainer intercepted the *H IV a* and repeatedly passed close enough to buffet the wing with his prop wash, forcing Opitz lower and lower over rough terrain. Just minutes from a forced landing, he spotted a powerful dust devil, flew directly into it, and began to circle tightly, climbing fast. The wing carried Opitz to the clouds, where he lost the harasser.[73] He went on to earn his Diamond C badge pin by flying 514 kilometers (319 miles) on day seven and finished the contest in seventh place. Opitz was the sixth pilot in the United States to earn this pin, and the tenth worldwide.[74]

It was a fine showing for the decade-old Horten wing, competing against newer sailplanes such as the all-metal Schweizer 1–23 (flown by the second, third, and fourth place finishers) and the RJ-5, flown by the contest winner, Richard H. "Dick" Johnson.[75] The RJ-5 had already begun to revolutionize the design of sailplanes after Johnson and codesigner Harland Ross, on advice from engineer Dick Lyon of the Hughes Aircraft Corporation, became the first to shape a sailplane wing to the new, low-speed laminar airfoils developed during the late 1940s by the NACA. Johnson had flown the RJ-5 to a world distance record in 1951, and his win at the nationals underscored the potential of the new airfoils to increase sailplane cruising speed, as Johnson demonstrated on day nine by setting a world speed record when he averaged 88 km/h (54.78 mph) over a triangular course.[76]

E. J. Reeves, president of the Texas Soaring Association, which had hosted the national contest, proclaimed that "the Horten wing has competed impressively against the best conventional designs—not as an exhibition or as a prototype show-piece, but as a work-a-day aircraft ready to stand or fall on its own revolutionary features." Opitz had indeed flown the wing day after day; in fact, its performance had at times exceeded his physical endurance and that of his ground crew. Opitz and the wing had, said Reeves, "convinced everybody of the tailless design's practical values and it appeared to us that this will hasten acceptance of flying wings, and thereby accelerate their further development in this country."[77]

Further development was important if Reimar's ideas were to keep pace with the new conventional sailplanes such as the Slingsby Sky, Ross-Johnson RJ-5, and Schweizer 1–23.[78] August Raspet of Mississippi State College (MSC) had monitored evaluations of the wing's performance at least since 1950, when he credited Reimar's "Praying Mantis" cockpit (a British term for the Horten semiprone couch) with lowering the *H IV a*'s drag coefficient to "0.01, a world's record for sailplane aerodynamics."[79] The consistently high standings earned by Opitz and the wing in the competitions at Dayton,

Toledo, and Grand Prairie had intensified Raspet's interest and before the nationals ended, he offered both cash and another sailplane to Hollis Button for the *H IV a*. Button accepted, and Opitz delivered the wing to MSC the month after the national contest in September 1952.[80]

Mods and Measurements at Mississippi State

What followed was a series of evaluations conducted on the wing by MSC staff over the remainder of the decade. These are the only independent scientific analyses conducted on a Horten wing after World War II. It is important to understand how MSC reached their conclusions because Reimar disputed their findings for the remainder of his life.

August Raspet had been studying soaring flight since the early 1940s. After earning a doctorate in physics in 1942, Raspet joined the Pratt Read Company, which built gliders for the U.S. Army and navy during the war. He published papers on soaring instruments, flight techniques, and the flight performance of sailplanes he had carefully measured using scientific methods. In early 1949, Harold Flinsch, dean of the MSC School of Engineering, asked Raspet to join his staff to boost its research output. Like Raspet, Flinsch was a sailplane pilot and enthusiast, and he believed that research into soaring flight could be profitable to MSC at a relatively modest cost. Flinsch created the Department of Aerophysics solely for Raspet to continue his work, and Raspet won a $16,000 grant from the Office of Naval Research to study the soaring characteristics of buzzards soon after arriving at MSC.[81]

In 1950, Raspet led a group that included student Dick Johnson in an effort to improve the sailplane Johnson had designed along with Harland Ross, the RJ-5. The team improved the L/D from about 30:1 to better than 40:1, and along the way, Raspet compiled the critical knowledge required to extract the full promise of the new low-speed, laminar-flow airfoil technology. "In particular," Raspet said, "the importance of wave-free contours was found to be of utmost importance by the series of profile drag measurements made in flight on the RJ-5."[82] By wave-free, Raspet meant a smooth wing surface with imperfections of less than .002 of an inch.

Dick Johnson flew the modified sailplane to victory at the USA national meet in 1950 and again in 1951. Later that year, he flew the RJ-5 a world record distance of 861 km (535 miles) that remained unbroken until 1964.[83] Grand Prairie was Johnson's third national title won in the RJ-5 in as many years. Raspet was eager to reprise the improvements made to the RJ-5 and release the potential for extraordinary performance that he saw trapped within the *H IV a*. "We believe," he said two months after Grand Prairie, that the Horten wing's glide ratio "can be increased to as much as 65 to 1."[84]

This represented an astonishing 23 percent increase over the RJ-5's 40:1 glide ratio. Almost forty years would pass before a practical production sailplane eventually exceeded 60:1 in the early 1990s. The ratio of 65:1 was quite unachievable in the 1950s, unless Raspet intended to transform the *H IV a* from a practical aircraft capable of flying every contest day into a temperamental experiment, which is precisely what the scientist envisioned. Raspet and Johnson had sealed gaps and smoothed surfaces on the RJ-5 to promote laminar airflow and improve glide ratio. In 1951, Raspet discovered that by using a pump to suck air into thousands of pin-size holes drilled into a wing surface, he could greatly reduce drag, lower stall speed, and boost glide ratio. This is the only technology that could have enabled Raspet to bring the Horten sailplane anywhere near 65:1.

The first step toward this towering goal was to establish performance benchmarks for the Horten wing to compare to the results of future work carried out on the aircraft. The task fell to Opitz, who carefully piloted the *H IV a* during a series of flights in the calm air of early morning, probably during late fall 1952 and winter 1953. He spoke to Raspet about the wing's flight-path oscillations, but the aerophysicist was more interested in boosting performance, not improving handling.[85] Raspet asked the German flyer to mentor MSC pilots on the idiosyncrasies of the *H IV a*, and Opitz obliged by tutoring Ray Parker and Lew Everett, but both left the program before they could carry out meaningful work.[86] Opitz moved on, too, and lacking pilots interested in flying the wing, the project lay dormant during the next four years until a new candidate stepped from the throng of Hungarian immigrants fleeing the Soviet crackdown in November 1956. Dezsö George-Falvy had been interested in aviation since his youth. He had built hundreds of flying models, learned to fly gliders and powered aircraft, and served in the Hungarian army during World War II.

After the war, George-Falvy earned a master's degree from the Technical University of Budapest and served on the faculty of that school. He earned another master's degree from the University of Colorado after arriving in the USA in 1956. George-Falvy's blend of engineering training, flight test experience, and considerable time piloting sailplanes led Raspet to hire him in about 1958.[87] Along with the other members of Raspet's team, George-Falvy had much work to do, in part because Opitz's baseline flights had generated some alarming data about the *H IV a*. The L/D was not 37, as many had expected, based on Reimar Horten's claims, but rather 32.[88]

To explain the discrepancy, Opitz recalled that the Horten wing was "quite old" by the time that MSC began studying it and besides, he said, "we did not care much for sealing" gaps between control surfaces and the wing, and in other areas, with tape to smooth airflow and reduce drag.[89] Wear and tear on the wing and failure to seal all the gaps might have cost some performance, but this hardly accounts for a drop of more than 13 percent; there must therefore be other factors to consider.[90] More than a decade and a half had passed since the wing emerged from the Göttingen workshop in April 1943. The Wilkinson report claimed that before the war ended, pilots had flown the wing for five hundred hours, including 1.5 hours in clouds. Repeated exposure to temperature and humidity fluctuations could have made the aircraft heavier, relaxed the alignment of the wooden structure, and degraded the carefully refined aerodynamic performance.

After hostilities ended, military personnel airlifted the wing from Germany to England and then moved it by sea to the USA in 1950. The ocean humidity would likely have penetrated the wing's wooden components. The air at Button's ranch in Valley City, North Dakota, was much dryer, but the wing remained there only a short time. In 1952, Opitz took it south to the hot and dry center of Texas and before year's end, he had flown the aircraft east into the sultry air of Mississippi. The long-term effects of several crashes and other structural damage must also be factored in. A. E. Slater mentioned a crash at the Royal Aircraft Establishment test center in Farnborough, England, and Reimar reported in *Nurflügel* a structural test at Farnborough that broke the wing. Even Opitz's considerable woodworking skills may not have been enough to restore the wing to the precise alignment it had during the war.[91]

Over the next several years, Raspet's team made extensive modifications to the *H IV a* aimed at increasing the L/D. Smoothing the external surfaces was perhaps the most important alteration made, due to its potential to significantly alter performance, and some of this work is visible in photographs, yet nothing is known of the materials or methods used. Raspet and Dick Johnson had used common putty compounds to fill imperfections on the RJ-5, which had a wing and fuselage skinned with wood. However, putty was unsuitable to smooth the fabric that covered parts of the outer wing sections. The group reshaped the hatch covering the pilot inside the semiprone cockpit and fabricated a streamlined housing to enclose the forward landing skid.[92] Whereas the Germans had applied a translucent clear dope to finish the wing prior to its first flight and Opitz had followed this practice when he repaired the glider, Raspet had the *H IV a* painted blue and orange.[93]

With funding and support from the U.S. Army Transportation Command and the U.S. Navy Office of Naval Research, Raspet started a new round of tests on the wing in 1959. In fact, it was part of a larger military effort to develop aircraft capable of vertical/short takeoffs and landings (VSTOL). Reducing or even eliminating runways had led the navy to sponsor Raspet's first research proposal at MSC in 1948 to "conduct viscous flow studies using sailplanes." Raspet continued the work through the 1950s, testing different ideas using a variety of production aircraft and also designing and building several experimental aircraft.[94]

Raspet's team planned a series of tests to first calibrate the airspeed indicator and ensure its accuracy and then measure glide performance, airfoil profile drag, boundary-layer transition, pitch-angle, and elevon deflections with the CG at three different positions. The team mapped the flow of air around various parts of the aircraft using tuft testing. Preferring to fly in the smooth air of early morning just before dawn, George-Falvy took the wing up thirty-five times and spent around sixty hours in the air to complete the tests.[95] In addition to his flying duties, it appears that George-Falvy led the investigation, and many conclusions reached about how the wing performed, and why, were his alone. The demands of developing other experimental VSTOL projects explains Raspet's absence while tests continued on the *H IV a*.

When using only the elevons during his tests, George-Falvy could execute coordinated turns "totally free" of the adverse yaw "which plagued many other tailless designs" because of the differential gearing that moved the elevons about twice as far up as down.[96] The drag rudders at each wing tip, he said, "were hardly ever needed, except for a very abrupt direction change."[97] He could maneuver the wing into a steep bank and begin to circle, the control yoke held fully aft without stalling, and fly a 360° revolution in as little as eight seconds.

During his tests, George-Falvy encountered the pitch oscillations noted by Opitz, assessing them harmless at low speeds but dangerous when flying faster than 145 km/h (90 mph) when the oscillations could "lead to dangerous flutter of the outer wing panels."[98] Further tests showed him that deploying one drag rudder simultaneously rolled the wing and pitched it down, which could become a safety issue when flying at low speed and altitude, and yet he easily countered the roll and dive with opposite drag rudder (the instinctive response of pilots flying conventional aircraft) or avoided it altogether by maneuvering with elevons alone. The spoilers permitted a steep approach at 70 to 75 km/h (43 to 47 mph) and touchdown on a precise spot followed by a ground roll distance as short as 30 m (98 ft.), thanks to the pilot's view from the semiprone cockpit plus the strong braking action provided by the skid landing gear.[99]

George-Falvy conducted most of his tests with the CG positioned 138 cm (54 in./4 ft. 6 in.) behind the nose. He tested other locations but came to regard this location as optimal because at this position the wing always stalled smoothly below 61 km/h (38 mph) with the wing tips remaining level throughout the maneuver, which George-Falvy described as an instant of "mushing" before the aircraft "gently pitched down and resumed normal flight."[100] He observed very rapid responses about the lateral axis to pitch control inputs, but he noted that many flying wings share this trait.[101]

Having documented the wing's good handling qualities, it must have disappointed the staff at MSU (Mississippi State University, renamed in 1957), and Raspet in particular, to discover the L/D had fallen yet again from 32 to 29.5, despite their efforts to smooth the surface to minimize drag and boost L/D.[102] George-Falvy pinned much of the drop on drag caused by excessive elevon deflection required to hold up the wing's nose.[103] At higher speeds, he found the data tracked closer to earlier estimates, but photographs from the tuft tests revealed strong drag from separated airflow and turbulence on the cockpit hatch behind the windscreen, caused after the team reshaped this area.[104] The wing's sensitivity to seemingly minor irregularities was proven during the 1990s when Reinhold Stadler calculated that the L/D lost a point for every 2 percent drag increase. Stadler also confirmed the negative effects of slight changes in CG.[105]

George-Falvy realized that excessive elevon deflection resulted if the center of gravity "was located considerably farther back in the [1943] DFS test, or if the two [*H IV a* wings] were not the same," but he never acknowledged his own size and weight as crucial factors in the performance drop.[106] The *H IV a* (Werk Nr., or airframe number 25) that he flew in 1959 weighed 40 kg (88 lb.) more than the *H IV a* (Werk Nr. 26) flown by Scheidhauer against the Darmstadt D 30 in May 1943. George-Falvy was markedly taller and heavier than Scheidhauer, whom Reimar had specifically designed the wing to accommodate. MSU staff enlarged the cockpit hatch to accommodate his extra height, disrupting the smooth contours that Reimar had carefully calculated, while George-Falvy's greater weight, forward of the CG, forced him to fly with significantly more elevon deflection than either Scheidhauer or Opitz had used (the two German pilots were nearly identical in size).[107]

George-Falvy presented his *H IV a* analysis in 1960 at the VIII Congress of the Organisation Scientifique et Technique du Vol á Voile (OSTIV) in Cologne, Germany, and it appeared in the January 1961 issue of the *Swiss Aero Revue*. Reimar acknowledged that 37:1 was probably impossible to achieve by

1959, but he believed that less than 30:1 resulted from flaws in testing and analysis that he regretted being unable to correct. The designer never met George-Falvy and believed the hatch modifications were a misguided attempt to improve streamlining. Ironically, George-Falvy's extra weight may have tilted the lift vector enough to create a bell lift distribution and make favorable yaw when turning. Whether this happened was never analyzed because Raspet's death in the spring of 1960 ended all scientific work on the *H IV a*.[108]

Reimar believed George-Falvy's report ended serious interest in Horten all-wing sailplanes within the sport soaring community in Europe and the USA, but in reality, conventional designs had rapidly overtaken and surpassed the wing's capabilities. In 1950, Dick Johnson proved the potential of laminar airfoils by surpassing an L/D of 40:1 while flying the RJ-5. In addition to improvements in airfoils, fiberglass-reinforced plastic was revolutionizing the construction of airframes in Europe and would soon engulf the USA. Reimar's high-performance wing had lost too much ground to innovations in airfoils and materials and became obsolete by 1960.

Club Wings

In 1951, several soaring clubs in Argentina asked for a sailplane equipped with side-by-side seating to use for training pilots. Reimar eagerly responded by designing the all-wing *I. Ae. 41 Urubú* (Vulture), similar in size, layout, and performance to the *H XII* motor-glider briefly tested during the war. If he could make the handling characteristics easy enough, Reimar hoped to encourage more pilots to fly the high-performance, single-seat wings such as the *I. Ae. 34 m*.

Reimar gave the *Urubú* 28 m^2 (301 ft.^{2}) of wing area across a wingspan of 18 meters (59 ft.), which he claimed to have designed to distribute lift in bell form. He understood that most clubs consisted of amateur members with neither skills nor desire to undertake building a complex and expensive aircraft. With this limitation in mind, he let simplicity guide his work, using only wood to construct the sailplane and employing just two elevon control surfaces that stretched across more than a quarter of each wing half. Drag rudders mounted near the wing tips gave additional yaw control when activated individually, or control of glide path and descent rate if activated in unison. Reimar mounted wheeled landing gear struts beneath the nose and the trailing edge of the center section, and provided extra capacity to absorb the shock and vibration of rolling over uneven ground. The pilot could retract the struts to lessen drag and boost performance.

Like many conventional training gliders, a pilot had to fly the tandem-seat *I. Ae. 34 a Clen Antú* training wing with either another occupant or equivalent ballast. Without a balanced load, the CG shifted dangerously. The ballast weight was essential for safe flight but awkward and inconvenient to handle, and doing away with it simplified operating the glider. Karl Nickel acknowledged that Reimar might have designed the wing for easy conversion to a motor-glider, as he had done with the *I. Ae. 34*.[109] Heinz Scheidhauer and Karl Nickel traveled to Argentina from Germany in August 1951, and Reimar immediately put them to work on the training wing. A year later, the project took on greater significance as a means to rekindle interest in the Horten wings, following Argentina's poor showing at the 1952 world meet.

Scheidhauer made a series of short hops to test the controls before he flew the prototype on December 4, 1953. Another *I. Ae. 41* flew in 1955, and craftspersons finished the third *Urubú* late in 1958. Reimar remembers that a club took one of the wings, one remained at the factory, and officials appropriated the third for display around the country, often outdoors with no concern for the fragile wooden airframe. This *Urubú* quickly deteriorated and became unsafe for flying. "It had gotten wet [in the mountains] and left in the hanger and begun to come apart" within about six months, Reimar remembered.[110]

Pilots flew each of the other two examples for a total of about one hundred hours. Reimar claimed they behaved well in turbulent air and during stalls, but these aircraft were not without vices. Victor Riviera contacted Nickel and Scheidhauer, and both agreed that the *I. Ae. 41* was "a wonderful plane but still very dangerous [to fly because] as soon as water drops, or spots of dirt (dead flies for example) [contaminated the wing, it became] very unstable."[111] Nickel also found it tail-heavy and there was a problem with flutter, but adding counterweights to balance the elevons eliminated the latter problem.[112] Reimar speaks of several record flights that Scheidhauer proposed to carry out while flying the *Urubú*. An attempt to set an endurance record by remaining aloft for at least twenty-four hours failed when the men assigned to support the flight fell asleep. Scheidhauer had asked them to light gasoline lamps to help him navigate safely in the dark and close to mountains where rising air currents were strong.[113]

Soaring the Andes

Another record attempt of sorts was more successful. In 1956, Scheidhauer piloted the training wing for the first soaring flight over the Andes Mountains into Chile. Argentine historian Ricardo Burzaco wrote that the directors of the

Dirección de Fomento de la Aviación Civil (Civil Aviation Public Relations) and the I. Ae. organized the flight. Civil Aviation PR procured a Slingsby Type 34 Sky (the same type flown by the British team when it won the 1952 world contest in Spain) and selected Claudio Dori to fly it. The agency also provided a tow airplane flown by Guido Rossi and a meteorologist named Podestá. The I. Ae. supplied Scheidhauer and an *Urubú*, a tow aircraft and pilot, and meteorologist Rubén Sanchez. Aircraft engineer Alfonso Zuárez managed both teams. Assigning meteorologists and towing apparatus to each team suggests a good-natured competition between rival government organizations.

Planners chose to launch near the city of San Carlos de Bariloche, nestled among the Andean foothills in the province of Río Negro, 1,125 km (698 miles) south-southwest of Córdoba and less than 120 km (75 miles) from Chile. Exceptional weather was required if the sailplanes were to clear even the lowest mountain elevations between Argentina and Chile at this latitude. Early in the morning on October 30, 1956, the meteorologists predicted a splendid combination of up-slope winds, thermals, and mountain waves along the proposed route of flight. So good was the weather that after taking off, the tow airplanes reached just 300 meters (980 ft.) when Scheidhauer and Dori released their tow lines and turned west toward the Andes.

They flew together until Dori had to wrench his Sky violently off course to avoid an enormous Andean condor that threatened to collide with him. Scheidhauer lost sight of Dori but continued on past Catedral Mountain and el Brazo Tristeza (the Arm of Sadness). He then flew over the Tronador Volcano, elevation 3,491 m (11,450 ft.), and entered Chilean airspace. Minutes passed as the pilot soared the wing over the southern boundary of Lake Frias and gazed upon the towering summit of Lanín Volcano. Scheidhauer eased the controls forward and began a steady descent over lush green forests and fields. After spending three hours in the air, he soon began spiraling down, set up his approach, and then landed on the eastern shore of Lake Llanquihue near Ensenada, 115 km (71 miles) from San Carlos de Bariloche. Dori landed twenty minutes later. Weather conditions did not permit safe return flights until fifteen days had passed. The tow airplanes then pulled the gliders back over the mountains to Argentina.[114]

Positioning the *Urubú* to begin the flight had been almost as challenging as the actual trans-Andes sortie—a tow pilot had to haul Scheidhauer and the wing over more than 1,000 km (620 miles) of unsettled country from Córdoba to San Carlos de Bariloche—and to what end? Although the project demonstrated that a very experienced pilot could soar the *Urubú* over substantial obstacles, Scheidhauer's flight went unnoticed in magazines catering to the soaring community, such as *Soaring* in the USA and the British magazine *Sailplane and Glider*. Even the respected journal *Flight,* specializing in international developments, remained silent. Coverage appears to have been limited to the Patagonia region, but perhaps the organizers aimed to raise public enthusiasm and increase government support in Argentina for soaring and civil aviation, which certainly needed a boost after Eduardo Lonardi had deposed Perón in 1955, and Pedro Aramburu replaced Lonardi a short time later. Historian Fernando Siarez has confirmed that unlike Perón, neither Lonardi nor Aramburu supported civil and sport aviation.

Reimar is vague about the flight's origins and purpose, but he certainly welcomed the chance to direct favorable attention on his work. Since their poor showing during the 1952 world contest in Spain, the Horten wings had remained unpopular in Argentina. Only one club had accepted an *Urubú* since the type was first flown in 1953, yet Horten had designed the airplane specifically for club use. Scheidhauer made a flight in an *I. Ae. 34 m Clen Antú* that covered a distance of 353 km (219 miles) during the Argentine national meet in February that year, but the deed earned just a half-sentence in the July 1953 issue of *Sailplane and Glider*. The magazine again covered the Argentine contest scene the following year, but no Horten wing was mentioned.[115]

CHAPTER 6

Winglets and More Wings

WHEN FRED HOINVILLE flew for the Australian team during the 1952 and 1954 world soaring championships, he noticed how the expense of high-performance sailplanes was denying many enthusiasts the opportunity to enjoy the sport. Hoinville reasoned that smaller aircraft with more modest performance capabilities should be less expensive to own and operate. They should require fewer ground crews and less expensive transport equipment, such as automobiles and trailers, and need less costly launch apparatus, such as winches and towing aircraft. With careful design, performance would probably remain good enough to satisfy most pilots. He argued that pilots could probably maneuver the smaller sailplane more nimbly in tight thermals and, therefore, climb more efficiently than would be possible in larger sailplanes. The category of sailplanes spanning 7.6 m (25 ft.) that Hoinville was seeking did not become popular until the late 1960s, but once it caught on, it took the world by storm as the lightweight, aluminum-framed hang glider.[1]

Unbeknownst to Hoinville, Reimar Horten had started developing the prototype for a small and simple sailplane in the late 1940s.[2] Reimar called it the Horten *H X*, but the Argentineans named it *Piernifero* ("leg-trained") because the pilot was supposed to support and maneuver the wing on the ground using his or her own legs. Argentine builder and *Piernifero* test pilot,

Rogelio Bartolini, described it as a "winglet" because of its small size and low weight when compared to other sailplanes of the day. Reimar speaks of discussing with Scheidhauer before the war the operational and performance aspects of a hypothetical foot-launched sailplane weighing about 30 to 40 kg (66 to 88 lb.); however, the idea for lightweight gliders dates to the earliest days of gliding as an organized sport.

Reinhold Platz, who had worked as chief designer for a Dutch entrepreneur of powered aircraft, Anthony Fokker, during and after World War I, designed a very light glider in 1922 while in Japan. The pilot's feet served as landing and takeoff gear for a "High Performance Hanging Glider" built and flown in 1937. Reimar speculated that pilots of his small soaring wings could scout thermals for the pilots flying larger, high-performance sailplanes, but what would attract amateur builders and pilots was the promise of lower costs to construct and operate these aircraft. Reimar pointed out that small sailplanes did not require airfields with paved runways, could be constructed nearly anywhere, and were beyond the regulatory control of government organizations.[3]

Hoinville had envisioned that such minimalist aircraft would span less than eight meters (26 ft.); however, Horten designed three different versions of the *Piernifero* that spanned from 7.5 m (24 ft. 7 in.) to 15 m (49 ft.). Scheidhauer constructed about 80 percent of a 10 m (32 ft. 10 in.) wing called the *H X b*, but he never flew it, and the 15 m *H X c* progressed no further than the drawing stage. The *H X a* weighed 37 kg (82 lb.), and each square meter of wing surface lifted 10 kg/m^2 (2 lb./ft.^{2}), or as much as the *H I* that Reimar had flown in 1933 (Figure 28).

The *H X a* wing loading was only slightly greater than the foot-launched Rogallo-wing hang gliders developed in Australia and the USA during the 1960s; however, the hang gliders had a distinctive advantage. One person could disassemble their aluminum-tube airframes and pack them into a bag approximately 6 m (20 ft.) long and less than .3 m (1 ft.) in diameter, which one or two people could then carry short distances. By contrast, the wooden *H X a* was built in one piece to reduce cost and speed construction time. Because of its weight and size, a trailer was required to move the wing any distance on the ground.[4]

Reimar also took a minimalist approach to designing the cockpit and control arrangements for the *Piernifero*. The pilot rested his chest on an inclined board, and after taking off, he swung his legs back and hooked each ankle on small, curved beams that doubled as two of the three wooden skids

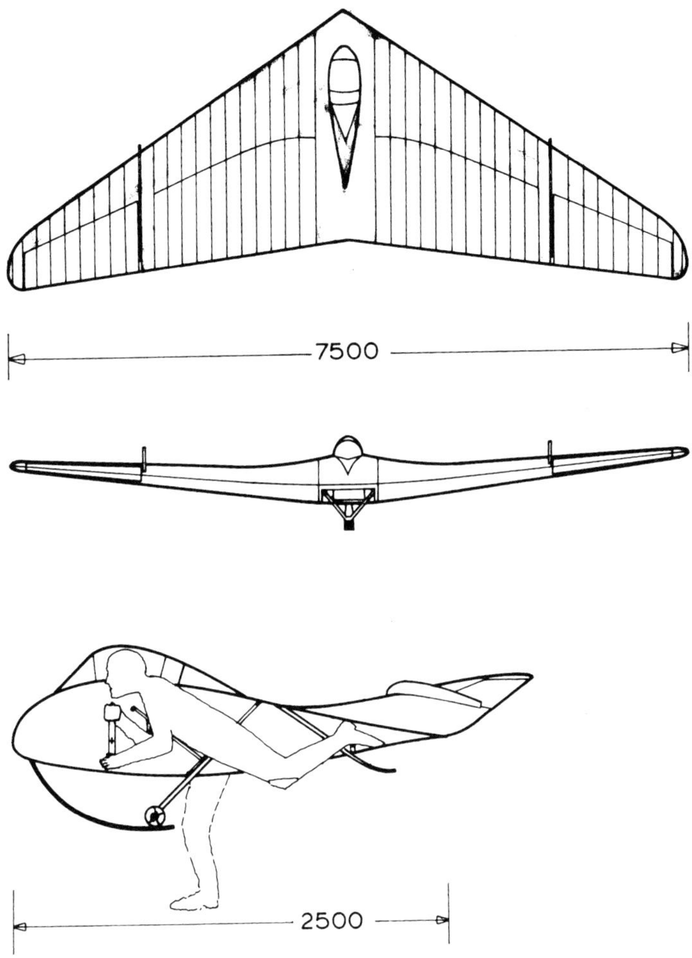

Figure 28. *H X a* winglet three-view drawing (Peter F. Selinger).

that supported the wing on the ground. Except for the dangling legs, the effect resembled the semiprone cockpits devised for the *H IV a* and *H VI*. A small canopy covered the pilot's head, allowing him a slight view over the top of the wing. The pilot manipulated just two elevons, and both surfaces had Frise noses to reduce the adverse yaw. Horten added wing fences along the inboard side of both elevons. Each fence extended from the leading edge across the top of the wing and nearly to the trailing edge. Their purpose was twofold: to generate vortices that impinged against the elevons and provided extra traction against the airflow to control pitch and roll when the wing was flown at higher angles of attack, and to provide a stabilizing effect about the yaw axis. Such devices violated Reimar's quest for purity in all-wing design, but they appear to have been necessary for good handling in the air.

Reimar sought to emulate the handling characteristics of a sailplane called the Grunau Baby. It was a shrewd choice because the airplane was a sport sailplane of conventional layout designed in Germany in 1931 that had become one of the most popular sailplanes in the world. Thousands were flying in many countries in Europe, Great Britain, and Australia by the late 1940s, and the aircraft had influenced the development of other popular sailplanes. Reimar understood that the Grunau Baby competed directly with the *H X a*, and to be popular, his new wing had to be easier to build and less expensive. This marks a rare example of Horten factoring market potential into his design work.[5]

Three men began building the *H X a* late in 1949 and finished the project four years later in January 1953, after about two thousand hours of work, a figure that compared well with the five thousand hours or so needed to build the Grunau Baby. Once they had the primary structure framed, they enclosed the leading edge with thin layers of wood and then covered the remainder of the wooden airframe with fabric. Only after finishing the prototype did the men discover that with certain modifications, they could have nearly halved the construction time. The group spent another few weeks building a trailer before moving the wing from the workshop to the airfield (Figure 29). Including the trailer, materials for the entire project cost around five thousand Argentine pesos, or about $357.[6]

Pilots made at least eight unsuccessful attempts to take off by running with the wing, which they had nicknamed *L'alita* (the little one), but the wind velocity rarely increased enough to lift the bulky aircraft into the air, and the men could hardly run fast enough to reach flying speed when it did. They installed a skid landing gear designed by Reimar to keep the wing from falling on the pilot's legs should he stumble during takeoff or landing. Several months passed before Rogelio Bartolini completed a short flight in tow behind an automobile. In May 1954, he also made the first flights in tow behind an aircraft. Bartolini and his two friends had prepared carefully for air towing, which entailed a significant, and somewhat risky, increase in the experimental wing's speed and altitude. They packed a parachute into the small cockpit, added a chin rest to ease strain on the pilot's neck, and installed a variometer

Figure 29. Enthusiasts prepare to unload the Horten *H X a Piernifero* from its transport trailer. Reimar fitted wing fences perpendicular to the span and extending from the leading edge almost to the wing's trailing edge to generate vortices that gave the elevons extra traction at higher angles of attack and to stabilize the wing about its yaw axis (National Air and Space Museum, Smithsonian Institution, SI 80-19275).

to accompany the airspeed indicator already fitted. Both car and plane vastly improved the takeoff performance compared to muscle power alone.[7]

Aircraft towing made it possible for the men to probe nearly the full range of aerodynamic qualities, which they found as good-natured as a Grunau Baby. This is a surprising claim for an all-wing sailplane equipped with such a simple control system. Bartolini explained that

> turns are very easy. All you have to do is move the stick [mounted right in front of the pilot] towards the direction of the turn [you wish to make] and immediately she banks and begins to turn herself. This synchronization obtained by the [Frise] effect of the lifting aileron has proved so good that so far it has not been necessary to modify it at all. In a rapid change of direction she yaws a little to begin with but at once settles down.[8]

He reported that the wing remained stable when circling at 60 km/h (37 mph) in a steady 30° bank. A complete turn under these conditions required a relatively slow twelve seconds, but Bartolini could turn no faster, nor induce the wing to stall, until he and his friends increased the elevon deflection angle. The wing tried to pitch down when Bartolini flew fast at 100 km/h (62 mph), but he maintained control by firmly holding the stick. He trusted the wing enough that during his second flight, he executed a loop following a dive to 110 km/h (68 mph), followed immediately by another loop from 120 km/h (74 mph). The maneuvers brought the aircraft low enough to the ground that Bartolini could begin his approach to land and found that by firmly swinging the stick from side to side, "the turbulence set up by the sudden movements increases the speed of descent and is thus equivalent to a sideslip."[9] This handy trait was accidental, but welcome, for it allowed the pilot to increase the descent angle while maintaining a relatively constant airspeed, permitting shorter and safer landings, and improving control in gusty conditions. Usually only pilots flying conventional aircraft equipped with rudders could exercise such flexible handling.

Bartolini's friends took turns flying a Grunau Baby while he flew the *H X a* alongside them. At speeds below 100 km/h (62 mph), the Grunau Baby showed less sinking speed, but there was parity when the pilots pushed the two gliders to speeds between 100 and 140 km/h (62 and 87 mph). These preliminary evaluations also revealed operational limitations. Soaring conditions were often better above the mountains, but pilots feared breaking their legs if they stumbled during a landing on the steep terrain. Scheidhauer told Reimar that the wing flew "like a little ship in the ocean . . . in air currents it was unstable." Consequently, a pilot would have strained to soar for extended periods as the wing pitched and bucked in turbulent thermals. In short, the wing was too small, and Horten believed this led pilots to shun the aircraft.[10]

Ancient One Reborn

A year after arriving in Argentina in 1948, Reimar was asked by the president of the Otto Ballod gliding club in Adolfo Gonzáles Chaves (almost 400 km, 248 miles, southwest of Buenos Aires) to design a new wing for his club. All-out performance again gave way to economy. With the requirement to be simple and inexpensive to build in mind, the group asked for a design with little or no metal components. The aircraft had to handle well in flight to compensate for the skills of pilots with little or no experience flying sailplanes. In 1950, Reimar gave the club plans to build his original *H I*, the same wing that the Horten brothers had built and flown in 1933 and then had to burn on the Wasserkuppe a year later. "I thought well, the ancient one [*H I*] has not been terminated after all," the designer recalled years later.[11] Reimar never clarified why he did not offer the club plans for the *H X a Piernifero*, but Otto Ballod may have wanted the enclosed cockpit, stronger airframe, and better performance available from the old design.

Although the *H I b* wingspan placed it outside Hoinville's class of "mini-midget" gliders, wing loading matched the newer *H X a*. Reimar simplified the design by eliminating the flaps fitted to the original *H I*, and the club expedited the construction when they opted to make the wing in one piece. Nonetheless, the difficulty of obtaining building materials, and the need for Reimar to travel for two days to consult in person, slowed construction. Four years passed before Horten watched a club pilot fly the wing for the first time on May 2, 1954. Reimar's faithful test pilot, Scheidhauer, later flew the wing and pronounced it so well balanced that it needed no adjustments. The wing proved stable and easy to fly, and an automobile could easily tow it into the air. Pilots likened the performance and handling to the popular Grunau Baby. Club Otto Ballod successfully operated the *H I b* for twenty-five years with no accidents and only the occasional attention of a carpenter was needed to maintain the wing.[12]

In 1950, the Cóndor soaring club near Buenos Aires asked Reimar for plans to build a wing with many of the *H X a Piernifero's* characteristics, particularly the small span, but better performance, namely higher speed, to enhance the wing's appeal to experienced pilots as well as students. The club members hoped that limiting the span would allow Reimar to design a one-piece wing that did away with the expensive steel fittings needed to join the two outer wing panels to the center section of bigger wings. These needs presented Horten with a design dilemma. Low induced drag, derived from a long span, high-aspect ratio wing with high wing loading gave more speed, yet there was a limit to how small a wing could be made and remain safe to fly. Reimar tried to meet these diverse requirements with the *H XVI Colibri* (Spanish for "hummingbird").

The *Colibri* spanned 12 m (39 ft.), slightly less than the *H I b*, but the aspect ratio more than doubled, and this drove the wing loading 56 percent higher, perhaps too high. Reimar included drag rudders to help a single pair of elevons control pitch and roll, but the wing's control architecture was otherwise uncomplicated. He seated the pilot in a standard upright position and enclosed him fully within a wooden fairing topped with a clear canopy. A skid landing gear provided support during landing and takeoff.[13]

The club built the wing in one piece to save cost and reduce construction time. In November 1952, the club workshop manager and flight instructor,

Waldemar Sturm, flew the *Colibri* at Merlo, near Buenos Aires, after an automobile hauled him into the air. He completed several flights at low altitude, guiding the wing through gentle turns, but he balked at flying higher behind a tow airplane. On January 20, 1953, Heinz Scheidhauer attempted the first flight in tow behind a Fieseler Storch light aircraft taking off from the Juárez Celman airfield in Córdoba province. After leaving the ground, Scheidhauer lost control and crashed from a height of about 15 m (50 ft.). He walked away from the wreck uninjured, but damage to the *Colibri* was beyond repair, and the club made no efforts to build another.

Reimar gave considerable thought to what caused the crash, and in a rare reproach of his favorite test pilot, Reimar implied that Scheidhauer became overconfident and bypassed the important interim step of flying at a lower altitude and less airspeed behind an automobile before ascending higher, and flying faster, in tow behind an aircraft. Horten believed the tow airplane's roiling wake upset the wing, but the *Colibri's* high wing loading had made it stall too fast for safe flight. Sturm had not stalled the wing because he remained in ground effect, where the stall speed was lower. Reimar came to realize that he had violated a basic design law governing sailplanes by making the *H XVI Colibri* too small and too heavy.[14] The minimum size and maximum weight for sailplanes, he said, was "about [13 m span, 43 ft.] and about 140 kilos [309 lb.]. This size, like the Grunau Baby, is good, but [if] it is bigger, then it is better. The H XVI was too small."[15] It had spanned just 12 m (39 ft.) and weighed 160 kg (354 lb.) with a pilot aboard.

Supersonic Research Resumes

In parallel with his low, slow, and inexpensive sailplane projects, Reimar resumed working at the other end of the performance spectrum. The end of World War II had halted development of the *H XIII* supersonic jet fighter, and Reimar had made no effort to resume the work after the war because the cost to design, build, and fly aircraft at these speeds required a level of support available only from national governments. Such funding was not forthcoming from the Perón administration until the early 1950s. Tank had refused to consider Horten's "friendly competition" with the Pulqui II in 1949, but a few years later, Reimar renewed this line of development during a burst of creativity that saw him working simultaneously on the large *I. Ae. 38* freighter wing, the *I. Ae. 34 a Clen Antú* and *H XVI Colibri* sailplanes, and the *H I b* club glider.

With authority and support from the FMA, a team led by Karl Nickel began conducting tests in 1953 using catapults propelled with large rubber bands to launch models fitted with delta wings from the shores of San Roque Lake into the water to recover and reuse. The models weighed from 5 to 20 kg (11 to 44 lb.), and Nickel and his men built each one to withstand the impact of launch speeds up to 200 km/h (124 mph). The group carried out more than two hundred tests and gathered valuable data, which they applied to refine such aerodynamic issues as airfoil profile shapes, lift distribution, and the CG. The model work allowed Karl Nickel to calculate the latter point with considerable accuracy.[16]

After technicians painted the models white and coated them with lamp-black mixed with kerosene, the tests revealed vortices emanating from the wing leading edges. These vortices grew larger at higher angles of attack, a finding particularly important for controlling delta aircraft during low-speed flying, such as during takeoff and landing. Reimar claimed that these studies helped the team to pinpoint the best sweep angle at around 60° and to settle on the form and location of the rudder, elevons, and air brakes. He later wrote that the team was probably the first to harness these vortices to control a delta wing airplane, following Dr. José Krasinski's observation of the phenomenon in the wind tunnel at Córdoba, but this is not true. Five years earlier in 1948, the U.S. Air Force had successfully flown a jet-propelled prototype called the Convair XF-92A, designed by engineers at the Consolidated Vultee Aircraft Corporation who had studied a 60° delta wing in 1945.[17]

Encouraged by the results of their tests, Horten's team began in 1954 to build a manned delta-wing glider designated the *I. Ae. 37-P* (*P* for *Planeador*, Spanish for "glider"). Like the XF-92A, the Planeador required a substantial vertical fin for yaw stability and control; it was therefore not an all-wing aircraft. The team built the airplane entirely of wood and fitted a semiprone cockpit. Captain Jorge Doyle completed the first flight on October 1 in tow behind a three-engine transport aircraft. Manuel Fentanes commented to historian Ricardo Burzaco that "his hair stood on end" while watching Doyle perform aerobatics in the sleek delta as he descended from 2,000 m (6,560 ft.) after riding the tow to that altitude.[18] Rogelio Balado remembered resting his chin in a quilted bracket during his own first flight in the glider and controlling the aircraft using a handlebar. As Reimar coached him by radio, Balado quickly grew accustomed to flying from the semiprone position, and like Doyle, he too finished his flight with acrobatics.

The pilots were most comfortable taking off and landing with the delta's nose raised about 25 to 30 degrees, and the glider touched down smoothly and rolled quickly to a stop at this angle. Reimar's craftspersons installed a nose wheel attached to a long strut to fix the glider's attitude at this angle. When

Scheidhauer flew the delta, he reported "negative stick force" at slow speed, suggesting that the powerful wing tip vortices were impinging on the elevons; nonetheless, Reimar claimed that "many" officers in the Argentine air force flew the airplane a total of about one hundred hours. The cockpit was reconfigured to a conventional upright seating position and then flown for another fifty hours. Satisfaction with both versions of the glider compelled Horten to begin building the second piloted prototype delta-wing aircraft, the *I. Ae. 37-P*, designed to fly on the jet propulsion provided by a single Rolls Royce Derwent V power plant, which Reimar believed would push the jet to Mach 1.

Work was well underway by 1955, and Reimar expected to finish and be ready for flight the following year when dissatisfaction with the Perón government led to a coup d'état in September. This so-called Liberating Revolution set about destroying all vestiges of Perón's regime, hitting government aircraft projects particularly hard because the Argentine army and navy supported the coup and saw the uprising as a chance to cripple or destroy the rival Argentine air force, then just ten years old, and an independent service organization. Government air minister and air force officer Cesar Ojeda remembered the revolution as "a great fracture in the Air Force," whose leadership was imprisoned or forced to quit, replaced by men who cared little for aeronautical work. The two administrations that followed the Liberating Revolution both continued to suppress the air force, and work did not resume on the *I. Ae. 37-P* until voters elected Dr. Arturo Frondizi in 1958.[19]

During the late 1950s, Reimar started designing the *I. Ae. 48* two-engine delta fighter for the navy and air force. He says he planned to apply the area rule developed by the NACA in the USA, narrowing the fuselage to resemble a coke bottle to improve performance at supersonic speeds. Ricardo Burzaco reports that the Argentine navy discussed with Reimar how to improve the jet's takeoff performance from an aircraft carrier because the catapults aboard the carrier bought by Argentina from the British Royal Navy in 1958 were suited only to launch propeller-driven airplanes. Both the *I. Ae. 37-P* and the *I. Ae. 48* were under construction when the FMA ordered all work stopped in 1960. Horten believed that the FMA leadership lost faith in the *I. Ae. 48* when they heard how several NATO countries, among them West Germany, had bought significant numbers of the Lockheed F-104 interceptor designed and built in the USA. The F-104 was unsuitable for carrier operations but much less expensive to purchase ready-to-fly for the air force compared to the cost to develop the Horten delta.[20]

The fate of the *I. Ae. 37-P* disappointed Reimar, since he had thought it safe from public exhibition because the aircraft was built in one piece and thus difficult to move. The authorities nonetheless ordered the finished aircraft moved to Córdoba where "thousands could see it and come and touch it," said Reimar. The move went well until the wing reached a street lined with trees leading to the center of town, but narrower than the aircraft's wingspan. To make it fit, "someone took a saw and cut the wing tips off the delta [glider]!"[21] The public display opened on schedule, but the damage had compromised the glider's structural integrity and it never flew again.

Fruit Hauler

Lack of efficient transportation infrastructure drove the development of the only powered tailless aircraft designed by Reimar in Argentina, the *I. Ae. 38.* In 2005, it remained the largest airplane built and flown in that country, according to historian and engineer Fernando Siarez. At the end of World War II, the process of distributing perishable goods from outlying provinces to the capital, Buenos Aires, remained difficult. The far northern and southern regions were particularly isolated. Oranges, for example, had to be imported from other countries because there was no method for moving this produce quickly enough from the Province of Misiones, where the fruit was harvested more than 650 km north of the capital. It was also impossible to get fresh fish from ports far to the south.

Early in 1950, Cesar Ojeda asked Reimar to design a large glider to fulfill this critical transportation requirement. Ojeda envisioned something similar to the Messerschmitt Me 321 Gigante transport with a cargo capacity of 18,000 kg (40,000 lb.), after hearing how the Luftwaffe had employed the aircraft for a short time to resupply the German army in North Africa during World War II. Crews would be landing and taking off from primitive airfields, which required good low-speed performance but high speed was required, too. Ojeda wanted an airplane that was easy to load, could lift 10,000 kg (22,100 lb.), and attain a speed of just "100 km/h (62 mph) in a straight line."[22] Horten studied the problem and further refined the requirements to solve it. The transport wing would have to haul 10 metric tons (22,000 lb.) of freight 1,000 km (620 miles), land on an unimproved runway, perhaps a clearing hacked into the jungle, load easily using a ramp at the rear of the airplane, and then take off again at maximum weight in less than 500 m (1,640 ft.).

Horten proposed a transport powered by four engines each developing 552 kW (750 hp). A large compartment suspended at the apex of the swept wing contained the cockpit and cargo cabin with a useable volume of 30 m^3 (1,059 $ft.^3$). Cargo was loaded through a pair of rear doors. One door doubled as an inclined ramp. Using a wind tunnel, Reimar verified that the crew could open these doors enough to drop their cargo by parachute. A team led by Karl Nickel built and tested flying models in order to measure the wing's skid-roll moment.

Reimar designed a special landing gear to distribute the aircraft's weight when flying from airfields softened by seasonal rains. Two large main wheels installed on each side of the cargo cabin supported most of the weight. Horten half-buried them in the base of the cabin to reduce their drag. Hydraulic oleo struts supported each main wheel strut to minimize bounced landings on rough airfields. Reimar engineered a system that allowed the pilot to determine the CG after the cargo was secure and shortly before taking off. If the load had altered the wing's angle relative to the ground, the pilot could operate a compressed air mechanism to adjust the ground angle to the optimum setting for takeoff. Without a tail, it was critical to set the right balance because the wing would have rolled much farther on the ground before lifting off if the ground angle were too high or too low. At such low speeds, the elevons probably were not powerful enough to compensate for a poor angle, particularly if the transport carried a heavy load. Up at the cabin nose, Reimar installed a tire and wheel at the end of a tall strut, which the pilot retracted after takeoff.[23]

Horten buried the four power plants inside the wing. Each engine turned a drive shaft connected to a two-blade propeller facing aft in a pusher configuration. He planned to install ventilator-type refrigeration turbines at the front of each engine to preserve the perishable cargo. He added wing fences to enhance the wing's handling but had trouble finding the hydraulic components needed to power the combination dive brakes and drag rudders used so often on Horten aircraft. As a stopgap measure, he grafted two vertical stabilizer and rudder assemblies onto the top of the *I. Ae. 38* near the wing tips to help control yaw. To construct the entire aircraft, the designer selected aluminum for its resistance to the heat and humidity found in the northern provinces near the equator. The wing was fundamentally a scaled-up version of the single-seat *H II* sailplane that spanned 32 m (105 ft.).[24]

Reimar said his team of engineers and draftspersons spent forty thousand hours designing the *I. Ae. 38* and making drawings before craftspersons labored for another 350,000 person-hours to build a prototype. These figures seem high, but there is no doubt the job was a massive one for the small government-managed aircraft industry. The airplane could have been finished sooner were it not interrupted by another coup d'etat in 1955. General Eduardo Lonardi ousted Perón, but Lonardi's government lasted less than two months before General Pedro Eugenio Aramburu took over. Only after the government changed again three years later when Arturo Frondizi came to power in 1958 did political stability return long enough for work to resume on the transport wing.[25] Construction was at an advanced stage when he discovered that the engine selected in 1950 to power the wing would not be ready to install when workers finished building the airframe. The only available substitute was the 253 kW (316 hp) El Gaucho, an engine that produced less than half the power of Reimar's first choice. The replacements could get the wing airborne, but they reduced significantly almost every important performance parameter.

Commander Rogelio Balado conducted a successful first flight in 1960, despite a miscalculation in the CG that made the event a hazardous milestone (Figure 30). Technicians corrected the problem, and three more flights quickly followed that confirmed, according to historian Ricardo Burzaco, the potential for the *I. Ae 38* to become a good cargo airplane. Reimar spoke of "outstanding low-speed characteristics and controllability," but whatever the wing's potential, other means were devised to move the goods from jungle to capital, and the wing was no longer needed after ten years of fitful development.[26]

The project was canceled; nonetheless, Burzaco argues that the *I. Ae. 38* was not obsolete. Installing turboprop engines, lighter and more powerful

Figure 30. Onlookers watch as technicians prepare the *I. Ae. 38* all-wing transport for flight in 1960 (Reimar Horten via Peter F. Selinger).

than the reciprocating type, would have breathed new life into Reimar's design. Burzaco further speculates that the wing could have successfully dropped airborne soldiers by parachute, or replaced the twenty-five-year-old Douglas DC-3 as the mainstay transport for Argentine military operations in the Antarctic. Successfully adapting the airplane to turboprop power might have increased load capacity, speed, and ceiling, but such an adaptation would have been no small undertaking.

Yet better engines could not make up for other operational weaknesses of an all-wing transport. Handling the *I. Ae. 38* on the ground would have been more difficult compared to the conventional DC-3 transport. The big Horten wing lacked a powerful aerodynamic fulcrum (fuselage and a tail) with which to maneuver on the ground when heavily loaded, although reversible-pitch propellers might have helped. The wing's cargo bay design, tall and narrow due to the restricted movement of the CG inherent in the all-wing layout, would have forced operators to stack cargo high rather than along the length of a conventional fuselage. Cargo at the bottom would have to be crated to withstand the weight of the material stacked above it. The *I. Ae. 38* carried less cargo than the DC-3 transport, yet the projected cruise speed was only slightly greater, not enough to justify the costs of further development. The Argentine military scrapped the prototype soon after it had deposed the Frondizi government in 1962.[27]

Final Projects

Reimar's fruit hauler was his last government or military all-wing aircraft, but long before the project ended in 1962, the designer was struggling to support himself and his family. He had joined the Escuela de Ingenieria Aeronáutica (School of Aeronautical Engineering) in 1954 where he taught aerodynamics courses, and published numerous articles on these subjects, but his professorship alone was insufficient to survive as the financial situation in Argentina worsened leading up to Perón's ouster in 1955.[28]

The new Aramburu government formed the Comisión Nacional de Investigaciones (National Commission of Investigations) and began to harass the German immigrant community. Many left Argentina and fled to other countries, among them Kurt Tank and some of his staff who relocated to India in February 1956. Reimar arranged to stay in Argentina, and the revolutionaries left him and his family alone.[29] He may have bargained with the new government to remain in Argentina if he curtailed designing military aircraft. In

1960, he was appointed an adviser to the Dirección Nacional de Fabricaciones e Investigaciones Aeronáuticas (DINFIA, the National Directorate of State Industries), a government agency that managed the automobile and aviation industries, including the FMA.

His finances remained tight, and with a family of four to support, Reimar travelled to Germany in 1962 looking for work but found nothing. Upon returning to Argentina, the Instituto Aerotécnico Córdoba (IAC, the Aeronautical Engineering Institute of Córdoba) offered to employ him without a contract. The firm would guarantee his pension if he worked only on small-scale conventional aircraft. Reimar must have agreed to the terms because he committed to finally settling in Argentina by purchasing a ranch south of Córdoba in 1964 and building a house there. The institute assigned him modest design tasks, such as modifying an existing light transport, and he continued to work there for more than fifteen years. Reimar also became an advisor to the Grupo de Desarrollo de Aviones (Airplane Development Group) in 1969, and in 1972 he joined the Córdoba branch of the Instituto de Investigaciones Aeronáuticas y Espaciales (Institute of Aeronautical and Space Investigations), again in an advisory role.[30]

Around 1965, a Venezuelan air force colonel asked Reimar if he would be willing to relocate to Venezuela, start a flying school, and teach students aircraft design, fabrication, and flight training by building and flying several *H VII* two-engine wings. The offer greatly excited Reimar. "We would have had all the support and experienced men needed to build a whole line of all-wing aircraft," he said, but Walter, who had saved the drawings needed to build the airplane, saw difficult problems with the proposal and refused to send them to Reimar.[31]

The plan was truly risky. Reimar knew little about the working conditions, living standards, or political stability of Venezuela, and yet he desperately wanted to escape from the stifling restrictions on intellectual curiosity and experimentation that seemed to prevail after Perón's final fall. The overall Argentine philosophy of aeronautical development was fundamentally different from what he had known in Germany. "The idea of testing was not supported" in Argentina, but in Germany, "we test to make better. [We were always] testing to find improvements."[32] Reimar spoke bitterly about wasting his mathematics PhD in Argentina after working for so long in Germany to earn it. Whether hobbled by a lack of resources, will, or expertise, Argentina seemed to follow other nations at a considerable distance rather than trying to compete equally, or even lead in aeronautical research:[33]

> If America would build a plane without a tail, then the Argentine's would do the same. They have the idea that they must copy . . . and one cannot go against it . . . If I mentioned to my superiors that you can read it here in this book, then they are satisfied. [But] If I go on to do something from my own mind, but it has not been published, then they would put it aside without interest.[34]

Reimar eventually retired from the IAC in 1978 after thirty years of service. From his five-hundred-acre ranch nestled between the mountains at Villa General Belgrano, 162 km (100 miles) southwest of Córdoba, he now looked forward to lending his expertise to "aircraft projects throughout the world."[35] However, Horten began to fear that the data and techniques he had discovered after decades of careful tests and evaluations, data necessary to stabilize and control the all-wing aircraft, would be forever lost. In 1980, airline pilot and vintage glider enthusiast Jan Scott announced to the Vintage Sailplane Association (VSA) that

> This remarkable man is now looking for someone to carry on his work. Apparently none can be found in Argentina. He asked if VSA would try to find a bright young student of Aerodynamics that would be willing to come and stay with the Hortens for as long as necessary to learn what he has to offer. Such a person should have access to facilities where an aircraft could be built later. Interested persons should contact [the VSA].[36]

Stealth Claims

After several years had passed with no reponse to Scott's appeal, Reimar began to discuss more openly a claim he had first raised in 1950 in the article, "Ala Volante Caza 'Horten IX'" (Flying Wing Fighter Horten IX), published in the May 1950 issue of *Revista Nacional de Aeronautica* (National Magazine of Aeronautics). After describing how speed, ceiling, and turn radius were among the important factors that determined the combat effectiveness of a fighter aircraft, Reimar added another factor he said was equally important: the degree to which the fighter was visible to radar. "Because the electrical wave reflection is good [from] metallic surfaces," said Horten, "however, in wood surfaces, the reflection of these waves is small, inasmuch as these are hardly visible [on] the radar."[37]

In 1983, Reimar wrote that he had planned to sandwich a mixture of sawdust, charcoal, and glue between the layers of wood that formed large areas of the exterior surface of the *H IX* jet wing to shield, as he said, the "whole airplane" from radar, because "the charcoal should absorb the electrical waves. Under this shield, then also the tubular steel [airframe] and the engines [would be] 'invisible' [to radar]."[38] Reimar was describing the process of reducing the radar energy reflected from an aircraft, lowering its radar cross-section, or RCS, so that it could remain undetected by an adversary and carry out its mission with greater stealth.

When interviewed in the mid-1980s, Reimar further claimed that he had specifically used wood to build a substantial portion of the airplane because the material "would not reflect" radar energy.[39] He recalled that technicians sandwiched 12 mm (½ in.) of charcoal, sawdust, and glue between two 1.5 mm (1⁄16 in.) thick sheets of plywood before using this composite material to skin the wings. Reimar even claimed to have sized the elevon hinge bolts and other metal hardware to measure less at its widest point than the radar wavelength published in a secret report prepared by the Telefunken electronics firm, which had based its findings on the analysis of a radar set removed from the wreck of an Allied aircraft.[40] Sizing the hardware reduced its RCS until it was undetectable, Reimar claimed. Asked to explain the background to these actions, the designer replied that "we made it of our own inspiration," without direction from the RLM, to mask the wing from detection during attacks on Allied ships equipped with air-search radar.[41]

This story is weak for several reasons. Allied ships were equipped with air-search radar; however, these targets were a low priority compared to the waves of heavy bombers that had pounded Germany day and night from the beginning of summer 1943. Germany needed vast numbers of interceptors capable of penetrating the screens of Allied fighters protecting the bombers, not long-range strike aircraft with low RCS. Oddly enough, neither the Horten brothers nor Franz Berger mentioned the RCS techniques applied to the *H IX* to Allied intelligence specialists immediately after the war, a peculiar omission in light of Reimar's interest in resuming his work. Finally, if the *H IX* was as speedy as Reimar estimated, why would the aircraft need to evade radar detection when it could outrun any Allied fighter?

Following the heightened interest in all-wing aircraft after the public unveiling of the USA's Northrop B-2 bomber on November 22, 1988, imaginative writers considered none of these questions as they extrapolated from the similarity of the B-2 to the *H IX* (both all-wing aircraft) to conclude that Reimar had designed the first stealth aircraft by purposefully and deliberately reducing the *H IX* jet wing's RCS. Lengthy tests on special radar ranges proved that Northrop engineers had reduced the B-2 RCS to an extraordinarily low value by employing numerous innovations in aerodynamics and structural materials. However, it was regrettable that so many writers concluded, based solely on Reimar's words, and without supporting documentary or physical evidence, that he had not only accomplished a similar result more than forty years earlier but also purposely designed the wing to evade radar detection. Examples of this misguided historical fiction are *Stealth Bomber: Invisible Warplane* (Motorbooks, 1989) by Bill Sweetman, and David Baker's article, "In Valleys of Shadow: The Black World of Stealth (Part One)."[42]

Final Collaborations

In 1987, Reimar began collaborating with Siegfried Panek and his partner, Bernhard Mattlener, to design an all-wing sport airplane, the Panek Ultra-Light with a nine-meter (29 ft. 6 in.) wing, designated the PUL-9. A single-seat prototype built near Bologna, Italy, first flew in 1991.[43] The design goals were easy construction and inexpensive operation. The PUL-9 was successful enough that the three men went on to design a larger and more powerful two-seat wing, the PUL-10, powered by a 48 kW (64 hp) two-cycle engine. Panek and Mattlener constructed a prototype, and a test pilot took it aloft for the first time in September 1992. He was soon able to push the wing to a cruise speed of 155 km/h (96 mph) and a maximum speed of 190 km/h (118 mph). Best climb rate was 4 m/sec (787 ft./min.), and the wing carried enough fuel to convey pilot and passenger about 800 km (496 miles).

In 1993, Edward Uden and aerospace engineer Reinhold Stadler joined the group after Reimar passed away on August 14 (Figure 31). Reimar's career fascinated both men, and they spent several years testing and evaluating his theories. Their hard work, which they generously shared with an international community of engineers, aerodynamicists, pilots, and enthusiasts via the Internet, led to greater understanding of how Horten brought stability and control to all-wing aircraft. The enlightenment process began with reports about flight characteristics. One writer flew the PUL-10 and pronounced it free of adverse yaw, a quality he credited to the Horten lift distribution. However, Reinhold Stadler observed "slight" adverse yaw during his own flight but noticed pilot Siegfried Panek applying opposite aileron before rolling into a turn, suggesting that the adverse yaw may have been stronger than reported.[44] The wing's overall performance encouraged Panek's team to plan for fitting a more powerful four-cycle engine developing 59 kW (79 hp), because Reimar had designed the wing to cruise comfortably at about 200 km/h (124 mph) using his own airfoil design.[45]

Figure 31. Reimar Horten in Córdoba the year before he died on August 14, 1993, at age seventy-seven (Reimar Horten via Peter F. Selinger).

The PUL series marked the first rendering of the classic Horten all-wing form in fiberglass-reinforced plastic. The graceful sweeping lines of these smooth composite airframes enchanted both pilots and the aviation press. Articles about the PUL-10 appeared in Germany, Italy, France, England, and the United States, and Panek formed a company, Nurflügel G.m.b.H., in Hirschberg, Germany, to market the wing. The firm produced a slick and colorful brochure touting the aircraft as an "original Horten design [that was] . . . safe, simple, economical, comfortable, practical," and backed by a rich historical legacy.[46]

Panek's company built a second PUL-10 equipped with retractable landing gear, but it crashed in 1995 during a test flight carried out by Panek in France to observe the handling qualities with the CG at the aft limit. The wing stalled and began to spin, but because Panek believed that the Horten bell distribution's favorable yaw characteristics rendered drag rudders unnecessary, he had not installed these devices on either the PUL-9 or PUL-10. Without these vital controls, Panek could not arrest the spinning wing and only just managed to bail out before impact. A third example began flying in Germany late in 1997, and in May 2005, Bernhard Mattlener announced a new wing, the H 3000, designed based on lessons learned from the PUL-10. He wrote enthusiastically that the legislation recently adopted in the USA to create a new Light Sport Aircraft (LSA) category had renewed interest in all-wing designs, but this was only a dream. Although the LSA category led to many new designs by the end of 2010, all were of conventional configuration.[47]

Questions and Conclusions

Although we understand Reimar Horten's work more completely, some key questions linger. How and when he applied the bell distribution remains controversial. Karl Nickel worked closely with Reimar and married his sister, and should therefore have detailed knowledge about the designer's methodology, yet Nickel remained convinced that Reimar knew nothing of the bell distribution until 1955.[48] Reimar's fear of losing control of his ideas helps explain why he never shared the bell distribution with Nickel during the years they worked together. Reimar once recalled a conference late in 1944 at Dessau, Germany, where the design of long-range bombers was discussed with men from Junkers, Heinkel, and Messerschmitt—around thirty to forty people—which should have presented Reimar with the ideal opportunity to demonstrate his expertise, but he kept silent because

> I did not want to give them all my ideas. You see, I had been working for the acceptance of all-wing designs since the mid-1930's and most of the time it was a difficult road to take. I still felt that I should protect some of my ideas so that the all-wing could be better accepted by Göring and others in the Air Ministry [RLM]. I just did not want to give them away in a conference and later receive no credit for them.[49]

Both Nickel and coauthor Michael Wohlfahrt in their 1990 book, *Tailless Aircraft in Theory and Practice*, denounced the Horten bell distribution for prohibitive drag, writing that the drag values for the Horten *H II*, *III*, *IV*, and *VI* exceeded "the theoretical possible optimum value by nearly 40%!"[50]

Nickel and Wohlfahrt failed to point out that their measure of efficiency applied only to wingspan. The Horten bell distribution yields the greatest benefits when plotting its effects on wing structural efficiency, a condition that Prandtl had first described in 1932. It is worth noting that in 1950 and without knowledge of his 1932 paper, an aeronautical engineer employed by the NACA, Robert T. Jones, confirmed Prandtl's theory that minimum spar bending moment and least induced drag were indeed byproducts of a parabolic lift distribution similar to the Horten bell. Put another way, "Prandtl found the minimum drag for a given wing SPAR, perhaps a more important solution than the minimum drag for a given wing SPAN."[51] Yet Prandtl's findings were virtually unknown, as aeronautical engineer Don Stackhouse wrote on December 14, 2003:

> very few people understand (or even have heard of) Prandtl's [parabolic] study [and this] is probably why we don't see it investigated more extensively. "Everybody" learns in aero engineering school that the elliptical distribution is "always best" (even though it isn't), and it's also probably the one that's easiest to design (mainly because that's what we're all used to), so that's what everybody uses. It's a sacred cow.

Resurging interest in Reimar's work during the 1990s led Edward Uden to construct a series of radio-controlled model wings, each equipped with either a Horten bell or elliptical lift distribution and different control surface placement, and to then fly them to observe their behavior. His experiments showed a favorable yawing force that became stronger as he moved the elevons closer toward the wing tips on wings designed to distribute lift in the Horten bell form.[52] Uden tested for only a single lift coefficient value; that is, he fixed the elevons at one position and flew the wing at one speed, but Al Bowers wondered if favorable yaw occurred at other deflections and speeds. He ran a series of calculations that showed favorable yaw from low to medium speeds, but as velocity increased, the effect weakened considerably before reverting to adverse yaw at very low lift coefficients (high speeds).[53]

To test his findings, Bowers worked with student Michael Allen from the Embry-Riddle Aeronautical University. Allen modified a radio-controlled, all-wing model called the Klingberg Wing, spanning 2 m (6 ft. 6 in.), to conform to the aerodynamic architecture necessary to create the bell lift distribution. The model displayed extreme adverse yaw when built according to the original plans, but the Horten treatment cured it, and the results were clearly visible from the ground when Bowers flew the wing "directly overhead and then made a full[-]deflection roll command."[54] The wing yawed toward the direction of bank. "The fact that [the bell] solves BOTH problems (minimum drag for a given structural weight, AND adverse yaw flight mechanics)," remarked Bowers, "is something of a major miracle to me."[55]

Bowers noted that Reimar also used the bell form to dampen both Dutch roll at high speed and pitch-up during a stall, which could cause the wing to tumble into a spin. Engineer Bill Hinote explained the effect on stall stability by pointing out that the bell distribution worked with wing sweep to create a stabilizing down-force that functioned much like the horizontal stabilizer on a conventional aircraft. Bowers marveled that Reimar deduced these aerodynamic mechanisms using primarily paper and pencil calculations, supplemented with flight tests. Even with the aid of modern calculation devices, Bowers needed the better part of five years to piece together the story of Horten's bell distribution.[56]

Reinhold Stadler contributed much to Bowers' task. In 1997 at the Deutsches Museum in Munich, Stadler studied the three-part Lippisch article, "Definition of the Lift Distribution along the Wingspan," published in *Flugsport*, and he identified Prandtl's December 1932 paper, "About Wings of Smallest Induced Drag," as the source of Lippisch's main ideas. Inspired by his findings, Stadler calculated different lift distributions with varying sweep angles for an *H IV a* type of wing design, and his results verified the conclusions in Prandtl's article, as described in "Der NURFLÜGLER, ein parametrischer Vergleich" (The Flying Wing, a Parametrical Comparison), which Stadler presented at the winter meeting of the Oskar-Ursinus-Vereinigung Wintertagung (Oskar Ursinus Conference) early in 1998.

In 2001, he prepared another paper titled "HORTEN Auslegung, Anmerkungen zur Vorauslegung" (HORTEN Interpretation, Preliminary Notes), wherein he analyzed how Reimar had applied his bell-shaped lift distribution to wing design. Stadler's analysis showed that Reimar did not always use the bell form, but rather employed whatever distribution best met the design goals of a particular wing. For example, he twisted the *H IX* jet wing to a simple linear geometry rather than the more complex nonlinear twist required by the bell form.[57]

In addition to determining how the bell lift distribution had evolved, it was important to pinpoint whether Reimar had actually demonstrated favorable yaw in flight and when this important event occurred. Reimar complicated this task by not talking about distributing lift in bell form, and the favorable yaw properties that resulted, until 1980.[58] Jan Scott wrote that "negative wing tip loading causes the aileron-yaw to turn the aircraft in the proper direction" after visiting Reimar in Argentina during the spring, and Horten wrote an article about the bell distribution theory that appeared in *Soaring* in 1981, unfortunately without mathematical computations or flight tests to bolster the concept.[59]

As noted in Chapter 2, Reimar wrote in 1983 that he experienced favorable yaw while flying the first *H II*, but the model experiments conducted by Uden, Allen, and Bowers showed that favorable yaw required a particular placement of the elevons that the *H II* did not have. Not until the foot-launched *H X c Piernifero 3* appeared in the early 1950s did Horten sketch a wing with the configuration required to make favorable yaw, but no one built this aircraft (Figure 32). It must have been a great source of frustration for Reimar to imagine this breakthrough in yaw control after it was too late to demonstrate it in flight using a piloted wing.

Until the Northrop design team flew the B-2 bomber in 1989, no one else had improved the stability and control of all-wing aircraft to a greater degree than Reimar Horten did. I believe that he ranks among the most accomplished aircraft designers for this reason. Lingering questions about when and how he employed the bell lift distribution to produce favorable yaw do not diminish his mastery of the configuration, which he demonstrated by building and flying nearly sixty wings, frequently with little outside support, and

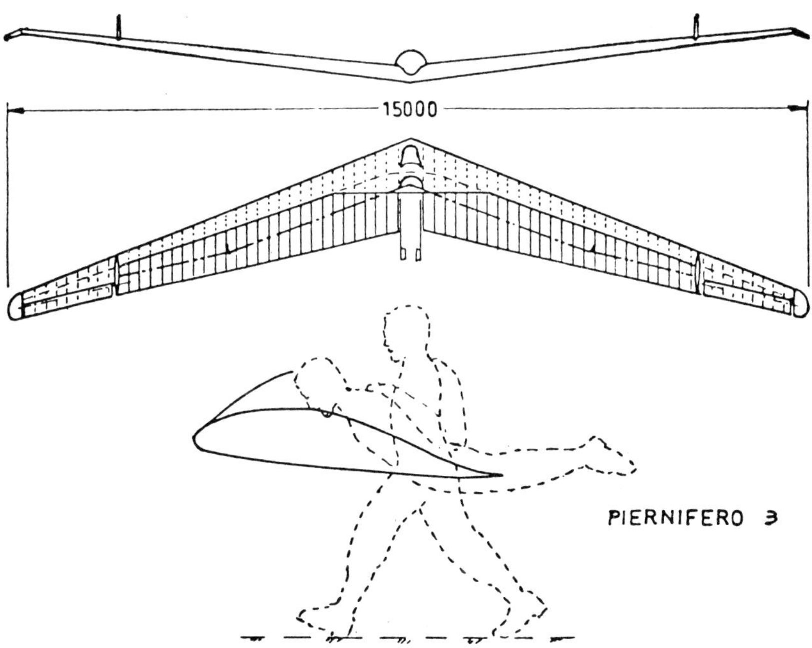

Figure 32. Reimar's *H X c Piernifero 3* sailplane, seen in this Jan Scott sketch, had the wing and elevon configuration necessary to make favorable yaw. No one built this aircraft, and the remarkable yaw properties of the Horten bell lift distribution remained unverified until Albion Bowers and Michael Allen demonstrated favorable yaw when they flew a modified radio-controlled model sailplane in 2002 (Jan Scott).

in austere, even primitive, working conditions. A complete appraisal of Reimar's work must also weigh his numerous other innovations. He was the first designer to build and then fly a successful all-wing aircraft (*H I*, 1933), the first to fly a supine cockpit (*H II*, 1935), the first to experiment with phenolic resin composites in aircraft construction (*H V a*, 1937), and the first designer to develop the semiprone pilot accommodations (*H IV a*, 1941). Horten also performed the first tests of a laminar airfoil for sailplanes (*H IV b*, 1944), and he designed the first all-wing jet (*H IX V2*, 1945).

In counterpoint to his successes, he neglected to address important aspects of all-wing design. He never devised an easy method for moving his wings on the ground, and at best could only offer unorthodox solutions to the pilot's restricted view out of the cockpit and to the often unusual in-flight handling characteristics.[60] His wings reached their highest levels of performance only when flown by experienced pilots who were capable of adapting to these unusual characteristics. Reimar and Walter agreed that a pilot should spend about fifty flying hours becoming acquainted "with the peculiarities of the [all-wing], and [to] master the aircraft."[61] Such familiarization required a commitment in time and money that could easily exceed both the will and the limited means of most casual flyers who frequented the sport flying clubs in large numbers after World War II. It was precisely to this group, however, whom Reimar needed to "sell" his wings if he hoped to gain broader acceptance and recognition as a successful designer.

Pilots already seasoned on conventional aircraft were accustomed to flying in a particular way using the control layout and functions that had become standard before World War I. With varying levels of training and experience, they desired airplanes with which to achieve useful flying goals, repeatedly and under varying conditions. They needed to feel comfortable, to sit upright and grasp and manipulate familiar controls in the cockpit, be able to see well ahead, down, above, and to the sides of an aircraft, and take off and land with relative ease. In short, Reimar had to move beyond the all-wing airplane as a successful experiment and make it more practical.

Because only a small group of pilots actually flew the Horten wings, the sampling rate was too low to mask differences in their flying styles and flight experience, which could adversely influence their assessments of handling qualities and performance. To make a simple level turn, for example, one pilot might think nothing of moving the controls abruptly while another might prefer smooth manipulations. As engineer Chris Bryant pointed out, British test pilot Eric Brown viewed the Horten *H IV a* through the lens of his vast experience flying hundreds of different *powered* airplanes when he cited the wing for objectionable characteristics that were actually common to many sailplanes and considered quite reasonable by the sport sailplane community.[62] Such inconsistencies hardly inspired confidence in pilots who frequently placed their greatest trust in each other to access accurately how an airplane behaved in the air. As historian Walter G. Vincenti said, "pilot opinion could make or break a machine."[63]

Inconsistent evaluations suggest another problem. The wings were fragile and quite sensitive to misalignment, particularly those with high aspect ratios. Ironically, these wings promised the best performance and elicited the bitterest disappointment when they did not meet performance goals. Special attention was required to tune them for optimum performance, but once the wings had moved beyond Reimar's careful scrutiny and care, handling and performance were sure to decline.

Then there is Reimar's own coveted independence, which he admitted was a key factor in his success. "You see," he once remarked, "it is the individual who creates the great ideas, not the large factories. Individuals working alone in university laboratories, in small rooms in their homes, etc."[64] He did not admit that his self-imposed isolation insulated him from the honest feedback necessary to correct several crippling limitations to the all-wing configuration. Rudi Opitz, who knew well the pilots who flew Horten aircraft before and during the war, among them Scheidhauer and Ziller, observed that Reimar acted "like people in politics, they surround themselves with [yes-men]."[65] The designer received only positive reports until the time came for public trials, when flaws appeared that he should have addressed much earlier; for example, when the soaring community in Argentina began to lose interest in all-wing aircraft after the 1952 contest in Spain.

Reimar also failed to tap the rich and promising byproduct of the wing in its purest form—a profound visual aesthetic that has resonated with generations of enthusiasts.[66] By insisting that no fin, strake, or other nonlifting appendage mar his designs, Reimar created a shape nearly devoid of spatial tension that projected the simple, streamlined essence of bird flight. Piloting the Horten *H IV* and *H VI*, for example, equipped with semiprone cockpits, approached "the ideal of the 'flying human being.'" The wings appeared to extend "out from [the pilot's] shoulders with a minimum of aircraft technology."[67] Intent on selling derivatives of the original Horten designs, one salesperson openly acknowledged the exhilarating grip on pilot ego exerted by this startling imagery:

> Let's be truthful. One reason for appearing at an airport in a flying wing is to create a sensation. Therefore, let us dare to admit that it would be fun to have your aircraft mistaken for something from a far more advanced planet. The enjoyment of fame and the attention you will attract is not something of which you should be ashamed.[68]

Fundamental weaknesses, such as poor visibility from the cockpit and difficult ground handling, continue to limit pure wings to performing only a few highly specialized roles that include military missions requiring a low RCS. From time to time, there will appear amateur pilots who possess the determination to endure these flaws and the skill and resources to design, build, and fly all-wing aircraft for sport and pleasure. Notwithstanding the significant improvements that Reimar Horten made in stability and control, all-wing aircraft remain on the fringes of aeronautical practice.

APPENDIX 1

Horten Aircraft Type Data

THUS FAR IN *Only the Wing*, I have described selected types of Horten aircraft and touched on the operational history of several specific airframes. Yet from 1933 until the late 1980s, Reimar led the design of 27 different aircraft types and collaborated on designing the 2 types of PUL sport wings, and craftspersons began or finished at least 65 examples of these aircraft types. Each main type, subtype, and specific airframe listed in Appendix 1, 2, and 3 is followed by certain physical characteristics, performance data, and operational events with the aim to provide an overall perspective on Reimar's body of work.

Note that several different designations refer to the same airframe before and after modification: the *H II* and *H II m, H III b* and *H III d, H V b* and *H V c*, and the *H IV a* and *H XIV*. In Germany in the early 1950s, Walter Horten attempted to revive interest in his brother's designs by using the *H III* as the basis to develop a sport wing equipped with a motor that he called the *Ho 33*. Walter devised a cockpit seating two in tandem beneath a long canopy and auxiliary landing gear that retracted into the wing near mid-span. He supervised the construction of two versions fitted with different engines; both appear in the appendices. They flew and remained active for several years but difficulty moving them about on the ground greatly limited their utility and no additional examples followed.

Table A1.1 shows a steady progression in wingspan, angle of leading edge sweep, wing area, and wing loading as Reimar made his wings larger and more complex until World War II ended, with one glaring exception. Pressured by Dynamit A. G., Reimar and Walter attempted to fly the *H V a* wing back in 1937, yet the design preceding it numerically, the *H IV a*, did not fly until 1941 (see Appendix 3, Table A3.2). Reimar began his career anew in Argentina with an advanced sport sailplane similar to the wartime *H IV a*, only to revisit the smaller and simpler designs reminiscent of the *H I* a few years later.

Table A1.1. Wing Comparison

Aircraft Name	Wingspan, Sweep Angle	Wing Area	Wing Loading
Horten H I	12.4 m (40 ft. 8 in.), 23°	21 m² (226 ft.²)	10 kg/m² (2 lb./ft.²)
Horten H II	16.5 m (54 ft. 2 in.), 29.5°	32 m² (344 ft.²)	9.4 kg/m² (1.9 lb./ft.²)
Horten H II m	16.5 m (54 ft. 2 in.), 29.5°	32 m² (344 ft.²)	14 kg/m² (2.9 lb./ft.²)
Horten H V a	14 m (45 ft. 11 in.), 25.5°/40°	34 m² (366 ft.²)	54 kg/m² (11 lb./ft.²)
Horten H II L	16.5 m (54 ft. 2 in.), 29.5°	32 m² (344 ft.²)	10.3 kg/m² (2.1 lb./ft.²)
Horten H V b	16 m (52 ft. 6 in.), 25.5°/36°	38 m² (409 ft.²)	38 kg/m² (7.8 lb./ft.²)
Horten H III a	20.4 m (67 ft. 1 in.), 24.3°	36 m² (388 ft.²)	8.3 kg/m² (1.7 lb./ft.²)
Horten H III c	20.4 m (67 ft. 1 in.), 24.3°	36 m² (388 ft.²)	8.3 kg/m² (1.7 lb./ft.²)
Horten H III b	20.5 m (67 ft. 3 in.), 24.3°	37.5 m² (404 ft.²)	8.8 kg/m² (1.8 lb./ft.²)

Aircraft Name	Wingspan, Sweep Angle	Wing Area	Wing Loading
Horten *Parabel*	12 m (39 ft. 4 in.), parabolic curve	33 m^2 (355 $ft.^2$)	5.15 kg/m^2 (1 lb./$ft.^2$)
Horten H III d	20.5 m (67 ft. 3 in.), 24.3°	37.5 m^2 (404 $ft.^2$)	11.2 kg/m^2 (2.3 lb./$ft.^2$)
Horten H III e	20.5 m (67 ft. 3 in.), 24.3°	37.5 m^2 (404 $ft.^2$)	12 kg/m^2 (2 lb./$ft.^2$)
Horten H III f	20.5 m (67 ft. 3 in.), 24.3°	37.5 m^2 (404 $ft.^2$)	9.6 kg/m^2 (1.9 lb./$ft.^2$)
Horten H III g	20.5 m (67 ft. 3 in.), 24.3°	37.5 m^2 (404 $ft.^2$)	12.2 kg/m^2 (2.5 lb./$ft.^2$)
Horten H IV a	20.3 m (66 ft. 7 in.), 20°	18.9 m^2 (203 $ft.^2$)	17.5 kg/m^2 (3.6 lb./$ft.^2$)
Horten H IV b	20.3 m (66 ft. 7 in.), 16.8°	18.9 m^2 (203 $ft.^2$)	18.5 kg/m^2 (3.8 lb./$ft.^2$)
Horten H VI	24.2 m (79 ft. 5 in.), 16.7°	17.8 m^2 (192 $ft.^2$)	23 kg/m^2 (4.7 lb./$ft.^2$)
Horten H V c	16 m (52 ft. 6 in.), 25.5°/36°	38 m^2 (409 $ft.^2$)	38 kg/m^2 (7.8 lb./$ft.^2$)
Horten H VII	16 m (52 ft. 6 in.), 40°	44 m^2 (474 $ft.^2$)	45.5 kg/m^2 (9.3 lb./$ft.^2$)
Horten H VIII	40 m (131 ft. 3 in.), 33.7°	146 m^2 (1,572 $ft.^2$)	55 kg/m^2 (11.3 lb./$ft.^2$)
Horten H IX V1	16 m (52 ft. 6 in.), 32.2°	46 m^2 (495 $ft.^2$)	43.5 kg/m^2 (8.9 lb./$ft.^2$)
Horten H IX V2	16.8 m (55 ft. 1 in.), 32.2°	52.8 m^2 (568 $ft.^2$)	130 kg/m^2 (26.6 lb./$ft.^2$)
Horten H XIII a	12 m (39 ft. 4 in.), 60°	36 m^2 (387 $ft.^2$)	9.2 kg/m^2 (1.9 lb./$ft.^2$)

Aircraft Name	Wingspan, Sweep Angle	Wing Area	Wing Loading
Horten H X (Ho XIII b)	7.2 m (23 ft. 7 in.), 70°	37.8 m² (407 ft.²)	18.5 kg/m² (3.8 lb./ft.²)
Horten H XI	12 m (39 ft. 4 in.), 31°	n/a	25 kg/m² (5.1 lb./ft.²)
Horten H XII	16 m (52 ft. 6 in.), 29°	38.5 m² (414 ft.²)	32.3 kg/m² (6.6 lb./ft.²)
Horten H XIV	16 m (52 ft. 6 in.), 18°	15.76 m² (170 ft.²)	14.6 kg/m² (3 lb./ft.²)
Horten H XVIII	40 m (131 ft. 3 in.), 24.3°	150 m² (1,615 ft.²)	213 kg/m² (43.6 lb./ft.²)
I. Ae. 34 a Clen Antú (Horten H XV a)	18 m (59 ft.), 21.8°	18.9 m² (203 ft.²)	22.7 kg/m² (4.6 lb./ft.²)
I. Ae. 34 m Clen Antú (Horten H XV b)	18 m (59 ft.), 21.8°	18.9 m² (203 ft.²)	18.5 kg/m² (3.8 lb./ft.²)
I. Ae. 41 Urubú (Horten H XV c)	18 m (59 ft.), 26.6°	27 m² (291 ft.²)	15.2 kg/m² (3.1 lb./ft.²)
INAV-1A	13.6 m (45 ft.)		
Horten Ho 33 V1	19.2 m (63 ft.), 25°	35.6 m² (383 ft.²)	19.7 kg/m² (4 lb./ft.²)
Horten Ho 33 V2	19.2 m (63 ft.), 25°	35.6 m² (383 ft.²)	19.7 kg/m² (4 lb./ft.²)
Horten H X a *L'alita*	7.5 m (24 ft. 7 in.), 33.7°	11.2 m² (121 ft.²)	10 kg/m² (2 lb./ft.²)
Horten H X b *Piernifero 2*	10 m (32 ft. 10 in.), 63.4°	17.5 m² (188 ft.²)	6.4 kg/m² (1.3 lb./ft.²)
Horten H I b	12.4 m (40 ft. 8 in.), 23°	21 m² (226 ft.²)	10 kg/m² (2 lb./ft.²)

Aircraft Name	Wingspan, Sweep Angle	Wing Area	Wing Loading
Horten H XVI *Colibri*	12 m (39 ft. 4 in.), 26.6°	9 m^2 (97 $ft.^2$)	17.8 kg/m^2 (3.6 lb./$ft.^2$)
I. Ae. 37	10 m (32 ft. 10 in.), 63.4°	48 m^2 (517 $ft.^2$)	25 kg/m^2 (5.1 lb./$ft.^2$)
I. Ae. 38	32 m (105 ft.), 36.5°	133 m^2 (1,432 $ft.^2$)	131 kg/m^2 (26.8 lb./$ft.^2$)
PUL-9	9 m (29 ft. 6 in.), n/a	11 m^2 (118 $ft.^2$)	22.3 kg/m^2 (4.5 lb./$ft.^2$)
PUL-10 I	10 m (32.8 ft.), 34°	15.4 m^2 (165.7 $ft.^2$)	37.6 kg/m^2 (7.7 lb./$ft.^2$)
PUL-10 II	(as above)	(as above)	37.6 kg/m^2 (7.7 lb./$ft.^2$)
PUL-10 III	(as above)	(as above)	35.7 kg/m^2 (7.3 lb./$ft.^2$)

Table A1.2. Weight Comparison

Aircraft Name	Aspect Ratio	Empty Weight	Maximum Flying Weight
Horten H I	7.3	120 kg (265 lb.)	210 kg (464 lb.)
Horten H II	8.5	220 kg (486 lb.)	300 kg (663 lb.)
Horten H II m	8.5	330 kg (729 lb.)	450 kg (995 lb.)
Horten H V a	6.75	1,600 kg (3,536 lb.)	1,840 kg (4,066 lb.)
Horten H II L	8.5	265 kg (586 lb.)	365 kg (807 lb.)
Horten H V b	6.75	1,360 kg (3,006 lb.)	1,600 kg (3,536 lb.)
Horten H III a	11.6	220 kg (486 lb.)	300 kg (663 lb.)
Horten H III c	11.6	220 kg (486 lb.)	300 kg (663 lb.)
Horten H III b	11.1	250 kg (553 lb.)	330 kg (729 lb.)

Aircraft Name	Aspect Ratio	Empty Weight	Maximum Flying Weight
Horten *Parabel*	4.4	90 kg (199 lb.)	170 kg (376 lb.)
Horten H III d	11.1	300 kg (663 lb.)	420 kg (928 lb.)
Horten H III e	11.1	340 kg (751 lb.)	450 kg (995 lb.)
Horten H III f	11.1	280 kg (619 lb.)	360 kg (796 lb.)
Horten H III g	11.1	300 kg (663 lb.)	460 kg (1,017 lb.)
Horten H IV a	21.8	250 kg (553 lb.)	330 kg (729 lb.)
Horten H IV b	21.8	270 kg (597 lb.)	350 kg (774 lb.)
Horten H VI	32.4	330 kg (729 lb.)	410 kg (906 lb.)
Horten H V c	6.75	1,440 kg (3,182 lb.)	1,600 kg (3,536 lb.)
Horten H VII	5.8	1,550 kg (3,426 lb.)	2,000 kg (4,420 lb.)
Horten H VIII	11	5,000 kg (11,050 lb.)	8,000 kg (17,680 lb.)
Horten H IX V1	5.6	1,900 kg (4,199 lb.)	2,000 kg (4,420 lb.)
Horten H IX V2	5.35	4,844 kg (10,705 lb.)	6,876 kg (15,196 lb.)
Horten H XIII a	4	250 kg (553 lb.)	330 kg (729 lb.)
Horten H X (Ho XIII b)	1.4	600 kg (1,326 lb.)	700 kg (1,574 lb.)
Horten H XI	n/a	200 kg (442 lb.)	300 kg (663 lb.)
Horten H XII	6.7	460 kg (1,017 lb.)	700 kg (1,574 lb.)
Horten H XIV	16.2	150 kg (332 lb.)	230 kg (508 lb.)

Aircraft Name	Aspect Ratio	Empty Weight	Maximum Flying Weight
Horten H XVIII	10.7	11,000 kg (24,310 lb.)	32,000 kg (70,720 lb.)
I. Ae. 34 a Clen Antú (Horten H XV a)	17.1	270 kg (597 lb.)	430 kg (950 lb.)
I. Ae. 34 m Clen Antú (Horten H XV b)	17.1	270 kg (597 lb.)	350 kg (774 lb.)
I. Ae. 41 Urubú (Horten H XV c)	12	250 kg (553 lb.)	410 kg (906 lb.)
INAV-1A			
Horten Ho 33 V1	10.4	460 kg (1,017 lb.)	700 kg (1,574 lb.)
Horten Ho 33 V2	10.4	460 kg (1,017 lb.)	700 kg (1,574 lb.)
Horten H X a *L'alita*	5	38 kg (84 lb.)	112 kg (248 lb.)
Horten H X b *Piernifero 2*	5.7	38 kg (84 lb.)	112 kg (248 lb.)
Horten H I b	7.3	120 kg (265 lb.)	210 kg (464 lb.)
Horten H XVI *Colibri*	16	80 kg (177 lb.)	160 kg (354 lb.)
I. Ae. 37	2.1	1,100 kg (2,431 lb.)	1,200 kg (2,652 lb.)
I. Ae. 38	7.8	7,800 kg (17,238 lb.)	17,500 kg (38,675 lb.)
PUL-9		172 kg (379 lb.)	250 kg (551 lb.)
PUL-10 I		364 kg (802 lb.)	580 kg (1,279 lb.)
PUL-10 II		342 kg (753 lb.)	(as above)
PUL-10 III		317 kg (698 lb.)	550 kg (1,212 lb.)

Table A1.3. Production Notes

Aircraft Name	Seats	Propulsion	Year Completed	Qty.
Horten H I	1		1933	1
Horten H II	1		1935	1
Horten H II m	1	Hirth HM 60R, 59 kW (79 hp)	1935	1
Horten H V a	2	(2) Hirth HM 60R, 59 kW (79 hp)	1937	1
Horten H II L	1		1937	3
Horten H V b	2	(2) Hirth HM 60R, 59 kW (79 hp)	1937	1
Horten H III a	1		1938	1
Horten H III c	1		1938	1
Horten H III b	1		1939	9
Horten *Parabel*	1		1939	1
Horten H III d	1	Walter Mikron, 33 kW (44 hp)	1943	1
Horten H III e	1	VW, 22 kW (30 hp)	1944	1
Horten H III f	1		1943	3
Horten H III g	2		1944	2
Horten H IV a	1		1941	4
Horten H IV b	1		1944	1
Horten H VI	1		1944	2
Horten H V c	1	(2) Hirth HM 60R, 59 kW (79 hp)	1942	1
Horten H VII	2	(2) Argus As 10, 176 kW (236 hp)	1944?	1

Aircraft Name	Seats	Propulsion	Year Completed	Qty.
Horten H VIII	3	(6) Argus As 10, 176 kW (236 hp)	not completed	1
Horten H IX V1	1		1944	1
Horten H IX V2	1	(2) Junkers 004B, 900 daN (1,989 lb. thrust)	1944	1
Horten H XIII a	1		1944	1
Horten H X (Ho XIII b)	1	n/a	not completed	1
Horten H XI	1		concept only	
Horten H XII	2	DKW 6 cylinder, 66 kW (90 hp)	1944	1
Horten H XIV	1		1945	1
Horten H XVIII	3	(6) Jumo 004B, 900 daN (1,989 lb. thrust)	not completed	1
I. Ae. 34 a Clen Antú (Horten H XV a)	2		1949	3
I. Ae. 34 m Clen Antú (Horten H XV b)	1		1951	2
I. Ae. 41 Urubú (Horten H XV c)	2		1953	
INAV-1A	1		1954	1
Horten Ho 33 V1	2	Zundapp Z9-092, 37 kW (50 hp)	1954	1
Horten Ho 33 V2	2	Porschemotor, 59 kW (79 hp)	1955	1
Horten H X a *L'alita*	1		1954	1

Aircraft Name	Seats	Propulsion	Year Completed	Qty.
Horten H X b *Piernifero 2*	1		not completed	
Horten H I b	1		1954	1
Horten H XVI *Colibri*	1		1953	
I. Ae. 37	1		1954	1
I. Ae. 38	2	(4) El Gaucho, 253 kW (316 hp)	1960	1
PUL-9	1	Rotax 582, 48 kW (64 hp)	1991	1
PUL-10 I	2	Rotax 582, 48 kW (64 hp)	1992	1
PUL-10 II	2	Rotax 912 UL, 60 kW (80 hp)	1994	1
PUL-10 III	2	(as above)	1997	1

Table A1.4. Notes on Composition

Aircraft Name	Composition (center section/ wing)	Remarks
Horten H I	wood	
Horten H II	steel tube/wood	
Horten H II m	steel tube/wood	First engine HM 60 (48 kW, 65 hp) overheated aft.er about fift.een hours of operation.
Horten H V a	steel tube/wood	
Horten H II L	steel tube/wood	
Horten H V b	steel tube/wood	

Aircraft Name	Composition (center section/ wing)	Remarks
Horten H III a	steel tube/wood	
Horten H III c		Fitted with Vorflugel, forewing.
Horten H III b		
Horten *Parabel*	steel tube/wood	Never flown, deteriorated, burned by Reimar Horten.
Horten H III d	steel tube/wood	A more powerful 48 kW (66 hp) Mikron engine installed near end of war.
Horten H III e	steel tube/wood	
Horten H III f	steel tube/wood	Flat-prone cockpit.
Horten H III g	steel tube/wood	Trainer.
Horten H IV a	steel tube/wood	Horten used same nose wheel assemblies for H III and H IV a gliders (Nickel and Wohlfahrt, *Tailless Aircraft.*, 214).
Horten H IV b	steel tube/wood, composite	
Horten H VI	steel tube/wood	
Horten H V c	steel tube/wood	Rebuilt Ho V b, flown by Stüper of Göttingen AVA, Knemeyer, etc.
Horten H VII	aluminum/wood	
Horten H VIII	steel tube/wood	Not completed, flying wind tunnel.

Aircraft Name	Composition (center section/ wing)	Remarks
Horten H IX V1	steel tube/wood	
Horten H IX V2	steel tube/wood	Crashed, destroyed, Ziller died.
Horten H XIII a	steel tube/wood	Testbed for supersonic design; renamed to foil RLM discovery.
Horten H X (Ho XIII b)	wood	Model tested, renamed to foil RLM discovery, 12 m (39 ft. 4 in.) length on p.153 dwg.
Horten H XI	steel tube/wood	Similar to H IV a but with only two elevons.
Horten H XII	steel tube/wood	H VII trainer, He 177 APU, Mustang airfoil, Scheidhauer flew as glider.
Horten H XIV	steel tube/wood	Modified Ho IV Wr. Nr. 24, shorter span, destroyed by U.S. troops.
Horten H XVIII	steel tube/wood	Jet bomber.
I. Ae. 34 a Clen Antú (Horten H XV a)	wood	
I. Ae. 34 m Clen Antú (Horten H XV b)	wood	m - Mono Plaza or single-seat.
I. Ae. 41 Urubú (Horten H XV c)	steel tube/wood	Two-seat trainer, 3-4 built (RHTT, 357-358).

Aircraft Name	Composition (center section/ wing)	Remarks
INAV-1A		
Horten Ho 33 V1	wood	H III modified by Walter Horten.
Horten Ho 33 V2	wood	Forty flying hours.
Horten H X a *L'alita*	wood	
Horten H X b *Piernifero 2*	wood	Scheidhauer built most of airframe but did not cover it, never flown.
Horten H I b	wood	Similar to 1934 H I but with elevons & bell distribution, flew 25 yrs.
Horten H XVI *Colibri*	wood	Sport sailplane for Club Cóndor in Buenos Aires, crashed second test flight, not repaired.
I. Ae. 37	wood	Glider testbed, prone pilot, about 150 flying hours.
I. Ae. 38	aluminum	2x H II with cargo pod, scrapped (Burzaco, *Las Alas de Perón*, 192-193).
PUL-9	steel-tube/fiber-glass-reinforced plastic	
PUL-10 I	fiberglass-rein-forced plastic	
PUL-10 II	fiberglass-rein-forced plastic	Fitted with retractable landing gear.
PUL-10 III	fiberglass-rein-forced plastic	

Unless otherwise stated, all data in Appendix 1, 2, and 3 from Horten and Selinger, *Nurflügel: Die Geschichte der Horten-Flugzeuge 1933–1960* (Graz, Germany: H. Weishaupt Verlag, 1983).

APPENDIX 2

Horten Aircraft Performance Characteristics

THESE FIGURES are based mainly on Reimar's performance calculations for wings built to his specifications and in top condition. The actual flight performance of most Horten aircraft was never measured in flight. Table A2.1 captures the essence of performance while soaring. Min. Sink @ Speed refers to the speed a pilot should fly to descend at the slowest rate possible and still maintain control of the aircraft such as when circling within a thermal. L/D @ Speed refers to the velocity a pilot should fly to minimize the height lost between sources of lift.

Maximum Level Flight Speed, Table A2.3, refers to powered wings. The 1,000 km/h (620 mph) listed as the Neer Exceed Speed for the *H IX V2* jet wing was estimated; the aircraft never came close to this figure in testing. Power Loading is the ratio of a unit of power available per unit of weight. Less weight per unit of power often indicates a capability for higher speeds and faster rates of climb.

Table A2.1. Speed Performance

Aircraft Name	Min. Sink @ Speed	L/D @ Speed
Horten H I	0.80 m/s @ 60 km/h (2.62 ft./s @ 37 mph)	21 @ 72 km/h (45 mph)
Horten H II	0.70 m/s @ 60 km/h (2.30 ft./s @ 37 mph)	24 @ 72 km/h (45 mph)
Horten H II m	0.70 m/s @ 60 km/h (2.30 ft./s @ 37 mph)	24 @ 72 km/h (45 mph)
Horten H V a		
Horten H II L	0.70 m/s @ 60 km/h (2.30 ft./s @ 37 mph)	24 @ 72 km/h (45 mph)
Horten H V b		
Horten H III a	0.48 m/s @ 45 km/h (1.57 ft./s @ 28 mph)	28 @ 60 km/h (37 mph)
Horten H III c	0.48 m/s @ 45 km/h (1.57 ft./s @ 28 mph)	28 @ 60 km/h (37 mph)
Horten H III b	0.50 m/s @ 45 km/h (1.64 ft./s @ 28 mph)	28 @ 60 km/h (37 mph)
Horten *Parabel*	0.80 m/s @ 45 km/h (2.62 ft./s @ 28 mph)	19.5 @ 61.5 km/h (38 mph)
Horten H III d	0.60 m/s @ 53 km/h (1.96 ft./s @ 33 mph)	25 @ 70 km/h (43 mph)
Horten H III e	0.60 m/s @ 53 km/h (1.96 ft./s @ 33 mph)	28 @ 70 km/h (43 mph)
Horten H III f	0.52 m/s @ 47 km/h (1.70 ft./s @ 29 mph)	28 @ 63 km/h (39 mph)

Aircraft Name	Min. Sink @ Speed	L/D @ Speed
Horten H III g	0.60 m/s @ 53 km/h (1.96 ft./s @ 33 mph)	28 @ 70 km/h (43 mph)
Horten H IV a	0.50 m/s @ 60 km/h (1.64 ft./s @ 37 mph)	37 @ 73 km/h (45 mph)
Horten H VI	0.45 m/s @ 70 km/h (1.48 ft./s @ 43 mph)	45 @ 83.5 km/h (52 mph)
Horten H IV b		Ca. 40
Horten H V c		
Horten H VII		
Horten H VIII		
Horten H IX V1		
Horten H IX V2		
Horten H XIII a	1.1 m/s @ 60 km/h (3.6 ft./s @ 37 mph)	16 @ 80 km/h (50 mph)
Horten H X (Ho XIII b)	2.0 m/s @ 70 km/h (6.5 ft./s @ 43 mph)	9 @ 97 km/h (60 mph)
Horten H XI	0.90 m/s @ 80 km/h (3 ft./s @ 50 mph)	22 @ 90 km/h (56 mph)
Horten H XII		
Horten H XIV	0.55 m/s @ 55 km/h (1.8 ft./s @ 34 mph)	30 @ 70 km/h (43 mph)
Horten H XVIII		
I. Ae. 34 a Clen Antú (Horten H XV a)	0.75 m/s @ 70 km/h (2.4 ft./s @ 43 mph)	28 @ 88 km/h (55 mph)

Aircraft Name	Min. Sink @ Speed	L/D @ Speed
I. Ae. 34 m Clen Antú (Horten H XV b)		30 @ 83 km/h (51 mph)
I. Ae. 41 Urubú (Horten H XV c)	0.85 m/s @ 65 km/h (2.8 ft./s @ 40 mph)	24 @ 81 km/h (50 mph)
INAV-1A		
Horten Ho 33 V1	0.80 m/s @ 75 km/h (2.6 ft./s @ 47 mph) @ 18.3 kg/m^2	26 @ 90 km/h (56 mph) @ 18.3 kg/m^2
Horten Ho 33 V2		26 @ 90 km/h (56 mph) @ 19.7 kg/m^2
Horten H X a *L'alita*	0.80 m/s @ 50 km/h (2.6 ft./s @ 31 mph)	18 @ 78 km/h (48 mph)
Horten H X b *Piernifero 2*	0.70 m/s @ 45 km/h (2.3 ft./s @ 28 mph)	20 @ 60 km/h (37 mph)
Horten H I b	0.80 m/s @ 60 km/h (2.6 ft./s @ 37 mph)	21 @ 72 km/h (45 mph)
Horten H XVI *Colibri*	0.65 m/s @ 70 km/h (2.1 ft./s @ 43 mph)	30 @ 77 km/h (48 mph)
I. Ae. 37		10 @ 149 km/h (92 mph)
I. Ae. 38		
PUL-9		n/a
PUL-10 I		n/a
PUL-10 II		n/a
PUL-10 III		n/a

Table A2.2. Speed Comparison

Aircraft Name	Stall Speed	Landing Speed	Cruise Speed
Horten H I	45 km/h (28 mph)	60 km/h (37 mph)	
Horten H II	40 km/h (25 mph)	40 km/h (25 mph)	
Horten H II m	49 km/h (30 mph)	49 km/h (30 mph)	180 km/h (112 mph)
Horten H V a	97 km/h (60 mph)	84 km/h (52 mph)	250 km/h (155 mph)
Horten H II L	42 km/h (26 mph)	42 km/h (26 mph)	
Horten H V b	82 km/h (51 mph)	70 km/h (43 mph)	230 km/h (143 mph)
Horten H III a	37 km/h (23 mph)	37 km/h (23 mph)	
Horten H III c	37 km/h (23 mph)	37 km/h (23 mph)	
Horten H III b	39 km/h (24 mph)	39 km/h (24 mph)	
Horten *Parabel*	30 km/h (19 mph)	30 km/h (19 mph)	
Horten H III d	44 km/h (27 mph)	44 km/h (27 mph)	130 km/h (81 mph)
Horten H III e	46 km/h (29 mph)	46 km/h (29 mph)	120 km/h (74 mph)
Horten H III f	41 km/h (25 mph)	41 km/h (25 mph)	
Horten H III g	46 km/h (29 mph)	46 km/h (29 mph)	

Aircraft Name	Stall Speed	Landing Speed	Cruise Speed
Horten H IV a	55 km/h (34 mph)	55 km/h (34 mph)	
Horten H VI	55 km/h (34 mph)	55 km/h (34 mph)	
Horten H IV b	60 km/h (37 mph)	60 km/h (37 mph)	
Horten H V c	60 km/h (37 mph)	60 km/h (37 mph)	230 km/h (143 mph)
Horten H VII	77 km/h (48 mph)	77 km/h (48 mph)	300 km/h (186 mph)
Horten H VIII	80 km/h (50 mph)	80 km/h (50 mph)	250 km/h (155 mph)
Horten H IX V1	75 km/h (47 mph)	75 km/h (47 mph)	
Horten H IX V2	150 km/h (93 mph)	130 km/h (81 mph)	900 km/h (558 mph)
Horten H XIII a			
Horten H X (Ho XIII b)	56 km/h (35 mph)	88 km/h (55 mph)	n/a
Horten H XI	66 km/h (41 mph)	66 km/h (41 mph)	n/a
Horten H XII	75 km/h (47 mph)	75 km/h (47 mph)	180 km/h (112 mph)
Horten H XIV			
Horten H XVIII	192 km/h (119 mph)	136 km/h (84 mph) @ 16 tons	750 km/h (465 mph)

Aircraft Name	Stall Speed	Landing Speed	Cruise Speed
I. Ae. 34 a Clen Antú (Horten H XV a)			
I. Ae. 34 m Clen Antú (Horten H XV b)			
Ae. 41 Urubú (Horten H XV INAV-1			
Horten Ho 33 V1	55 km/h (34 mph)	60 km/h (37 mph)	ca. 175 km/h (109 mph)
Horten Ho 33 V2	55 km/h (34 mph)	ca. 60 km/h (37 mph)	ca. 200 km/h (124 mph)
Horten H X a *L'alita*			
Horten H X b *Piernifero 2*			
Horten H I b			
Horten H XVI *Colibri*			
I. Ae. 37			
I. Ae. 38	133 km/h (82 mph)	133 km/h (82 mph)	235 km/h (146 mph)
PUL-9		50 km/h (31 mph)	
PUL-10 I		68 km/h (42 mph)	
PUL-10 II		64 km/h (40 mph)	
PUL-10 III		60 km/h (37 mph)	

Table A2.3. Speed Notes

Aircraft Name	Maximum Level Flight Speed	Never Exceed Speed	Power Loading	Remarks
Horten H I		228 km/h (141 mph)		Flown to 170 km/h (105 mph)
Horten H II		230 km/h (143 mph)		
Horten H II m	200 km/h (124 mph)	230 km/h (143 mph)	7.6 kg/kW (12.5 lb./hp)	
Horten H V a	280 km/h (174 mph)	350 km/h (217 mph)	15.6 kg/kW (25.6 lb./hp)	
Horten H II L		230 km/h (143 mph)		Flown to 450 km/h (279 mph)
Horten H V b	260 km/h (161 mph)	350 km/h (217 mph)	13.5 kg/kW (22.2 lb./hp)	
Horten H III a		210 km/h (130 mph)		
Horten H III c		210 km/h (130 mph)		
Horten H III b		210 km/h (130 mph)		
Horten *Parabel*		164 km/h (102 mph)		
Horten H III d	160 km/h (99 mph)	210 km/h (130 mph)	12.7 kg/kW (20.9 lb./hp)	
Horten H III e	140 km/h (87 mph)	210 km/h (130 mph)	20.4 kg/kW (33.5 lb./hp)	
Horten H III f		210 km/h (130 mph)		

Aircraft Name	Maximum Level Flight Speed	Never Exceed Speed	Power Loading	Remarks
Horten H III g		210 km/h (130 mph)		
Horten H IV a		200 km/h (130 mph)		
Horten H VI		200 km/h (124 mph)		
Horten H IV b		200 km/h (124 mph)		
Horten H V c	260 km/h (161 mph)	350 km/h (217 mph)	13.5 kg/kW (22.2 lb./hp)	
Horten H VII	350 km/h (217 mph)	550 km/h (341 mph)	5.7 kg/kW (9.4 lb./hp)	
Horten H VIII	280 km/h (174 mph)	430 km/h (267 mph)	7.6 kg/kW (12.5 lb./hp)	
Horten H IX V1		1,000 km/h (620 mph)		
Horten H IX V2	960 km/h (595 mph)	1,000 km/h (620 mph)		
Horten H XIII a		180 km/h (112 mph)		
Horten H X (Ho XIII b)		500 km/h (310 mph)		
Horten H XI		350 km/h (217 mph)		
Horten H XII	200 km/h (124 mph)	300 km/h (186 mph)	10.6 kg/kW (17.4 lb./hp)	
Horten H XIV		250 km/h (155 mph)		

Aircraft Name	Maximum Level Flight Speed	Never Exceed Speed	Power Loading	Remarks
Horten H XVIII	820 km/h (508 mph)	900 km/h (558 mph)	5.93 kg/daN (13 lb./lb.)	
I. Ae. 34 a Clen Antú (Horten H XV a)		250 km/h (155 mph)		
I. Ae. 34 m Clen Antú (Horten H XV b)		250 km/h (155 mph)		
I. Ae. 41 Urubú (Horten H XV c)		200 km/h (124 mph)		
Horten Ho 33 V1	ca. 185 km/h (115 mph)	240 km/h (149 mph)	18.9 kg/kW (31 lb./hp)	
Horten Ho 33 V2	ca. 220 km/h (136 mph)	240 km/h (149 mph)	11.9 kg/kW (19.5 lb./hp)	
Horten H X a *L'alita*		170 km/h (105 mph)		
Horten H X b *Piernifero 2*		150 km/h (93 mph)		
Horten H I b		228 km/h (141 mph)		
Horten H XVI *Colibri*		200 km/h (124 mph)		
I. Ae. 37		250 km/h (155 mph)		

Aircraft Name	Maximum Level Flight Speed	Never Exceed Speed	Power Loading	Remarks
I. Ae. 38	252 km/h (156 mph)			
PUL-9	185 km/h (115 mph)			
PUL-10 I	175 km/h (108 mph)			Fixed landing gear.
PUL-10 II	220 km/h (136 mph)			Retractable landing gear.
PUL-10 III	230 km/h (143 mph)			Retractable landing gear.

APPENDIX 3

Horten Aircraft Airframe Data

CONFIDENCE IS high that historians have identified the numerous different types of Horten aircraft, but too many records were lost at the end of World War II to be certain of the total number of airframes built. Reimar reused major components, such as wing panels and center sections, to build new aircraft, and someone other than the Horten brothers may have built additional examples in Germany. Despite these uncertainties, we know that workers finished or began work on at least 65 different airframes, including one conventional sailplane, the INAV-1A designed and built in Argentina. Appendix 3 lists the 71known Horten airframe designations (including the four airframes twice given new designations, see above) along with a brief history of each airframe when it is known.

There was no large-scale mass production of one type or version. Rather, Reimar's career output of airplanes resembles one continuous work-in-progress. Even within a series of airframes of the same basic type, nearly every one differed in some significant way from the others. For example, Reimar used different types and placement of control surfaces on all four *H IV a* sailplanes. Of course, something interesting usually happened to each airframe, too, as the Remarks column in Table A3.1 suggests.

Table A3.1. Aircraft History

Aircraft Name	Registration	Remarks
1. Horten H I	D-Hangwind	Burned to ashes on the Wasserktüppe early August 1934.
2. Horten H II	D-Habicht	New W. Nr. and designation H II m after motor installed.
3. Horten H II m	D-Habicht	This H II m, Werk Nr. 7, in Krakow, Poland, ca. 1994 (Chris Wills January 3, 1994, correspondence, hereafter Wills corr.).
4. Horten H V a	---	Destroyed in 1937 crash during first flight.
5. Horten H II L	D-10-125	Flown by Reimar Horten in 1937 Rhön contest, modified to flight-test shape of Ho IX V3 engine intakes and several versions of Ho IX V6 nose section with dummy engines. Transferred from Wright Field, OH, to NASM; restored by, and displayed at, Deutsches Technikmuseum Berlin in 2006.
6. Horten H II L	D-10-131	Flown by Walter Horten in 1937 Rhön contest.
7. Horten H II L	D-11-187, D-13-387, '42'	Scheidhauer flew 240 km, Hanna Reitsch evaluated November 1938, Kurt Hieckmann killed during aerobatics March 1939.
8. Horten H V b	---	Rebuilt to H V c in 1942.
9. Horten H III c	D-11-347, D-12-347	Fitted with Vorflügel, D-12-347 flown in thunderstorm by Werner Blech in 1938 Rhön, wing destroyed and Blech killed.

Aircraft Name	Registration	Remarks
10. Horten H III a	D-11-348, D-12-348/'43' (Myhra, DFS 228, 37)	D-12-348 flown into thunderstorm by Scheidhauer 1938 Rhön, wing destroyed and Scheidhauer injured.
11. Horten *Parabel*	---	Not flown. Succumbed to the elements, burned 1940.
12. Horten H III b	D-4-681, '21'	Sponsored by Mr. Helbig, Chief Government Counsel at Ministry of Education and Culture.
13. Horten H III b	D-4-682	
14. Horten H III b	D-4-683	
15. Horten H III b	D-4-684	
16. Horten H III b	D-4-685	
17. Horten H III b	D-4-686	
18. Horten H III b	D-4-687	
19. Horten H III b	---	
20. Horten H III b	WL-VI-45?	
21. Horten H IV a	D-10-1359, LA-AA	Prototype, square wing tips, span 20.1 m (65 ft 11 in), push-rod and cable controls, wing twist 9°, destroyed during war (Reinhold Stadler, "Horten IV, Vergleich der Leistungsdaten," January 25, 1998, unpublished, 2).

Aircraft Name	Registration	Remarks
22. Horten H III d	DV+LK	Finished and flown at Göttingen in 1943 H III b with engine added (Nickel, December 20, 2005, correspondence).
23. Horten H IV a	D-10-1450, LA-AB	Rounded wing tips, span 20.3 m (66 ft 7 in), push-rod control only, wing twist 6°, tips shortened—became H XIV, destroyed during war (Stadler, "Horten IV, Vergleich der Leistungsdaten," 2).
24. Horten H IV a	D-10-1452, LA-AC, BGA 647, N79289	Rounded wing tips, span 20.3 m (66 ft 7 in), push-rod control only, wing twist 6° (Stadler, "Horten IV, Vergleich der Leistungsdaten," 2). To England via Kronfeld, U.S. via Button, Opitz 1952 U.S. Nationals, MSU, Planes of Fame.
25. Horten H IV a	D-10-1451, LA-AD	Rounded wing tips, span 20.3 m (66 ft 7 in), push-rod control only, wing twist 6° (Stadler, "Horten IV, Vergleich der Leistungsdaten," 2). Scheidhauer fly-off w/ Zacher in D 30, May 23–26, 1944; ninety-three flying hours before wing tip smashed; to RAF BAFO postwar, tested bridle tow-line configuration for winch launching; Wills coor.); wings only to Deutsches Museum Oberschleissheim in Munich for restoration (Stadler, "Horten IV Restoration," T.W.I.T.T. website, April 3, 2002).
26. Horten H V c	PE+HO (***Pe***schke and ***Ho***rten)	Rebuilt from H V b, tested at Göttingen.
27. Horten H III f	LA-AF	

Aircraft Name	Registration	Remarks
28. Horten H VII	---	Destroyed at Minderheide (Wills coor.).
29. Horten H III f	LA-AH	
30. Horten H III h	LA-AI	Allies collected at Nellingen near Stuttgart, along w/ H III f Werk Nr. 32, October 1947 to Northrop Technical Institute from U.S. Air Force, ca. 1960 to NASM, displayed at the Steven F. Udvar-Hazy Center in 2005 ("C. I. O. S. Trip No. 377 & 412, Report on Development . . .," 4, 8; Hall, "The Horten Tailless Sailplanes," 2).
31. Horten H III f	---	Allies collected at Nellingen near Stuttgart, along w/ H III h Werk Nr. 31, October 1947 to Northrop Technical Institute from U.S. Air Force, ca. 1960 to NASM, displayed at the Steven F. Udvar-Hazy Center in 2005 ("C. I. O. S. Trip No. 377 & 412, Report on Development . . .," 4, 8; Hall, "The Horten Tailless Sailplanes," 2).
32. Horten H VI V1	LA-AK	Scheidhauer flew as much as ten hrs./five to ten flights (WROH CD 24b), set afire outside Göttingen in summer 1945 by RAF lieutenant (RHTT, 177).
33. Horten H VI V2	---	Not flown, Allies collected at Hersfeld, October 1947 to Northrop Technical Institute from U.S. Air Force, to NASM ca. 1960, DTM Berlin finished restoring in 2005, displayed at the Steven F. Udvar-Hazy Center (RHTT, 177; Hall, "The Horten Tailless Sailplanes," 2).

Aircraft Name	Registration	Remarks
34. Horten H III e	LA-AE, DV+LL	VW engine, construction began at Peschke. New center section with H III b wings, then another set modified to waggle tips, scrapped 1945 (Wills coor.).
35. Horten H III g	LA-AG	
36. Horten H III g	LA-AJ	
37. Horten H IX V1	---	Glider prototype; moved to Brandeis from Oranienburg early April 1945 as Russians closed on Oranienburg, destroyed (Walter Horten, WROH CD 3a (April 21, 1988)).
38. Horten H IX V2	---	Crashed February 18, 1945, during third flight, killing pilot Lt. Erwin Ziller.
39. Horten H IX V3 (Ho 229 V3)		Allies recovered incomplete at Friedrichsroda April 1945, to Farnborough, England, then to Wright Field, ca. 1960 to NASM.
40. Horten H IV b		Laminar airfoil copied from crashed P-51D Mustang, crashed February 18, 1945, Hermann Strebel killed.
41. Horten H VIII		Half-finished when war ended.
42. Horten H XII		
43. Horten H XIII a		New center section plus H III b wings, scrapped Göttingen 1945 (Wills coor.)
44. Horten H X (H XIII b)		

Aircraft Name	Registration	Remarks
45. Horten H XIV		H IV modified with shorter wing tips after September 6, 1944, crash at Klippeneck.
46. I. Ae. 34 a Clen Antú (H XV a)		
47. I. Ae. 34 a Clen Antú (H XV a)		
48. I. Ae. 34 a Clen Antú (H XV a)		
49. I. Ae. 34 m (H XV b)	LV-XXA	Built to compete in 1952 world soaring meet in Spain.
50. I. Ae. 34 m (H XV b)	LV-XXB	
51. I. Ae. 41 Urubú (H XV c)		Scheidhauer first to fly sailplane over Andes land in Chile, October 30, 1956, displayed at Museo Nacional de Aeronautica.
52. I. Ae. 41 Urubú (H XV c)		
53. I. Ae. 41 Urubú (H XV c)		
54. I. Ae. 41 Urubú (H XV c)		
55. Horten H XV c	D-9003, HB-SAA	Walter Kirschsieper added motor, scrapped 1980s (*T.W.I.T.T. Newsletter*, October 2002).

Aircraft Name	Registration	Remarks
56. Horten H XV c	D-1109	
57. Horten H XV c		
58. Horten H XV c		
59. INAV-1A		Claudio Dori tested April 1954 with good results (Burzaco, *Las Alas de Perón*, 226–227).
60. Horten Ho 33 V1	D-5332, D-EJUS	Scrapped ca. 1960 (Wills coor.).
61. Horten Ho 33 V2	D-EGOL	Displayed at Wasserkuppe ca. 1994 (Wills coor.).
62. Horten H X a *L'alita*		Foot-launch hang glider.
63. Horten H X b *Piernifero 2*		Airframe constructed by Scheidhauer, not flown.
64. Horten H I b	LV-X-17	
65. Horten H XVI *Colibri*		Crashed, destroyed 1953 (Wills corr.).
66. I. Ae. 37		Technicians installing more powerful engine when project canceled.
67. I. Ae. 38		
68. PUL-9		Flight tests at NIKE Aeronautica Airfield, Italy (www.nurflugel.com/…H_3000).

Aircraft Name	Registration	Remarks
69. PUL-10 I		Flight tests Italy 1993.
70. PUL-10 II		Crashed during 1995 test at Sarrebourg Airfield, France, with aft center of gravity.
71. PUL-10 III		Flight tested 1997 Breitscheid Airfield, Germany.

Table A3.2. Production History

Aircraft Name	Werk Nr.	Builder/ Location	First Flight or Date Completed
Horten H I	1	Bonn	Jul-33
Horten H II	2	Bonn	May/Jun-35
Horten H II m	3 (2)	Bonn-Essen	Sep-35
Horten H V a	5	Troisdorf	May-37
Horten H II L	6	Lippstadt	1937
Horten H II L	7	Lippstadt	1937
Horten H II L	8	Lippstadt	1937
Horten H V b	9	Cologne	1937
Horten H III c	10	Cologne	7-May-38
Horten H III a	11	Cologne	1938
Horten *Parabel*	12	Cologne	1939
Horten H III b	13	Berlin	11-Jun-39
Horten H III b	14	Berlin	1939
Horten H III b	15	Berlin	1939
Horten H III b	16	Fürth	1939

Aircraft Name	Werk Nr.	Builder/ Location	First Flight or Date Completed
Horten H III b	17	Fürth	1939
Horten H III b	18	Fürth	1939
Horten H III b	19	Giebelstadt	1939
Horten H III b	20	Minden	1941
Horten H III b	21	Minden	1941
Horten H IV a	22	Königsberg	May-41
Horten H III d	23 (see remarks)	Göttingen	1943
Horten H IV a	24	Göttingen	11-Feb-43
Horten H IV a	25	Göttingen	28-Apr-43
Horten H IV a	26	Göttingen	20-Jun-43
Horten H V c	27 (9)	Peschke/ Minden	26-May-42
Horten H III f	28	Göttingen	10-Aug-43
Horten H VII	29	Peschke/ Minden	1944?
Horten H III f	30	Göttingen	25-Nov-44?
Horten H III h	31	Göttingen	1944
Horten H III f	32	Göttingen	1944
Horten H VI V1	33	Bonn	24-May-44
Horten H VI V2	34	Bonn	1944
Horten H III e	35	Göttingen	25-Jan-44
Horten H III g	36	Göttingen	1944
Horten H III g	37	Göttingen	10-Jul-44
Horten H IX V1	38	Göttingen	1 or 5-Mar-44

Aircraft Name	Werk Nr.	Builder/ Location	First Flight or Date Completed
Horten H IX V2	39	Göttingen	2-Feb-45
Horten H IX V3 (Ho 229 V3)		Friedrichsch-roda	not completed
Horten H IV b	40	Hersfeld	1944
Horten H VIII	41	Göttingen	not completed
Horten H XII	42	Kirtorf	1944
Horten H XIII a	43	Hersfeld	27-Nov-44
Horten H X (H XIII b)			not completed
Horten H XIV	44 (24)	Göttingen	1945
I. Ae. 34 a Clen Antú (H XV a)	1	Córdoba	20-Jun-49
I. Ae. 34 a Clen Antú (H XV a)	2	Córdoba	1949
I. Ae. 34 a Clen Antú (H XV a)	3	Córdoba	1949
I. Ae. 34 m (H XV b)	1	Córdoba	19-Oct-51
I. Ae. 34 m (H XV b)	2	Córdoba	9-Feb-52
I. Ae. 41 Urubú (H XV c)	P	Córdoba	4-Dec-53
I. Ae. 41 Urubú (H XV c)	2	Córdoba	7-Mar-55
I. Ae. 41 Urubú (H XV c)	3	Córdoba	17-Dec-55
I. Ae. 41 Urubú (H XV c)	4	Córdoba	
Horten H XV c	---	Tübingen	20-May-52

Aircraft Name	Werk Nr.	Builder/ Location	First Flight or Date Completed
Horten H XV c	---	Pappenheim	2-May-53
Horten H XV c	---	Wangen	
Horten H XV c	---	Mannheim	
INAV-1A		Córdoba	1954
Horten Ho 33 V1	---	Bonn	1954
Horten Ho 33 V2	---	Bonn	28-Sep-55
Horten H X a *L'alita*	---	Córdoba	May-54 (1st air tow)
Horten H X b *Piernifero 2*	---	Córdoba	not completed
Horten H I b	---	Gonzales-Chavez	2-May-54
Horten H XVI *Colibri*	10	Buenos Aires	20-Jan-53
I. Ae. 37	P	Córdoba	1-Oct-54
I. Ae. 38	---	Córdoba	1960
PUL-9		Bologna, Italy	1991
PUL-10	I	Bologna, Italy	Sep-92
PUL-10	II	Bologna, Italy	1994
PUL-10	III	Bologna, Italy	1997

Notes

Introduction

1. After consulting curators and other experts, I decided to change the designation from *H IX V3* to *Ho 229 V3* . . . In the first edition of *Only the Wing*, I called the flown powered versions of the jet wing Horten IX the *H IX V2* ("nine-vee-two") and *H IX V3*, but I now believe the designations *Ho 229 V2* and *Ho 229 V3* to be more accurate. Except for the revised Introduction, the old designation using a Roman numeral remains in this edition of *Only the Wing*.

2. Bowers to Lee email, March 31, 2015.

3. Bowers to Lee email, March 31, 2015.

4. Bowers to author, January 8, 2020.

5. Bowers to author, January 8, 2020. The article is Reimar Horten, "Lift Distribution on Flying Wing Aircraft," translated by Jan Scott, *Soaring*, June 1981, 40–42.

6. Technicians repaired the Prandtl-d 3 and NASA transferred the glider to the California Science Center, Bowers email to author, January 8, 2020.

7. Jay Levine, "New Wing Shape Tested in Wind Tunnel," 21 September 2015, NASA Armstrong Flight Research Center, https://www.nasa.gov/feature/new-wing-shape-tested-in-wind-tunnel (accessed 11/26/19).

8. Bowers to author, January 8, 2020.

9. Bowers to author January 24, 2020.

10. Bowers to author, January 8, 2020.

11. Bowers to Steel et al, 19 February 2016; Albion H. Bowers, Oscar J. Murillo, Robert "Red" Jensen, Brian Eslinger, Christian Gelzer, "On Wings of the Minimum Induced Drag: Spanload Implications for Aircraft and Birds," NASA/TP-2016-219072, 3 March 2016.

12. Bowers et al, "On Wings of Minimum Induced Drag," 4–5.

13. Bowers et al, "On Wings of Minimum Induced Drag," 16.

14. Cunningham to Dorney, 28 January 1946, RG 342, Box 3485, file "386.3 – Captured Enemy Equipment – Freeman Field 1945–46. Lt. Col. Richard J. Cunningham was chief of the Photographic Division, Intelligence, Air Technical Service Command, at Wright Field, Ohio. Col. H. C. Dorney commanded Freeman Field, Indiana.

15. Ibid.

16. Horten and Selinger, *Nurflügel*, 136? 1983, Russ Lee translation.

17. https://airandspace.si.edu/collections/horten-ho-229-v3/stealth/

18. Lauren Horelick, https://airandspace.si.edu/collections/horten-ho-229-v3/stealth/ (accessed 1/2/2020).

19. Dobrenz, Spadoni, and Jorgensen, "Aviation Archeology of the Horten 229 V3 Aircraft," AIAA 2010-9214, 4–5, http://arc.aiaa.org, downloaded by Russell Lee on June 10, 2014.

20. Dobrenz et al, "Aviation Archeology," 1.

21. Dobrenz et al, "Aviation Archeology," 5.

22. Dobrenz et al, "Aviation Archeology," 6.

23. Dobrenz et al, "Aviation Archeology of the Horten 229 V3 Aircraft," 9.

24. A. R. Weyl, "Tailless Aircraft and Flying Wings—A Study of Their Evolution and Their Problems," *Aircraft Engineering and Aerospace Technology*, December 1944, 340. Writing in another well-read periodical, *The Aeroplane*, Weyl published 21 articles from December 1944 to April 2, 1948, about all-wing, tailless, and semi-tailless aircraft, and aircraft with low-aspect ratio wings. His study of unusual configurations is detailed and comprehensive.

25. In July 1995 at the Harris Hill airfield near Elmira, New York, the author watched a pilot, during attempts to take off in his semi-tailless sailplane, repeatedly ground loop because of a mild crosswind that would not have hindered the pilot of a conventional sailplane.

26. Krauss email to author, March 15, 2005. In 1999, Krauss published the sixth edition of his bibliography, *Tailless Aircraft: An Extensive Bibliography of Subsonic Types* (Cleveland Heights, OH: Serge Krauss, 1999).

27. The U.S. Air Force used the term *stealth aircraft* in 1988 to refer to the Lockheed F-117 bomber. The F-117 was designed to absorb and reflect little or no electromagnetic energy and thus elude or delay detection by radar.

28. R. Horten, "Das Nurflügelflugzeug," *Luftfahrt und Schule*, August 1938, 244–245.

29. Records of U.S. Air Force Commands, Activities, and Organizations, accessible at the National Archives, College Park, MD.

30. Donald J. Bush, *The Streamlined Decade* (New York: George Braziller, 1975), 4.

31. Bush, *The Streamlined Decade*, 1, 5–6.

32. Matt Walker, "Vine Seeds become 'giant gliders', Earth News, BBC, http://news.bbc.co.uk/earth/hi/earth_news/newsid_8391000/8391345.stm (accessed 11/14/19); Leon Wang, "Seeds with efficiently shaped wings glide slowly to earth, Javan cucumber," Biomimicry Institute, https://asknature.org/strategy/seeds-with-efficiently-shaped-wings-glide-slowly-to-earth/#.Xc2yiFdKjb0 (accessed 11/14/19).

33. Ernst H. Hirschel, Horst Prem, Gero Madelung, *Aeronautical Research in Germany–from Lilienthal until Today*, (Berlin: Springer, 2004) 29–30; n. a., "Flying Wings, Aviation and Aeromodeling – Interdependent Evolutions and Histories," http://www.ctie.monash.edu.au/hargrave/etrich.html (accessed 11/26/19); E. T. Wooldridge, *Winged Wonders: The Story of the Flying Wings* (Washington: Smithsonian Institution, 1983), 10–11; Tom D. Crouch, *The Bishop's Boys* (New York: W. W. Norton, 1989), 366, 379–394.

34. Harold Penrose, *British Aviation: The Pioneer Years* (London: Putnam, 1967), 72–73; Wooldridge, *Winged Wonders*, 14; C. H. Barnes, *Short Brothers Aircraft since 1900* (London: Putnam, 1989), 439.

35. Penrose, *British Aviation*, 88, 487; Wooldridge, *Winged Wonders*, 14; A. R. Weyl, "Tailless Aircraft and Flying Wings: A Study of Their Evolution and Their Problems," *Aircraft Engineering and Aerospace Technology*, January 1945, 9; Weyl, "Tailless Aircraft and Flying Wings," *Aircraft Engineering and Aerospace Technology*, May 1945, 142.

36. N. a., https://commons.wikimedia.org/wiki/File:Burgess-Dunne_AH-7_seaplane_US_Navy.jpg, accessed 11/15/19); https://en.wikipedia.org/wiki/Dunne_D.8 (accessed 11/15/19).

37. Woodridge, *Winged Wonders*, 19–21.

38. Walter and Reimar Oral History Interviews, CD 19a, Smithsonian Institution National Air and Space Museum (NASM) Archives, Washington, D. C., hereafter referred to as WROH, CD number, and recording date where known.

39. Gary R. Pape and John M. Campbell, *Northrop Flying Wings: A History of Jack Northrop's Visionary Aircraft* (Athlen, PA: Schiffer, 1995), 11–12; René J. Francillion, *Lockheed Aircraft since 1913* (Annapolis, MD: Naval Institute, 1987), 7–9, 52, 495–497; Reimar Horten and Peter F. Selinger, *Nurflügel: Die Geschichte der Horten-Flugzeuge 1933–1960* (Graz, Germany: Herbert Weishaupt Verlag, 1983), 229–230.

40. Reimar Horten Oral History Tape Transcript, 317–318, hereafter RHTT and page number.

41. Pape and Campbell, *Northrop Flying Wings*, 13.

42. Francillion, *Lockheed Aircraft since 1913*, 10; Pape and Campbell, *Northrop Flying Wings*, 13–17.

43. René J. Francillion, *McDonnell Aircraft since 1920: Volume I* (Annopolis, MD: Naval Institute, 1979), 21, 23; Maurer Maurer, *Aviation in the U.S. Army 1919–1939* (Washington: Office of Air Force History, 1987), 437; Pape and Campbell, *Northrop Flying Wings*, 32, 34.

44. Pape and Campbell, *Northrop Flying Wings*, 33.

45. On Horten's claim of access denied to wind tunnels, see David Myhra interviews with Reimar, RHTT, 25; and WROH [Myhra], CD 27a.

46. Pape and Campbell, *Northrop Flying Wings*, 258.

47. Horten and Selinger, *Nurflügel*, 57–58; Pape and Campbell, *Northrop Flying Wings*, 36, 52, 250.

48. Quoted in Pape and Campbell, *Northrop Flying Wings*, 40.

49. Horten and Selinger, *Nurflügel*, 55–56, author translation; my thanks to Christina Nutting for searching newspapers using ProQuest internet software.

50. Pape and Campbell, *Northrop Flying Wings*, 77, 83, 93–94, 258; https://planesoffame.org/news/181/0/Statement-on-Accident-April-22-2019 (accessed 11/18/19).

51. Bill Norton, *U.S. Experimental & Prototype Aircraft Projects: Fighters 1939–1945* (North Branch, MN: Specialty, 2008), 247; Pape and Campbell, *Northrop Flying Wings*, 98–103.

52. Pape and Campbell, *Northrop Flying Wings*, 75, 118–123, 130–131, 151–152, 161.

53. Pape and Campbell, *Northrop Flying Wings*, 79.

54. Pape and Campbell, *Northrop Flying Wings*, 83.

55. Pape and Campbell, *Northrop Flying Wings*, 77.

56. Carl E. Reichert, "Summation of All-Wing Aircraft," 4, Aircraft Laboratory, Engineering Division, U.S. Army Air Forces Materiel Command, March 22, 1948, File "452.1—All-Wing Aircraft, 1948," Box 3720, R and D Case Files, RG 342, National Archives at College Park (NACP).

57. Pape and Campbell, *Northrop Flying Wings*, 125–132, 149, 156–158, 183–185, 262; https://en.wikipedia.org/wiki/Boeing_B-47_Stratojet#Engines_and_performance (accessed 11/18/19).

58. Pape and Campbell, *Northrop Flying Wings*, 195–196, 200; Bill Sweetman, *Inside the Stealth Bomber* (Osceola, WI: MBI, 1999), 125–126.

Chapter 1

1. Reimar Horten biography in Royal Aeronautical Society pamphlet, n. d. but ca. 1993, and Walter Horten, "Eidesstattliche Erklärung" (Personal Statement), December 8, 1961. David Myhra, *The Horten Brothers and Their All-Wing Aircraft* (Atglen, PA: Schiffer, 1998), 7.

2. Author interviews with Dr. Karl Nickel and his wife, Gunilde (nee Horten), October 4–8, 2005, tape 7, October 7, 2005, hereafter KGNI, tape number, and interview date. Karl Nickel correspondence to author, May 6, 2002, and June 11, 2002. RHTT, 86. Max Horten's scholarship continued to influence the study of Islamic history and philosophy in 2005.

3. RHTT, 88–89; KGNI 1, October 4–5, 2005.

4. WROH CD 18b, 1982; KGNI 1, October 4–5, 2005.

5. Peter Riedel, *Über sonnige Weiten: Erlebte Rhöngeschichte 1933–1939* (Stuttgart: Motorbuch Verlag, 1985), 73, translated by the author. Elizabeth Horten tried to interest her sons in music. The boys received lessons in classical cello (Wolfram), violin (Walter), and piano (Reimar), but aviation pushed it all aside, to mother Horten's regret, in Myhra, *The Horten Brothers*, 8.

6. RHTT, 15, 61.

7. RHTT, 113. Reimar may have underestimated both his piloting skill and its value to his work.

8. RHTT, 90.

9. RHTT, 78.

10. KGNI 1, October 4–5, 2005; RHTT, 79; Walter Horten to Jan Scott, September 5, 1978, author trans.; RHTT, 117; WROH CD 14a.

11. Peter Fritzsche, *A Nation of Fliers: German Aviation and the Popular Imagination* (Cambridge, MA: Harvard University Press, 1992), 119.

12. Fritzsche, *A Nation of Fliers*, 120.

13. Fritzsche, *A Nation of Fliers*, 121.

14. Peter Riedel, *Über sonnige Weiten*, 73; RHTT, 97.

15. Bob Wander, *Badge Soaring: The A B C Badges . . . Made Easy!*, Number Eight in the Gliding Made Easy Series (Soaring Books & Supplies, 2000), 17, 19.

16. Many pilot veterans of World War I stepped forward to educate the new generation of German air enthusiasts, see Fritzsche, *A Nation of Fliers*, 111.

17. Riedel, *Über sonnige Weiten*, 73; WROH CD 17b, 19a (1982). Prandtl's scientific standing in John D. Anderson, *A History of Aerodynamics and Its Impact on Flying Machines* (Cambridge: Cambridge University Press, 1998), 257.

18. Günter Schmitt, *Hugo Junkers and His Aircraft* (Berlin, VEB Verlag für Verkehrswesen, 1988), 19.

19. Ibid., 19–22; Reimar Horten, "Das Nurflügelflugzeug," *Luftfahrt und Schule*, August 1938, 244–246; Reimar Horten and Peter F. Selinger, *Nurflügel: Die Geschichte der Horten-Flugzeuge 1933–1960* (Graz, Germany: Herbert Weishaupt Verlag, 1983), 17, translated by Jan Scott unless noted otherwise; WROH CD 24a.

20. WROH CD 34b.

21. V. R. Horton, *The Wing (Selected Articles)*, FTD-ID(RS)T-0870-83, translated by Leo Tanner Association, Foreign Technology Division, U.S. Air Force Systems Command, July 25, 1983 (Translation Division, Foreign Technology Division, Wright-Patterson Air Force Base, OH), 17–21, 23–24. Junkers G.38 specifications and history from Kenneth Munson, *Airliners between the Wars 1919–1939* (London: Blandford Press, 1972), 145–147. The visit to Bonn-Hangelar Airfield is noted in Horten and Selinger, *Nurflügel: Die Geschichte der Horten-Flugzeuge*, 16, 19.

22. Hermann Blenk and Werner Schulz, *Jahrbuch 1956: Der Wissenschaftlichen Gesellschaft für Luftfahrt E. V.* (Braunschweig, Germany: Friedrich Vieweg & Sohn, 1957), 123. First flight of the F.13 occurred in 1919.

23. KGNI 1, October 5, 2005.

24. KGNI 1, October 4–5, 2005; KGNI 3, October 5, 2005; Stadler e-mail to author, July 31, 2005.

25. Blenk and Schulz, *Jahrbuch 1956*, 123. Nickel quoted from his written correspondence to the author, May 6, 2002.

26. RHTT, 1–2, quote on 3.

27. Horten and Selinger, *Nurflügel*, 19; Horten, "Das Nurflügelflugzeug," 245. Horten refers to the 1930 aircraft by a temporary nickname, *Hermann Köhl*, in this article. Groenhoff's demonstrations in the Delta I made a positive impression on many observers, see A. R. Weyl, "Tailless Aircraft and Flying Wings, A Study of Their Evolution and Their Problems," *Aircraft Engineering and Aerospace Technology*, December 1944, 348.

28. Alexander Lippisch, *The Delta Wing: History and Development* (Ames, IA: Iowa State University Press, 1981), 5, 116, quote on 17.

29. WROH CD 6a (1982).

30. WROH CD 19a (1982), CD 34a.

31. Giorgio Zanrosso, *History and Evolution of the Glider, Volume I (1773–1914)* (Vicenza, Italy: Egida, 1996), 119; Edwin W. Teale, *The Book of Gliders* (New York: E. P. Dutton, 1930), 163–164.

32. Fritzsche, *A Nation of Fliers*, 109.

33. Walter Horten to Jan Scott, September 5, 1978, Jan Scott trans.; WROH CD 6b (1982).

34. RHTT, 20.

35. Horton, *The Wing*, 26; WROH CD 6b (1982).

36. RHTT, 19.

37. Ibid., 9; WROH CD 6b (1982). Reimar wrote that "we decided in 1928 to go into the construction of all-wing airplanes," see Gebruder Horten, "Zehn Jahre Entwicklung am Nurflügel-Schnellkampfflugzeug," Lilienthal-Gesellschaft für Luftfahrtforschung, April 1943, Reel 2630, Frame 145, Captured German and Japanese Technical Documents Collection, NASM Archives, Smithsonian Institution.

38. RHTT, 19; Horten, "Das Nurflügelflugzeug," 245; WROH CD 6b (1982).

39. RHTT, 15.

40. RHTT, 17.

41. RHTT, 14–15; WROH CD 34a.

42. WROH CD 6b (1982).

43. Horten Brothers, "Ten Years Development of the Flying-Wing High Speed Fighter," April 14, 1943, Lilienthal Gesellschaft, Berlin, Report No. LGB 164, 1, translated by Chance Vought Aircraft, July 18, 1948.

44. WROH CD 6b (1982).

45. Horton, *The Wing*, 26.

46. Ken Stewart, *The Glider Pilot's Manual* (Shrewsbury, England: Airlife, 2001), 199.

47. Horton, *The Wing*, 26–27.

48. RHTT, 20–21; WROH CD 6b (1982); Martin Simons, *Sailplanes 1920–1945* (Königsburg, Germany: Eqip, 2002), 99, 65.

49. WROH CD 6b (1982) and Oskar Ursinus, "Modelle," *Flugsport*, June 25, 1930, vol. 13, 231. These sources disagree on the distance flown; the shorter figure published in *Flugsport* is used here.

50. Horton, *The Wing*, 26. One of the largest models that Reimar and Walter tested was a giant spanning 4 m (13 ft.) and weighing 15 kg (33 lb.), one of the smallest spanned 1.2 m (4 ft.).

51. WROH CD 6b (1982), 34a.

52. In the early 1980s, Reimar still valued flying models as tools to develop sophisticated piloted aircraft, see Horton, *The Wing*, 30.

53. RHTT, 19; WROH CD 6b (1982); Horton, *The Wing*, 33.

54. WROH CD 34b.

55. Ferenc A. Vajda and Peter Dancey, *German Aircraft Industry and Production 1933–1945* (Warrendale, PA: Society of Automotive Engineers, 1998), 12.

56. Fritzsche, *A Nation of Fliers*, 191.

57. Richard J. Evans, *The Coming of the Third Reich* (New York: Penguin, 2004), 331–333, 348.

58. Evans, *The Coming of the Third Reich*, 380–381.

59. Membership cards for Reimar, Walter, who joined in February, and other family members are in Foreign Records Seized Collection, Berlin Document Center Microfilm,

NSDAP Zentralkartei (the central record file of the Nazi Party), A3340-MFKL-J082 and NSDAP Ortsgruppenkartei, A3340-MFOK-I066, RG 242, National Archives at College Park, MD (NACP). My thanks to Dr. Michael Neufeld for pointing out this important collection. See RHTT, 300–303, for Reimar's denial of Nazi Party membership.

60. Walter Horten penned this account in 1979 and Peter Riedel quoted it in *Über sonnige Weiten*, 73, author trans. The relative ease of achieving lengthy flights using unsophisticated flying models led many early designers to underestimate the difficulties of developing successful piloted all-wing aircraft, see A. R. Weyl, "Stability of Tailless Aeroplanes," *Aircraft Engineering and Aerospace Technology*, April 1945, 103.

61. RHTT, 22.

62. KGNI 2, October 5, 2005. The *H I* was slightly smaller in dimensions and wing loading. Compare drawings and characteristics of the two aircraft in Horten and Selinger, *Nurflügel*, 34, and Lippisch, *The Delta Wing*, 20.

63. KGNI 2, October 5, 2005; KGNI 4, October 7, 2005. Horten applied the same formula for wing chord thickness to the *H II* and *H III* sailplanes.

64. WROH CD 6b (1982); RHTT, 24; Myhra, *The Horten Brothers*, 40–41; Horton, *The Wing*, 33; WROH CD 35a; Horten and Selinger, *Nurflügel*, 26; Horton, *The Wing*, 34.

65. Horten and Selinger, *Nurflügel*, 34.

66. Horton, *The Wing*, 33.

67. WROH CD 6b (1982); n. a., "Full-Sized Wind Tunnels," *The Aeroplane*, January 29, 1930, 167, discusses applying the Reynolds calculations to scale models. KGNI 2, October 5, 2005; Peter Garrison, "Size Really Does Matter," *Flying*, September 1998, 116. My thanks to Dr. John Anderson for suggesting this line of inquiry.

68. Horton, *The Wing*, 34; KGNI 2, October 5, 2005. Nickel credited the drag rudder configuration to Reimar.

69. Horton, *The Wing*, 33; Horten and Selinger, *Nurflügel*, 26.

70. Simons, *Sailplanes 1920–1945*, 13; Oskar Ursinus, "Russicher Segelflug-Wettbewerb in der Krim 7–25 September 1924," *Flugsport*, November 29, 1924, Vol. 22, 426–428, and "Vom Krim-Segelflug-Wettbewerb," *Flugsport*, November 25, 1925, Vol. 23, 457–459; and E. T. Wooldridge, *Winged Wonders: The Story of the Flying Wings* (Washington, DC: Smithsonian, 1983), 28–31.

71. WROH CD 6b (1982). In 1929, Max Horten helped his son join a group of university students building a sailplane called the *Professor* in a workshop at the Akaflieg Bonn, see WROH CD 34a; KGNI 2, October 5, 2005.

72. Walter Horten quoted in Riedel, *Über sonnige Weiten*, 73–74, author trans. Reimar first experimented with this thin 1 mm plywood on flying models before using it to cover the *H I*, RHTT, 16, 18.

73. WROH CD 19a (1982).

74. RHTT, 16, 18, 22; WROH CD 32a, 34a, 34b; KGNI 1, October 4, 2005; Myhra, *The Horten Brothers*, 52–57.

75. KGNI 1, October 4–5, 2005.

76. Riedel, *Über sonnige Weiten*, 74, translated by the author; KGNI 1, October 4, 2005.

77. WROH CD 36a.

78. Horton, *The Wing*, 39–40.

79. Horton, *The Wing*, 40–41. Horten probably designed the control system with rapid adjustments in mind.

80. Horton, *The Wing*, 41.

81. On the interaction of different controls, see Weyl, "Tailless Aeroplane Control Systems," *Aircraft Engineering and Aerospace Technology*, May 1945, 133. Horten, *The Wing*, 41–42.

82. Horten, *The Wing*, 43, 44–45; WROH CD 39a.

83. Horton, *The Wing*, 45. The reduction in the size of the aileron was apparently not enough to void the enlargements Reimar had made to satisfy the contest technical committee.

84. WROH CD 19b (1982); Horton, *The Wing*, 46; Selinger, *T.W.I.T.T. Newsletter*, November 1993, 8.

85. WROH CD 35b; Horten and Selinger, *Nurflügel*, 29; Riedel, *Über sonnige Weiten*, 76; WROH CD 17a, quote from WROH CD 19b (1982).

86. Lippisch, *The Delta Wing*, 28; WROH CD 17a, CD 20a. Lippisch had hired Jacobs in 1927 to help him design tailless aircraft and in 1934, both men had transferred to the new DFS, see Simons, *Sailplanes 1920–1945*, 104; Horton, *The Wing*, 36–37.

87. Myhra, *The Horten Brothers*, 54; Horten and Selinger, *Nurflügel*, 33–34, author trans.

88. WROH CD 17a; KGNI 1, October 5, 2005; Myhra, *The Horten Brothers*, 54–57; WROH CD 17b, 34a, 18a.

89. Ibid., CD 30b.

90. RHTT, 1, 3–4, 55–56; KGNI 2, October 5, 2005.

Chapter 2

1. RHTT, 73.

2. RHTT, 117, 118; Reimar Horten discusses improvements to the *H I* in WROH CD D 20a and 35b.

3. KGNI 3, October 5, 2005. Nickel and Wohlfahrt show the *H II* mixer in *Tailless Aircraft in Theory and Practice*, 123. Drawing of Lippisch mixer design for his Delta I aircraft appears in Lippisch, *The Delta Wing*, 19.

4. NASM staff discovered the handholds and lifting handles in 1994 as they prepared to crate and ship the Horten *H II L* to the DTM Berlin for restoration.

5. WROH CD 17b.

6. RHTT, 72; KGNI 6, October 8, 2005; KGNI 1, October 4–5, 2005. Wingspan and other dimensions from Horten and Selinger, *Nurflügel*, 43, author trans. and calculations. Nickel and Wohlfahrt, *Tailless Aircraft in Theory and Practice*, 185.

7. The prototype *H II* no longer exists, but drawings of the *H II L* made by DTM staff in 1996 show a reflex airfoil profile extending to the elevon, see "Inventarisations und Restaurierungszeichnungen Flugzeugtypen (Inventory or Aircraft Drawings): Horten II, Zeichn.-Nr.: 2.00–B11, June 11, 1996, and Nr.: 4.08.009-020," dated July 8, 1996.

8. Weyl, "Stability of Tailless Aeroplanes," *Aircraft Engineering and Aerospace Technology*, March 1945, 77.

9. Bowers e-mail to author, March 22, 2002.

10. Bowers e-mail to nurflugel@yahoogroups.com, March 11, 2002, Digest Number 856, 3, emphasis in original note.

11. Bowers e-mail to nurflugel@yahoogroups.com, June 25, 2003, Digest Number 1143; Ludwig Prandtl, "Über Tragflügel kleinsten induzierten Widerstandes," *Zeitschrift für Flugtechnik und Motorluftschiffahrt*, Nr. 24, November 1933, 305–306, Lars Urban, trans.

12. For example, rules governing several classes of competition sailplanes restrict wingspan, see Bowers e-mails to author, May 20, 2002, and January 9, 2006.

13. Reinhold Stadler traced the transmission of these ideas from Prandtl to Lippisch, and then to Horten, see Stadler e-mail to Internet discussion group, nurflugel@telelists.com, January 21, 1999. Both Stadler and Bowers have commented on the brevity and clarity of Prandtl's paper.

14. Al Bowers e-mail to author, May 20, 2002. Dogma quote from Bill and Bunny Kuhlman, "Twist Distributions for Swept Wings, Part 1," *On the Wing . . . #161*, 5, web article available at www.B2Streamlines.com. Prandtl was an eminent fluid dynamicist whose work profoundly influenced the course of twentieth-century aerodynamics, see John Anderson, *A History of Aerodynamics*, 244–264.

15. H. H. Hurt, *Aerodynamics for Naval Aviators* (Aviation Maintenance Publishers, 1965, revised ed.), 76.

16. Alexander Lippisch, "Bestimmung der Auftriebsverteilung längs der Spannweite," (Definition of the Lift Distribution along the Wingspan), part I, *Flugsport*, December 20, 1933, 549; part II, *Flugsport*, January 10, 1934, 8; and "Bestimmung der Auftriebs-verteilung längs der Spannweite, Zusammenstellung der Ergebnisse" (Definition of the Lift Distribution along the Wingspan, Summary of Results), part III, *Flugsport*, February 7, 1934, 49–51; quote from p. 50, all articles trans. by Lars Urban. On p. 49 of part III, fig. 7 depicts the triangular distribution at 0.0 Ca (lift coefficient).

17. WROH CD 17b.

18. WROH CD 20a; Horton, *The Wing*, 50; RHTT, 2, 16.

19. Horton, *The Wing*, 55; RHTT, 120; WROH CD 20a, 17b; KGNI 3, October 5, 2005. Schrödter paid for welding supplies and steel tubing, and Horten paid a local flight instructor to do the welding.

20. WROH CD 20a; RHTT, 123; Horton, *The Wing*, 47–50; Harald J. von Beckh, "The Development and Airborne Testing of the PALE Seat," Report No. NADC-81200-60, June 20, 1981, Naval Air Development Center, Warminster, PA, 56.

21. Octave Chanute, *Progress in Flying Machines* (New York: The American Engineer, 1894), 286.

22. Marvin W. McFarland, ed., *The Papers of Wilbur and Orville Wright Including the Chanute-Wright Letters and Other Papers on Octave Chanute Volume One: 1899–1905* (New York: Arno, 1953), 40–50. I am indebted to retired Smithsonian curator Dr. Howard Wolko for pointing out Lilienthal's influence on the Wright brothers.

23. RHTT, 80; von Beckh, "Development and Airborne Testing of the PALE Seat," 56.

24. WROH CD 17b; RHTT, 123, 380.

25. Reimar said May in Horton, *The Wing*, 50–51, and June in WROH CD 17b.

26. Weyl, "Wing Tips for Tailless Aeroplanes," *Aircraft Engineering and Aerospace Technology*, September 1945, 259–266, quote on 264.

27. Jan Scott, "Reimar Horten Visited," *Bungee Cord*, Vol. VI, No. 2, June 1980, 6, contains the first printed reference to favorable yaw from a Horten wing; Reimar Horten, "Lift Distribution on Flying Wing Aircraft," *Soaring*, June 1981, 41.

28. Horton, *The Wing*, 54–55; RHTT, 123–124; Horton, *The Wing*, 51.

29. Horten and Selinger, *Nurflügel*, 36; Horton, *The Wing*, 56; WROH CD 20a; RHTT, 81, 118.

30. Reinhold Stadler, "Die Auslegung der Horten-Nurflügelflug-zeug" (The Interpretation of the Horten Only-Wing Aircraft), presented at the twelth Nurflügelsymposium, Scheidegg, Germany, October 2–3, 1999, author's office files, NASM; Stadler e-mail to author, March 5, 2000, and May 30, 1997; Edward Uden letter to author, October 31, 1995.

31. WROH CD 20a; Horton, *The Wing*, 56; RHTT, 82; Günter Brinkmann and Hans Zacher, *Die Evolution der Segelflugzeug* (Bonn: Bernard & Graefe, 1992), 91; von Beckh, "Development and Airborne Testing of the PALE Seat," 56.

32. Horton, *The Wing*, 57.

33. WROH CD 20a.

34. Horten and Selinger, *Nurflügel*, 36; RHTT, 81–82; WROH CD 16a, quote from 19b (1982).

35. Lippisch paraphrased in Weyl, "Stability of Tailless Aeroplanes," *Aircraft Engineering and Aerospace Technology*, 103.

36. WROH CD 16b; KGNI 3, October 5, 2005. Ferenc Vajda and Peter Dancey, *German Aircraft Industry and Production 1933–1945* (Warrendale, PA: Society of Automotive Engineers, 1998), 274, describes Udet's responsibilities and failures.

37. Reinhold Stadler correspondence with author, October 29, November 1, and November 18, 1995; WROH CD 14a.

38. Edwin Teale, *The Book of Gliders*, 99.

39. WROH CD 15b.

40. RHTT, 135–136.

41. WROH CD 1a (April 22, 1988).

42. KGNI 3, October 5, 2005.

43. Horten and Selinger, *Nurflügel*, 51, 53, 55; RHTT, 136, 138; WROH CD 19b (1982); Fritzsche, *A Nation of Fliers*, 190–200; Brinkmann and Zacher, *Die Evolution der Segelflugzeug*, 91.

44. KGNI 3, October 5, 2005; Horten and Selinger, *Nurflügel*, 51–52, 69, author trans., flight distance from German text, 52. Brinkmann and Zacher, *Die Evolution der Segelflugzeug*, 91.

45. Vajda and Dancey, *German Aircraft Industry*, 185.

46. Kenneth G. Wilkinson, *The Horten Tailless Aircraft*, "R. A. E. Report No. F. A. 259/1, Technical Note No. Aero. 1703," (Royal Aircraft Establishment, Farnborough Hants, England: October 1945), 43–44, hereafter "Aero. 1703." Reitsch dated her report November 17, 1938.

47. "Aero. 1703," 44.

48. "Aero. 1703," 44.

49. The statements about countering adverse yaw with the bell distribution appear in Horten and Selinger, *Nurflügel*, 38, author trans.; RHTT, 124; WROH CD 18a, 20a, and other records.

50. KGNI 3, October 5, 2005; WROH CD 46a; Dennis Piszkiewicz, *From Nazi Test Pilot to Hitler's Bunker: The Fantastic Flights of Hanna Reitsch* (Westport, CT: Praeger, 1997), 2.

51. "Aero. 1703," 36.

52. Nickel and Wohlfahrt, *Tailless Aircraft*, 356.

53. Nickel and Wohlfahrt, *Tailless Aircraft*, 142.

54. KGNI 3, October 5, 2005.

55. "Aero. 1703," 45.

56. Nickel and Wohlfahrt, *Tailless Aircraft*, 185.

57. Both the air speed and turn-and-bank instruments were prone to ice up and fail in clouds, see G. Brütting, "Hofmann's Flight through Four Local Thunderstorms," 108, in Wolf Hirth, *The Art of Soaring Flight*, Naomi Heron-Maxwell trans. (Stuttgart: Vaihingen-Stuttgart, 1938).

58. Nickel and Wohlfahrt, *Tailless Aircraft*, 204.

59. Hurt, *Aerodynamics for Naval Aviators*, 86–87; WROH CD 36a; Reinhold Stadler has done the best recent research, see his paper, *Die Auslegung der HORTEN-Nurflügelflugzeuge*, XII Nurflügelsymposium, Scheidegg, Germany, October 2–3, 1999, author office files, NASM; KGNI 8, October 8, 2005.

60. "Aero. 1703," 43–44.

61. RHTT, 127.

62. RHTT, 126–127; Brothers Horten, "Plastic, All-Wing Horten Aircraft," *Flugsport*, Vol. 35, No. 16, December 15, 1943, 249–251, H. I. Lewenz, trans. (R. T. P. 3 Translation No. 2337, issued by the Ministry of Aircraft Production, London); WROH CD 20b, 20a; N. a., Internet Wikipedia, articles "Phenolic resin," "Phenol formaldehyde resin," and "Bakelite"; Horten and Selinger, *Nurflügel*, 45.

63. WROH CD 20a, 20b.

64. Lewenz, "Plastic, All-Wing Horten Aircraft," 1. Horten covered most of the wing with conventional fabric; see photographs in Myhra, *The Horten Brothers*, 72–73.

65. WROH 37b.

66. RHTT, 130, 134. William Green summarizes the fighter destroyer concept in *Warplanes of the Third Reich* (Garden City, NY: Doubleday, 1972), 573–574.

67. RHTT, 131–132; Horten and Selinger, *Nurflügel*, 50, author trans. and calculations, 56.

68. Lewenz, "Plastic, All-Wing Horten Aircraft," 1–2. Trade name Lignofol identified in original *Flugsport* article, 251.

69. Quote from WROH CD 19b (1982), 20b; RHTT, 132–133, 149, 381; D. Appleyard and M. A. Biot, *The Horten Tailless Aircraft* (Combined Intelligence Objectives Subcommittee, G-2 Division, SHEAF [Rear] APO 413: May 1945), 4–5.

70. Lewenz, "Plastic, All-Wing Horten Aircraft," 2; RHTT, 133; Horten and Selinger, *Nurflügel*, 45.

71. WROH CD 31a; RHTT, 373–374; Horten and Selinger, *Nurflügel*, 56; Giorgio Zanrosso, *History and Evolution of the Glider* (Vicenza, Italy: La Serenissima, 2000), Vol IV, 15, 53.

72. WROH CD 43a (August 24, 1986).

73. RHTT, 132, 374.

74. L. M. Harris and J. T. Grey, "A Survey of German Chemical and Wood Industries Producing Plastic and Wood Materials for Aircraft Tooling and Fabrication," FIAT Final Report No. 466, U.S. Technical Industrial Intelligence Committee, Aeronautics Subcommittee, July 27–August 1945, 1.

75. Horten and Selinger, *Nurflügel*, 56; *The Green Guide: Germany* (Michelin Travel Publications, Watford, Herts, U.K.: 2000), 288.

76. RHTT, 374.

77. WROH CD 2a (April 20, 1988); Horten and Selinger, *Nurflügel*, 69, author trans.

78. Horten and Selinger, *Nurflügel*, 56, 71, quote on 117; RHTT, 374.

Chapter 3

1. RHTT, 140.

2. Airfoil profile information from Peter F. Selinger e-mail to nurflugel@onelist.com, June 9, 1999; statistics from Horten and Selinger, *Nurflügel*, 55, 59, 65, author calculations and trans.; WROH CD 20b; "Aero. 1703," 9.

3. KGNI 1, October 4–5, 2005; Günter Brinkmann and Hans Zacher, *Die Evolution der Segelflugzeuge* (Bonn: Bernard & Graefe, 1992), 91–92, Heinz Gerstle, trans.; Horten and Selinger, *Nurflügel*, 229.

4. Peter Bowers, *Modern Soaring Guide* (Blue Ridge Summit, PA: TAB Books, 1979), 78–79.

5. WROH CD 20b; Horten and Selinger, *Nurflügel*, 59; RHTT, 140–141.

6. Ibid., 143–145; WROH CD 20b; Horten and Selinger, *Nurflügel*, 59; KGNI 6, October 8, 2005; Karl Nickel handwritten comments to draft manuscript, October 8, 2005, author office files, NASM; Riedel, *Uber sonnige Weiten: Erlebte Rhöngeschichte 1933–1939* (Stuttgart, Germany: Motorbuch, 1985), 129, 167.

7. RHTT, 145–146; WROH CD 20b.

8. KGNI 6, October 8, 2005. Stadler correspondence to author, August 7, 1994.

9. RHTT, 141, 142, 144–146; Horten and Selinger, *Nurflügel*, 59; Riedel, *Über sonnige Weiten*, 221.

10. Ibid., 234, 241; Brinkmann and Zacher, *Die Evolution der Segelflugzeuge*, Heinz Gerstle, trans., 92.

11. Horten and Selinger, *Nurflügel*, 59; Riedel, *Über sonnige Weiten*, 234. Horten recalled in an interview Blech's preference to fly with his cockpit safety harness unfastened, see WROH CD 15b.

12. Riedel, *Über sonnige Weiten*, 239, 241; Horten and Selinger, *Nurflügel*, 59; WROH CD 15b, 20b, 22b; RHTT, 149–150, 185.

13. KGNI 6, October 8, 2005; Howard Siepen, "On the Wings of the Wind," *National Geographic Magazine*, June 1929, 759.

14. Reimar Horten, "The 'Horten-Parabola,'" *Sailplane and Glider*, August 1951, Vol. XIX No. 8, 186, G. S. Neumann, trans.

15. Hans Helbig, "Zwei Preise," *Luftfahrt und Schule*, August 1938, 245; WROH CD 23a; RHTT, 130.

16. Horten and Selinger, *Nurflügel*, 66, author trans.; RHTT, 150; WROH CD 15b. In 1934–1935, Helbig had worked with the DLV and the Education Ministry to implement a November 1934 Nazi decree that boy's schools begin teaching courses in constructing flying model airplanes, see www.segelflugmuseum.de/index.php?id=4538&psid=59fcaac1945b97e5f5b4131838b26111.

17. Brinkmann and Zacher, *Die Evolution der Segelflugzeuge*, Heinz Gerstle, trans., 92. Comparison data from Zanrosso, Vol. II, 178, 224; Betsy Woodward and Ales Stirnemann, *The World's Sailplanes* (OSTIV & Schweizer Aero-Revue, 1958), 72, 124.

18. WROH CD 15b, 16b; RHTT, 151–154, 382.

19. RHTT, 380.

20. KGNI 1, October 5, 2005.

21. RHTT, 381.

22. WROH CD 22b.

23. RHTT, 331.

24. Reidel, *Über sonnige Weiten*, 257; RHTT, 154–156; Horten and Selinger, *Nurflügel*, 66, 67, author trans.

25. Reimar Horten, "The 'Horten-Parabola,'" 186.

26. Horten and Selinger, *Nurflügel*, 66.

27. Galland correspondence to author, September 27, 1995.

28. RHTT, 157; WROH CD 21b, 33a, 44a; Horten and Selinger, *Nurflügel*, 71, 76, Jan Scott and author, trans.; Vajda and Dancey, *Germany Aircraft Industry*, 51; Bill Gallagher, "Charlie Spratt and Rudolph Opitz Honored during 40th Annual United States Soaring Hall of Fame Induction Ceremonies at the National Soaring Museum," *NSM Historical Journal*, Vol. 16, No. 2, 1994, 4–5.

29. WROH CD 44a; Horten and Selinger, *Nurflügel*, 94; Myhra, *The Horten Brothers*, 150; Adolf Galland, *The First and the Last* (New York: Ballantine, 1954), 75, 81; Vajda and Dancey, *German Aircraft Industry*, 108.

30. Horten and Selinger, *Nurflügel*, 22–24, Jan Scott and author, trans.

31. RHTT, 156.

32. Horten and Selinger, *Nurflügel*, 76; WROH CD 21b; RHTT, 157–158.

33. Fritz von Hermann, "Prone Flying," *Sailplane and Glider*, Vol. XX, No. 6, June 1952, 5.

34. Horten and Selinger, *Nurflügel*, 98, author trans.; RHTT, 156; WROH CD 18a.

35. Flight Lt. R. C. Forbes correspondence to Hollis Button, May 21, 1950; n. a., "Horten IV and VI High Performance Glider Aircraft," *LUFTFAHRT International,* 1849, 1861; Reinhold Stadler e-mail to nurflugel@onelist.com, August 31, 1999.

36. KGNI 6, October 8, 2005.

37. WROH CD 18a; A. R. Weyl, "Tailless Aircraft and Flying Wings: A Study of Their Evolution and Their Problems," *Aircraft Engineering and Aerospace Technology*, January 1945, 10.

38. "Aero. 1703," 46; Horten and Selinger, *Nurflügel*, 99, author trans.

39. Reimar Horten, "Flying Wing Pilot Position and Design Options," *Soaring*, Jan Scott, trans., August 1980, 22; "Aero. 1703," 13; WROH CD 27a; KGNI 4, October 7, 2005; Edward Uden personal interview, September 15, 1994.

40. Nickel and Wohlfahrt, *Tailless Aircraft*, 353.

41. Bertrand Handiwerk, "The Modified Horten IV of the AHQ BAFO Club," *Soaring*, November–December 1950, 9.

42. WROH CD 27a; Horten and Selinger, *Nurflügel*, 100, author trans.; "Aero. 1703," 13; Ulrich Albrecht, "Military Technology and Nation Socialist Ideology," *Science, Technology and National Socialism*, Monika Renneberg and Mark Walker, eds. (Cambridge: Cambridge University Press, 1994), 88–125; Hans Amtmann, *The Vanishing Paperclips* (Boylston, MA: Monogram, 1988); Crawford to Craigie, July 29, 1946, and August 19, 1946, File "740-*Prone Position in Aircraft*, Confidential 1946," Research Development Project Files, Engineering Division, Materiel Command, Box 3595, RG 342, NACP. Von Beckh's "The Development and Airborne Testing of the PALE Seat," cited in Chapter 2, places Horten's semiprone cockpit within the larger context of experiments conducted with pilot seating.

43. WROH CD 21b; RHTT, 158–159.

44. WROH CD 44a.

45. Discussion among Al Bowers, Bill and Bunny Kuhlman, and Doug Holverson late in February 2005 on the nurflugel@yahoo-groups.com discussion list generated key insights into difficulties that Horten faced in aerodynamically tuning his wings for maximum performance.

46. Lehmann, May 21, 1943, Reel 3743, Frame 650, Captured German and Japanese Technical Documents on Microfilm, NASM Archives; n. a., "The Horten IV: An Amazing German Tailless All-wing Glider," *Flight*, August 5, 1943, 139; von Hermann, "Prone Flying," 5; "Aero. 1703," fig. 11.

47. Reimar Horten, "Von der 'Ho-IV' zur 'Ho-VI,'" *Thermik*, Vol. 5, No. 3, March 1952, 33, Heinz Gerstle, trans.; "Aero. 1703," 10–11, fig. 8.

48. "Aero. 1703," 11, fig. 7.

49. Horten and Selinger, *Nurflügel*, 108, author trans.; "Aero. 1703," fig. 11; Karl Nickel written correspondence to *T.W.I.T.T. Newsletter*, December 7, 2001; Karl Nickel manuscript comments, October 8, 2005.

50. RHTT, 331.

51. WROH CD 21b, 44a.

52. Horton, *The Wing*, 57; Horten and Selinger, *Nurflügel*, 96–98, 229, author trans.

53. Bernd Ewald, "Will We Have a Horten Flying Wing in the Air Again?," *VGC News*, Winter 2001, 21; WROH CD 31a.

54. Horten and Selinger, *Nurflügel*, 98, 101, 102, author trans.; RHTT, 229.

55. Horten and Selinger, *Nurflügel*, 100, author trans.

56. Lehmann, May 21, 1943, Reel 3743, Frame 650, Captured German and Japanese Technical Documents on Microfilm, NASM Archives.

57. Horten and Selinger, *Nurflügel*, 102, author trans.; Reimar Horten, "Von der 'Ho-IV' zur 'Ho-VI,'" 33.

58. WROH CD 2a, April 20, 1988.

59. Al Bowers e-mail to author, April 25, 1997.

60. Martin Simons, *Sailplanes 1920–1945* (Königswinter, Germany: Eqip, 2001), 147–148.

61. Hans Zacher, "Ergebnisse der Leistungsmessung und Flugeigenschaftsprüfung des Segelflugzeuges D.30 'Cirrus,'" *Mitteilungen der Flugtechnischen Fachgruppen und Arbeitsgemeinschaften*, Folge 6, 1. Heft 1944, 2–23; Horten and Selinger, *Nurflügel*, 102, 104, 105, author calculations and trans.; n. a., "Horten IV and VI High Performance Glider Aircraft," *Luftfahrt International*, No. 12, November–December 1975, 1847–1870, Lucian Heichler, trans.

62. Forbes Aird, *Racer's Encyclopedia of Metals, Fibers & Materials* (Osceola, WI: Motorbooks, 1994), 18.

63. Horten and Selinger, *Nurflügel*, 104, author trans.

64. Reimar Horten, "Von der 'Ho-IV' zur 'Ho-VI,'" *Thermik*, 33; Horten and Selinger, *Nurflügel*, 104, author trans.

65. Nickel and Wohlfahrt, *Tailless Aircraft*, 383, 443–444, quotes on 443.

66. Jochen Ewald, "News from the Research Students: Idaflieg Summer Meeting 2005," *Quiet Flyer Magazine*, August 2006, 76.

67. Horten and Selinger, *Nurflügel*, 229; Reimar Horten, "Von der 'Ho-IV' zur 'Ho-VI,'" *Thermik*, 33.

68. WROH CD 44b; Horten and Selinger, *Nurflügel*, 87, 229; Nickel and Wohlfahrt, *Tailless Aircraft*, 352.

69. All quotes Nickel and Wohlfahrt, *Tailless Aircraft*, 353.

70. Nickel and Wohlfahrt, *Tailless Aircraft*, 353–354.

71. Nickel and Wohlfahrt, *Tailless Aircraft*, 354.

72. Nickel and Wohlfahrt, *Tailless Aircraft*, 351; KGNI 4, October 7, 2005.

73. Nickel and Wohlfahrt, *Tailless Aircraft*, 465.

74. Nickel correspondence to author, December 20, 2005.

75. Reimar Horten, "Von der 'Ho-IV' zur 'Ho-VI,'" *Thermik*, 34.

76. Nickel and Wohlfahrt, *Tailless Aircraft*, 354–355.

77. "Aero. 1703," 13.

78. D. C. Appleyard and M. A. Biot, "The Horten Tailless Aircraft—CIOS Target Number 25/157" (London: His Majesty's Stationary Office, May 1945), 22.

79. Horten and Selinger, *Nurflügel*, 94, 138; RHTT, 33–35, 168, 175–176.

80. WROH CD 2a (April 20, 1988).

81. N. a., "Reimar Horten," *VGC News*, No. 80, Winter 1993/4, 23–24; Karl Nickel manuscript notes, October 8, 2005.

82. Reimar Horten, "Von der 'Ho-IV' zur 'Ho-VI,'" *Thermik*, 33, 35, Heinz Gerstle, trans.; Martin Simons, *The World's Vintage Sailplanes 1908–1945* (Melbourne: Kookaburra Technical Publications Pty. Ltd., 1986), 163–164. *H VI* dimensions from Horten and Selinger, *Nurflügel*, 109, author calculations and trans; "Aero. 1703," 12, 17; Forest Products Laboratory, *Wood Handbook: Wood as an Engineering Material*, reprint ed. (Madison, WI: Forest Products Society, 1999), 19–20.

83. "Aero. 1703," 6.

84. Reimar Horten. "Flying Wing Position and Design Options," *Soaring*, Jan Scott, trans., August 1980 23; Reimar Horten, "Von der 'Ho-IV' zur 'Ho-VI,'" *Thermik*, 33, Heinze Gerstle, trans.

85. Reimar Horten, "Von der 'Ho-IV' zur 'Ho-VI,'" *Thermik*, 33, Heinz Gerstle, trans.

86. WROH CD 21b; von Hermann, "Prone Flying," 7; Horten and Selinger, *Nurflügel*, 94.

87. Selinger e-mail to author, December 4, 1998; "Aero. 1703," 11.

88. At the time, Reimar was attending Bonn University twice a week, WROH CD 41a, August 24, 1986.

89. Copied page from Heinz Scheidhauer's flight logbook via anonymous source, author trans.

90. Karl Nickel manuscript notes, October 8, 2005.

91. RHTT, 170–171, 176, 371; WROH CD 2b (April 20, 1988), 24b, 37b; Reimar Horten, "Von der 'Ho-IV' zur 'Ho-VI,'" *Thermik*, 35, 36, translated by Heinz Gerstle; Horten and Selinger, *Nurflügel*, 105, author trans.; "Aero. 1703," 17.

92. WROH CD 25b.

93. RHTT, 370.

94. Nickel and Wohlfahrt, *Tailless Aircraft*, 333.

95. "Aero. 1703," 17.

Chapter 4

1. WROH CD 43b (August 24, 1986), 44a.

2. RHTT, 102.

3. WROH CD 7b, 44a; RHTT, 159.

4. WROH CD 44a; RHTT, 159.

5. Horten and Selinger, *Nurflügel*, 71; WROH CD 44a.

6. RHTT, 267.

7. WROH CD 4b (April 21, 1988), 15b, 22a; RHTT, 159. Walter and Sabina married in May 1943.

8. RHTT, 184.

9. RHTT, 159; Horton, *The Wing*, 57; WROH CD 21b, 41a (August 24, 1986), 44a; RHTT, 160, 219.

10. WROH CD 31a; RHTT, 161, 374.

11. WROH CD 44a. Albion Bowers, NASA aerodynamicist at the Dryden Flight Research Center in 1997, validated this thinking in an e-mail to the discussion list at nurflugel-l@teleport.com, July 4, 1997, 1.

12. WROH CD 31a.

13. WROH CD 44a; RHTT, 225, 226. Author calculations.

14. WROH, 160; Karl Nickel manuscript comments, October 8, 2005.

15. RHTT, 262.

16. David Myhra, *The Horten Brothers and Their All-Wing Aircraft* (Atglen, PA: Schiffer, 1998), 150, 152; Horten and Selinger, *Nurflügel*, 71, 123, author trans.; WROH CD 26a, 57a (April 18, 1986); Karl Nickel manuscript comments, October 8, 2005; RHTT, 100, 162, 185.

17. RHTT, 188.

18. RHTT, 263.

19. RHTT, 206.

20. RHTT, 379.

21. RHTT, 324.

22. WROH CD 44a.

23. RHTT, 388.

24. WROH CD 45a.

25. Karl Nickel manuscript comments, October 8, 2005; Vajda and Dancey, *German Aircraft Industry*, 246; RHTT, 268, 275, 322, 323–324.

26. Ibid., 258–259, 324; WROH CD 23a; CD 41a (August 24, 1986); CD 44b.

27. Vajda and Dancey, *German Aircraft Industry*, 60, 68, 77.

28. Records of the Generalflugzeugmeister (director-general of air armament), Federal German Military Archives, Freiburg, quoted in Horst Boog, "The Strategic Air War in Europe and Air Defense of the Reich, 1943–1944," in *Germany and the Second World War, Vol. VII, The Strategic Air War in Europe and the War in the West and East Asia 1943–1944/5*, ed. Militärgeschichtliches Forschungsamt (Oxford: Clarendon, 2006), 407.

29. Owen Thetford, *Aircraft of the Royal Air Force since 1918* (London: Putnam, 1962), 164; Horten and Selinger, *Nurflügel*, 134; WROH CD 26a.

30. Ibid., CD 23b, 44b; RHTT, 163, 28; Andrei Shepelev and Huib Ottens, *Horten Ho 229, Spirit of Thuringia: The Horten All-Wing Jet Fighter* (Hersham, Surrey, England: Classic Publications, 2006), 41.

31. Horten and Selinger, *Nurflügel*, 56, 71, quote on 117; RHTT, 374.

32. "Aero. 1703," 17; WROH CD 23a.

33. "Aero. 1703," 15–17.

34. Reimar and Walter Horten, "Stabilitätsbetrachtungen an Pfeilflügeln," 164/ Vorbericht: *Flugzeugkonstruktion. Bericht über die Sitzung Nurflügelflugzeuge*, Lilienthal-Gesellschaft für Luftfahrtforschung, am 14 April 1943, Berlin (Considerations on the Stability of Arrow Wings, Preliminary Report 164: Airplane Construction, Report of the Flying Wings Meeting, sponsored by the Lilienthal Society for Aeronautics Research, April 14, 1943, Berlin), Reel 2630, Frame 126–139, Captured German and Japanese Technical Documents on Microfilm, NASM Archives. Historical summary, "Zehn Jahre Entwicklung am

Nurflügel-Schnellkampfflugzeug" (Ten Years Development of the Flying Wing High-Speed Fighter), Reel 2630, Frame 144–155, Captured German and Japanese Technical Documents on Microfilm, NASM Archives.

35. Horten and Selinger, *Nurflügel*, 118, author trans.

36. RHTT, 313–314; WROH CD 36a.

37. RHTT, 376–377.

38. WROH 42a (August 24, 1986).

39. RHTT, 290, 174, 284; Horten and Selinger, *Nurfügel*, 165, 128.

40. Horten and Selinger, *Nurflügel*, 78. Witness accounts of the altitude vary. Reimar remembers approximately 50 m (164 ft.), WROH CD 1b (April 22, 1988). Walter recalled a higher altitude, WROH CD 36a, and Nickel remembered about 5 m (16 ft.) in his draft manuscript comments, October 8, 2005.

41. Nickel to the author, May 6, 2002, 3; quote from Nickel and Wohlfahrt, *Tailless Aircraft*, 185.

42. WROH CD 36a; KGNI 4, October 7, 2005.

43. WROH CD 38b.

44. WROH, 18a, 24b, 36a, 45a; RHTT, 312–314.

45. WROH, 315.

46. Shepelev and Ottens, *Horten Ho 229*, 42.

47. WROH CD 44a; Vajda and Dancey, *German Aircraft Industry*, 80; Boog, "The Strategic Air War in Europe," 46–50.

48. RHTT, 28, 163–164, 264; WROH CD 44a, 44b; KGNI 4, October 7, 2005; Horten and Selinger, *Nurflügel*, 135, author trans.

49. RHTT, 264.

50. WROH CD 44b.

51. RHTT, 327.

52. Ibid., 166, 322; Karl Nickel correspondence to author, October 19, 2005; WROH CD 44b; Horten and Selinger, *Nurflügel*, 74, author trans.

53. RHTT, 29, 214–215, 219–220; Eric Schatzberg, *Wings of Wood, Wings of Metal: Culture and Technical Choice in American Airplane Materials, 1914–1945* (Princeton, NJ: Princeton University, 1999), 206.

54. Horten and Selinger, *Nurflügel*, 134; WROH CD 43b (August 24, 1986), 44b.

55. WROH CD 38b.

56. Horten and Selinger, *Nurflügel*, 134; KGNI 5, October 7, 2005; "Aero. 1703," 7; WROH CD 24b, 26a, 44b; RHTT, 266–269, quote from 269.

57. RHTT, 222; Green, *Warplanes of the Third Reich*, 634.

58. Author calculations based on J. Richard Smith and Eddie J. Creek, *Me 262 Volume Two* (West Sussex, England: Classic Publications, 1998), 442; Horten and Selinger, *Nurflügel*, 151, author trans. and calculations; WROH CD 8b, 24a, 47b.

59. Horten and Selinger, *Nurflügel*, 139; WROH CD 8a.

60. RHTT, 214, 218; Horten and Selinger, *Nurflügel*, 134.

61. My thanks to Smithsonian volunteer Alex Lampros for explaining in November 1994 some the challenges facing Horten when redesigning the center section.

62. Reinhold Stadtler e-mail to nurflugel@telelists.com, March 2, 1999; WROH CD 47b.

63. Horten and Selinger, *Nurflügel*, 141; WROH CD 45a; Nickel and Wohlfahrt, *Tailless Aircraft*, 358–359; WROH CD 53a; Reinhold Stadtler e-mail to nurflugel@telelists.com, March 2, 1999.

64. Peter M. Bowers, "Interesting Gliders," *Soaring*, January-February 1959, 23.

65. P. E. Peabody, "Information of development on 'Flying Wing' type planes, obtained from the brothers Horten, German pioneers in that field. British source. Received in Britain 28 June 1945," July 10, 1945, file "Captured Personnel Material B-800–826," 2, Box 431, G-2 Library "P" file, RG 165, NACP.

66. WROH CD 44a; Green, *Warplanes of the Third Reich*, 625.

67. Horten and Selinger, *Nurflügel*, 139.

68. Reinhold Stadler e-mail to author, July 17, 2005.

69. Horten and Selinger, *Nurflügel*, 134; RHTT, 168.

70. RHTT, 29.

71. Horten and Selinger, *Nurflügel*, back dust cover, 138, 144–145.

72. WROH CD 2a (April 20, 1988), 4b (April 21, 1988), 28b; "Aero. 1703," 36; Karl Nickel manuscript comments, October 19, 2005; Horten and Selinger, *Nurflügel*, unpublished manuscript, 186, 188.

73. WROH CD 3a, 45a.

74. Horten and Selinger, *Nurflügel*, 142, author trans.; WROH CD 25a, 27a, 45a; RHTT, 244–246.

75. WROH CD 50a.

76. WROH CD 16a; RHTT, 247–249. Reimar and Walter claimed that Ziller made additional flights in the *H IX V2* that he did not record in his personal logbook, see WROH CD 50a, 58a.

77. RHTT, 247, 266; WROH CD 26a.

78. WROH CD 16a.

79. "Aero. 1703," 5–6.

80. Ralf Schabel, *Die Illusion der Wunderwaffen: Die Rolle der Düsenflugzeuge und Flugabwehrraketen in der Rüstungspolitik des Dritten Reiches* (München, Germany: R. Oldenbourg, 1994), 282, Heinz Gerstle, trans.; Chris Wills, January 3, 1994, correspondence to author.

81. David Myhra, *DFS 228* (Atglen, PA: Schiffer Military History, 2000), 28.

82. Lippisch, *The Delta Wing*, 85.

83. Vajda and Dancey, *German Aircraft Industry*, 83, 116, 117; Alexander Werth, *Russia at War 1941–1945* (New York: Carroll & Graf, 1964), 229–230, 442, 900–901; WROH CD 37a.

84. Ulrich Albrecht, "Military Technology and Nation Socialist Ideology," *Science, Technology and National Socialism*, Monika Renneberg and Mark Walker, eds. (Cambridge: Cambridge University Press, 1994), 101. Albrecht assessed the Volksjäger on pp. 105–113. Horten and Selinger, *Nurflügel*, 150–153, author trans.

85. Karl Nickel correspondence to author, October 19, 2005; RHTT, 166, 325–326; Horten and Selinger, *Nurflügel*, 142; Myhra, *The Horten Brothers*, 137.

86. The only source available is contradictory on glide performance, see Horten and Selinger, *Nurflügel*, 159 and 162.

87. WROH CD 26b.

88. Horten and Selinger, *Nurflügel*, 158.

89. RHTT, 210; WROH CD 23a, 25a; Horten and Selinger, *Nurflügel*, 161, author trans.

90. WROH CD 25a, 57b; Edward Uden personal communication, September 4, 1995, proposal attached. British intelligence officers made a drawing of the *H XIII b* in 1945 or 1946, Helmut Walther, "Die vergessenen Horten-Deltas: Vom Nurflüler zum Deltajäger (Teil 1)," *Flugzeug Classic*, July 2004, 59, author trans.; RHTT, 253.

91. Horten and Selinger, *Nurflügel*, 150–151, 153, author trans.; RHTT, 253–254; WROH CD 38b. In his August 6, 2005, e-mail to the author, Reinhold Stadler reported finding no evidence that Horten officially proposed a Volksjäger design.

92. WROH CD 22b.

93. RHTT, 202; WROH CD 15b, CD 18b; Linda Hunt, *Secret Agenda: The United States Government, Nazi Scientists, and Project Paperclip, 1945 to 1990* (New York: St. Martin's Press, 1991), 82–83; Tom Bower, *The Paperclip Conspiracy: The Battle for the Spoils and Secrets of Nazi Germany* (London: Michael Joseph, 1987), 239; Michael H. Kater, "The Burden of the Past: Problems of a Modern Historiography of Physicians and Medicine in Nazi Germany," *German Studies Review*, Vol. 10, No. 1 (February 1987), 51.

94. RHTT, 210–211, 332; Horten and Selinger, *Nurflügel*, 111; n. a., "Reimar Horten," *VGC News*, No. 80, Winter 1993/4, 23. Reinhold Stadler analyzed the *H IV b* airfoil profiles described in "Aero. 1703" and reported his findings via e-mail to nurflugel@onelist.com on September 2, 1999.

95. Horten and Selinger, *Nurflügel*, 111.

96. WROH CD 3a.

97. RHTT, 312; WROH CD 28b; Karl Nickel manuscript comments, October 8, 2005.

98. Horten and Selinger, *Nurflügel*, 71; RHTT, 228.

99. Horten and Selinger, *Nurflügel*, 229–230; WROH CD 23b, 25b, 26a, 26b.

100. WROH CD 39a.

101. Horten and Selinger, *Nurflügel*, 134; WROH CD 3a (April 21, 1988), 39a, 39b.

102. Myhra, *DFS 228*, 44.

103. Al Bowers e-mail to nurflugel-l@teleport.com, July 4, 1997.

104. Myhra, *DFS 228*, 44.

Chapter 5

1. RHTT, 307. British test pilot Eric Brown wrote in his book, *Wings on My Sleeve* (Shrewsbury, England: Airlife: 1961), 82, that a British team picked up Reimar.

2. WROH CD 28a.

3. WROH CD 28a. Max passed away on June 2 after Reimar and Walter returned to England, see Nickel letter to author, December 20, 2005.

4. RHTT, 86–87, 160, 309–311, quote from 330.

5. D. C. Appleyard and M. A. Biot, *The Horten Tailless Aircraft*, May 1945, CIOS Target Number 25/157, Combined Intelligence Objectives Sub-Committee G-2 Division, SHAEF (Rear) APO 413, 4; "Aero. 1703," 4; RHTT, 308, 315–330.

6. "Aero. 1703," 6.

7. "Aero. 1703," 7.

8. RHTT, 320–321.

9. "Aero. 1703," 1.

10. KGNI 6, October 8, 2005; Nickel letter to author, December 20, 2005.

11. "Daily Activity Report—Technical Intelligence—1 June 1945," Reel 5720, Frame 0819, Air Force History Office collections, Bolling Air Force Base, MD.

12. Daily Activity Report, Frame 0580, June 15, 1945.

13. Daily Activity Report, Frame 1763, June 26, 1945.

14. Peter M. Bowers, *Boeing Aircraft since 1916* (Annapolis: Naval Institute Press, 1989), 381–382.

15. Walter J. Boyne, *The Aircraft Treasures of Silver Hill: The Behind-the-Scenes Workshop of Our Nation's Air Museums* (New York: Rawson Associates, 1982), 9.

16. RHTT, 123.

17. RHTT, 307.

18. RHTT, 301–310, 379; H. A. Taylor, *Fairey Aircraft since 1915* (Annapolis, MD: Naval Institute Press, 1974), 426; "Aero. 1703," 25, 26; Horten and Selinger, *Nurflügel*, 73, author trans., 167; Nickel letter to author, December 20, 2005.

19. Central Intelligence Agency, *The World Factbook 2003*, (Central Intelligence Agency, 2003), 23; José Fernández, ed., *The 21st Century World Atlas Millenium Edition* (Naples, FL: Trident, 2000), 420.

20. Uki Goñi, *The Real Odessa: Smuggling the Nazis to Perón's Argentina* (London: Granta Publications, 2002), 16, 63–65, 124.

21. James D. Rudolph, ed., *Argentina, A Country Study*, Foreign Area Studies, area handbook series (Washington: The American University, 1985), 52–53; Goñi, *The Real Odessa*, 122, 124, 137, 295.

22. Rudolph, *Argentina, A Country Study*, 290; Goñi, *The Real Odessa*, 128, 135; RHTT, 332, 334–35.

23. WROH CD 43a (August 24, 1986), 48b (August 1986); RHTT, 332–333; Ricardo Burzaco, *Las Alas de Perón—Aeronautica Argentina 1945/1960* (Buenos Aires: Artes Gráficas Morello, 1995), 44, author trans.

24. WROH CD 47a (August 1, 1986), 48b (August 1986).

25. Burzaco, *Las Alas de Perón—Aeronautica Argentina 1945/1960* (Buenos Aires: Artes Gráficas Morello, 1995), 99; Rudolph, *Argentina, A Country Study*, 57–58.

26. RHTT, 354.

27. WROH CD 47a (August 1, 1986).

28. RHTT, 340.

29. RHTT, 387.

30. Burzaco, *Las Alas de Perón*, 44; RHTT, 341, 368–369; WROH CD 31a.

31. Horten and Selinger, *Nurflügel*, 108, 171–174, 178, trans. and calculations by author. Horten applied the unofficial designation, *H XV a*, to the *I. Ae. 34 a*.

32. Horten and Selinger, *Nurflügel*, 174.

33. Horten and Selinger, *Nurflügel*, 171, 173, Jan Scott, trans.; n. a., "Gliding Reorganized in Argentine—Research Institute Founded," *Sailplane and Glider*, Vol. XIX, No. 11, November 1951, 262; Horten and Selinger, *Nurflügel*, 173, author trans.; Santiago Rivas and Juan C. Cicalesi, "Argentina's Jet Fighter Dreams—Pulqui, Pulqui II and IA-37/48," *International Air Power Review*, 2008, vol. 25, 164.

34. Horten and Selinger, *Nurflügel*, 172, author trans.

35. RHTT, 372.

36. n. a., "Argentine National Gliding Contest," *Sailplane and Glider*, Vol. XVIII, No. 5, November 1950, 119; Horten and Selinger, *Nurflügel*, 174–175, author trans.

37. RHTT, 377.

38. RHTT, 387.

39. Nickel correspondence to NASM volunteer John F. Fitzpatrick, January 15, 1994.

40. Arcier to Commanding General, Air Material Command, June 6, 1946, File "452.1—Airplanes—German (General) 1933–1948," Box 4035, RD 3888, RG 342, NACP.

41. TSNAC-4/BLE/emn/Intelligence to AAF OIC, Northrop Aircraft Inc., Hawthorne, CA, August 29, 1947, Cross Reference and Suspense Record, File "452.1—Gliders—Sailplanes—German 42–48," Box 4025, RD 3886, RG 342, NACP. The origin of the request to study the Horten gliders is not clear.

42. Stanley H. Hall, "Horten Tailless Sailplanes—Types III and VI," Report No. NM-14, August 30, 1948 (Northrop Aircraft Inc., Hawthorne, CA).

43. Pape and Campbell, *Northrop Flying Wings*, 246, 248. I am indebted to Dr. Richard P. Hallion for identifying other operational weaknesses in the Northrop program.

44. n. a., "Gliding Reorganized in Argentine—Research Institute Founded," *Sailplane and Glider*, Vol. XIX, No. 11, November 1951, 253; Horten and Selinger, *Nurflügel*, 175, 177, author trans.; Fernando Siarez e-mail to author, September 7, 2005; Ernst H. Hirschel, Horst Prem, and Gero Madelung, *Aeronautical Research in Germany: From Lilienthal until Today* (Berlin: Springer, 2004), 62–63.

45. KGNI 7, October 7, 2005; Fernando Siarez e-mail to author, September 7, 2005; RHTT, 338, 377.

46. "Gliding Reorganized in Argentine—Research Institute Founded," 262.

47. Horten and Selinger, *Nurflügel*, 177, 179, author trans.; n. a., "International Gliding Competition, 1952," *Sailplane and Glider*, Vol. XX, No. 6, June 1952, 8; Burzaco, *Las Alas de Perón*, 156.

48. Veronica Platt, "Olympic Impressions," *Sailplane and Glider*, Vol. XX, No. 8, August 1952, 3; Vernon Blunt, "Another Viewpoint," *Sailplane and Glider*, Vol. XX, No. 8, August 1952, 7; n. a., "International Gliding Competition, 1952," *Sailplane and Glider*, Vol. XX, No. 6, June 1952, 8.

49. Ann Welch, *The Story of Gliding*, 2nd ed. (London: John Murray, 1980), 164.

50. Chris Wills to the editor in "Mailsack . . . ," *Bungee Cord*, Vol. III #2, June 1977, 14.

51. Fred Hoinville, "Castillian Skies through Friendly Australian Eyes," *Sailplane and Glider*, September 1952, Vol. XX, No. 9, 16.

52. Ricardo Burzaco letter to author, January 25, 1996.

53. Horten and Selinger, *Nurflügel*, 179, author trans.; O. W. N., "Editorial," *Sailplane and Glider*, April 1953, Vol. XXI, No. 4, 1.

54. Veronica Platt, "Olympic Impressions," *Sailplane and Glider*, Vol. XX, No. 8, August 1952, 6; Vernon Blunt, "International Contests: Another Viewpoint—continued," *Sailplane and Glider*, Vol. XX, No. 9, September 1952, 2.

55. n. a., "Argentina Fifth National Contest," *Sailplane and Glider*, Vol. XXI, No. 7, July 1953, 20.

56. "Reimar Horten," *VGC News*, No. 80, Winter 1993/4, 23–24; Chris Wills to the editor in "Mailsack . . . ," *Bungee Cord*, Vol. III, No. 2, June 1977, 14; Reimar Horten, "The 'INAV-1,'" *Sailplane and Glider*, Vol. XXI, No. 8, August 1953, 10; Burzaco, *Las Alas de Perón*, 226.

57. Vernon Blunt, "Editorial," *Sailplane and Glider*, Vol. XX, No. 4, April 1952, 1.

58. n. a., "All-Wing Sailplane," *Flying*, November 1952, 50.

59. Simine Short, "Stamps That Tell a Story," *Gliding and Motorgliding International* (webzine), December 2002, 3; n. a., "A Memorial to Robert Kronfeld," *Aeroplane*, April 30, 1948, 494; Button letter to Disma, April 4, 1993.

60. Short, "Stamps That Tell a Story," 2; Rudi Opitz, videotape interview by author, Al Robinson videographer, July 18, 1997, National Soaring Museum, Elmira, New York, NASM author files, hereafter OVI.

61. n. a., "All-Wing Sailplane," *Flying*, November 1952, 50; Bill Gallagher, "Charlie Spratt and Rudolph Opitz Honored during 40th Annual United States Soaring Hall of Fame Induction Ceremonies at the National Soaring Museum," *NSM Historical Journal*, Vol. 16, No. 2, 1994, n. p.; n. a., "Fifth Mid-West Soaring Contest," *Soaring*, July–August 1952, 8.

62. Fred Hoinville, "The 19th American National Contest," *Sailplane and Glider*, Vol. XX, No. 10, October 1952, 10, 17.

63. E. T. Lange, "Weather at the 19th," *Soaring*, September–October 1952, 22.

64. OVI; Fred Hoinville, "The 19th American National Contest," *Sailplane and Glider*, Vol. XX, No. 10, October 1952, 11, 13; E. J. Reeves, "Texas Spirals: Bulletin of the Texas Soaring Association," September 1952, No. 55, 4; E. J. Reeves, "Soaring Surprise," *Flight Magazine*, October 1952, 20.

65. Rudi Opitz Lecture, Al Robinson videographer, July 17, 1997, Flying Wing Symposium, National Soaring Museum, Elmira, New York, hereafter OVL, NASM author files.

66. OVI.

67. OVI; n. a., "The 19th National," *Soaring*, September–October 1952, 3.

68. Quoted in Kim Rosenlof, "Horten Flying Wing: Germany's Tailless Gliders Led the Way," *Flight Journal*, June 2000, 62.

69. Lehmann, "H IV Flight Test, Brandenburg, 21 May 1943, Lehmann," Reel 3743, Frame 650, Captured German and Japanese Technical Documents on Microfilm, NASM Archives Division, Lars Urban, trans., NASM author files; OVL.

70. OVI.

71. OVI.

72. E. J. Reeves, "Soaring Surprise," *Flight Magazine*, October 1952, 20.

73. Gallager, "Charlie Spratt and Rudolph Opitz," 5; Reeves, *Texas Spirals*, 7.

74. Simine Short, "Stamps That Tell a Story," *Gliding and Motorgliding International* (webzine), February 23, 2004, 6.

75. N. a., "Here's Your Complete Contest at a Glance," *Soaring*, September–October 1952, 3.

76. Fred Hoinville, "The 19th American National Contest," *Sailplane and Glider*, Vol. XX, No. 10, October 1952, 18; David Raspet e-mail to author, July 30, 2004; Martin Simons, *Sailplanes 1945–1965* (Königsburg, Germany: Eqip, 2002), 144.

77. E. J. Reeves, "Soaring Surprise," *Flight Magazine*, October 1952, 21.

78. Vernon Blunt, "Editorial," *Sailplane and Glider*, December 1952, Vol. XX, No. 12, 1.

79. Quotes from August Raspet, "The Horten IV—Another Sailplane Catalogue Item," *Soaring*, November–December 1950, 8.

80. OVI; Button letter to Disma, April 4, 1993.

81. *Development and Procurement of Gliders in the Army Air Forces 1941–1944*, No. 47 of the USAF Historical Studies (Kansas State University: MA/AH Publishing, n. d.), 108; n. a., "Personal History of August Raspet," *Soaring*, August 1960, 11; David H. Bridges, "Of Aeronautics, Aerophysics, and Aerospace: A History of Aerospace Engineering at Mississippi State University," in *Aerospace Engineering Education during the First Century of Flight*, Barnes McCormick et al., eds. (Reston, VA: AIAA, 2004), 332–333.

82. August Raspet and Dezsö György-Falvy, "Boundary Layer Studies on the Phoenix Sailplane," *Schweizer Aero-Revue*, March 1961, 151.

83. Giorgio Zanrosso, *History and Evolution of the Glider, Volume III (1945–1998)* (Venezia, Italy: Il Cardo, 1998), 131–133. Bob Said, ed., "Ross RJK-5," *Soaring 1983 Sailplane Directory*, November 1983, 60; Martin Simon, *Sailplanes 1945–1965* (Königswinter, Germany: Eqip, 2002), 146.

84. E. J. Reeves, "Soaring Surprise," *Flight Magazine*, October 1952, 21.

85. David Bridges, "Of Aeronautics, Aerophysics, and Aerospace," 335; OVI; OVL.

86. Bruce Carmichael letter to author, October 15, 1993.

87. OVI; Dezsö George-Falvy, "I Flew the Horten IV," Web Memorial at www.dezso.org/writings/horten.html (website no longer exists).

88. OVI.

89. OVL.

90. Author calculation.

91. "Aero. 1703," 10; A. E. Slater, "Gliding Notes," *Aeroplane*, May 19, 1950, 596; Horten and Selinger, *Nurflügel*, 106, author trans.; Button letter to Disma, April 4, 1993; Reimar Horten, "Lift Distribution on Flying Wing Aircraft," *Soaring*, June 1981, 42.

92. Rosenlof, "Horten Flying Wing," 62.

93. OVI.

94. Dezsö George-Falvy, "Performance Analysis of the 'Horten IV' Flying Wing," presented at the VIII Congress of the Organisation Scientifique et Technique Internationale

du Vol a Voile, Cologne, Germany, June 1960, Acknowledgements, hereafter George-Falvy, *Performance Analysis*; Bridges, "Of Aeronautics, Aerophysics, and Aerospace," 335, 337.

95. Dezsö George-Falvy, "Memoirs—I flew the Horten IV," www.dezso.org/writings /horten.html (website no longer exists), September 28, 2005, 2, hereafter George-Falvy, "Memoirs."

96. George-Falvy, "Memoirs," 2–3.

97. Dezsö George-Falvy, "A Pilot's Notes on Flying the Horten IVa," www.dezso.org /writings/horten.html (website no longer exists), September 28, 2005, 1, hereafter George-Falvy, "Pilot's Notes."

98. George-Falvy, "Memoirs," 3.

99. George-Falvy, "Pilot's Notes," 2.

100. Quote from George-Falvy, "Pilot's Notes," 1.

101. George-Falvy, "Memoirs," 3.

102. George-Falvy, "Memoirs," 2.

103. George-Falvy, "Memoirs," 2, and *Performance Analysis*, 12.

104. George-Falvy, *Performance Analysis*, 8.

105. Reinhold Stadler, *Horten IV, Vergleich der Leistungsdaten* (Comparison of Performance Data), unpublished, January 25, 1998, 38, author trans.; Stadler letter to author, May 30, 1997.

106. George-Falvy, *Performance Analysis*, 3.

107. *H IV a* flying weights in Zacher, "Ergebnisse der Leistungsmessung . . . ," 20; and George-Falvy, *Performance Analysis*, Main Dimensions of the Horten IV. Button listed Scheidhauer as 5 ft. 3 in. tall, see Forbes letter to Button, May 21, 1950, 2. Author noted difference in sizes and weights after meeting all three pilots.

108. George-Falvy, "Memoirs," 3; Bridges, "Of Aeronautics, Aerophysics, and Aerospace," 336.

109. Horten and Selinger, *Nurflügel*, 182–183, author trans.; Karl Nickel, October 26, 2005, author telephone interview.

110. Horten and Selinger, *Nurflügel*, 182, 229, author trans.; RHTT, 358.

111. Victor Riviera, "Letters to the Editor," *T.W.I.T.T. Newsletter*, October 2002, 10.

112. KGNI 4, October 7, 2005.

113. WROH CD 31a; RHTT, 359–360.

114. Burzaco, *Las Alas de Perón*, 196–197, 198.

115. Horten and Selinger, *Nurflügel*, 182; Fernando Siarez e-mail to author, September 7, 2005; N. a., "Argentina Fifth National Contest," *Sailplane and Glider*, Vol. XXI, No. 7, July 1953, 20; José Cuadrado, "Argentine News," *Sailplane and Glider*, May/June 1954, Vol. XXII, No. 3, 5.

Chapter 6

1. Al Backstrom, "Small Is Beautiful—Midget and Submidget Sailplanes," *Soaring*, June 1980, 20. During the early 1950s, Backstrom developed a line of all-wing sport sailplanes without taper or sweep and often called "plank" wings, which nicely fit Hoinville's minimidget class.

2. Martin Simons, *Sailplanes 1945–1965*, 20, 178.

3. WROH CD 30a; RHTT, 345–346. Reinhold Platz, "A Novel Sailplane," *Sailplane and Glider*, April 1954, Vol. XXI, No. 4, 17. The Japanese hang glider appears in the same issue of *Sailplane and Glider*, 13; Backstrom, "Small Is Beautiful," 21.

4. Rogelio Bartolini, "The Flying Winglet," *Sailplane and Glider*, September/October 1954, Vol. XXII, No. 5, 20.

5. Horten and Selinger, *Nurflügel*, 216–221; Fernando Siarez e-mail to nurflugel@yahoogroups.com, October 29, 2004, Digest Number 1481, 3; Martin Simons, *Sailplanes 1920–1945*, 93.

6. Dollar value derived from www.measuringworth.org/exchangeglobal/.

7. Burzaco, *Las Alas de Perón*, 223; Rogelio Bartolini, "The Flying Winglet," 18; Rogelio Bartolini, "Test Flights of the Horten-X," *Sailplane and Glider*, March/April 1955, 15.

8. Bartolini, "Test Flights," 15.

9. Bartolini, "Test Flights," 15.

10. RHTT, 350, 371.

11. RHTT, 355.

12. RHTT, 355, 354–356; Horten and Selinger, *Nurflügel*, 212; Burzaco, *Las Alas de Perón*, 224.

13. RHTT, 374–375; Horten and Selinger, *Nurflügel*, 189–191, author trans.

14. Horten and Selinger, *Nurflügel*, 189, author trans.; Burzaco, *Las Alas de Perón*, 224; RHTT, 356–357; WROH 31b.

15. RHTT, 375. The original transcript reads "30 m," 98 ft., but this is clearly a typographical error.

16. Burzaco, *Las Alas de Perón*, 178–179, 181; Horton, *The Wing*, 65; KGNI 8, October 6, 2005.

17. Horton, *The Wing*, 66; Burzaco, *Las Alas de Perón*, 178–179; Horten and Selinger, *Nurflügel*, 198–199, author trans.; John Wegg, *General Dynamics Aircraft and their Predecessors* (Annapolis: Naval Institute, 1990), 199–201.

18. Burzaco, *Las Alas de Perón*, 179.

19. Burzaco, *Las Alas de Perón*, 180, quote on 191, 239.

20. Horten and Selinger, *Nurflügel*, 199; Burzaco, *Las Alas de Perón*, 180, 181.

21. Both quotes in RHTT, 359.

22. Fernando Siarez e-mail to author, November 18, 2005; Burzaco, *Las Alas de Perón*, 190.

23. Horten and Selinger, *Nurflügel*, 198, 207–208, 209, author trans.; Burzaco, *Las Alas de Perón*, 191, author trans.

24. Horten and Selinger, *Nurflügel*, 206, author trans.

25. Lisa North, *Civil-Military Relations in Argentina, Chile, and Peru* (Berkeley, CA: University of California, 1966), 69–70; Burzaco, *Las Alas de Perón*, 191–192.

26. Horten and Selinger, *Nurflügel*, 208, quote on 209.

27. *Pilot Training Manual for the C-47*, 84, reprinted in Len Morgan and G. G. O'Rourke, *Morgan's Aviation Hall of Fame* (New York: Arco Publishing, 1970), 97; Rudolph, *Argentina, A Country Study*, 59; Fernando Siarez e-mail, September 7, 2005.

28. n. a., "Se jubila el inventor del avión a chorro sin cola," *Los Principios*, Friday, September 1, 1978, no p. #; Peter S. Selinger, "Letter to the Editor," *T.W.I.T.T. Newsletter*, December 1994, 7.

29. Burzaco, *Las Alas de Perón*, 239, 240.

30. "Se jubila el inventor . . . ;" Etel Solingen, *Industrial Policy, Technology, and International Bargaining: Designing Nuclear Industries in Argentina and Brazil* (Stanford, CA: Stanford University, 1996), 37.

31. RHTT, 344–345, 367–368, 385.

32. RHTT, 376.

33. WROH CD 42a (August 24, 1986); RHTT, 379.

34. RHTT, 376.

35. RHTT, 365.

36. Jan Scott, "Reimar Horten Visited," *Bungee Cord,* Vol. VI, No. 2, June 1980, 6.

37. Reimar Horten, "Ala Volante Caza 'Horten IX,'" *Revista Nacional de Aeronautica*, May 1950, 19, cited in Shepelev and Ottens, *Horten Ho 229*, 116.

38. Horten and Selinger, *Nurflügel*, 136, author trans.

39. WROH CD 23b.

40. A British bomber crashed near Rotterdam and supplied the Germans with a functional H2S navigation radar, Vajda and Dancey, *German Aircraft Production*, 78.

41. Quote from WROH CD 45b; Horten and Selinger, *Nurflügel*, 134.

42. David Baker, "In Valleys of Shadow: The Black World of Stealth (Part One)" *Air International*, September 1995, Vol. 49, No. 3, 147–148.

43. Bernhard Mattlener, "The Company," 1, posted December 10, 2005, www.nurflugel.com.

44. Stadler e-mail to author, June 16, 2002.

45. Heiko Müller, "PUL-10: Hortens letzter Enwurf—Renaissance für den Nurflügel als UL," *Aerokurier*, March 1994, 64–65, author trans.; Bernhard Mattlener, "The PUL-10," *T.W.I.T.T. Newsletter*, June 1997, 7–8; KGNI 5, October 7, 2005; Miles McCallum, "Fly the Wing," *Kitplanes*, February 1993, 75–76; Stadler e-mail to author, November 28, 1997.

46. PUL-10 sales brochure, Nurflügel Flugzeugbau G.m.b.H. (aircraft construction company), ca. 1992.

47. Bernhard Mattlener, "The Company," 1–2, "Products," 3, posted December 10, 2005, www.nurflugel.com; KGNI 5, October 7, 2005; Douglas Bullard e-mail to www.nurflugel.com discussion list, May 25, 2005.

48. Karl Nickel, "Statements about some claims of Reimar Horten in his book Horten/Selinger: *Nurflügel*," September 2005, 2.

49. RHTT, 297.

50. Nickel and Wohlfahrt, *Tailless Aircraft*, 234–235, see also 47, 79, and 192 for explanation of critical gaps in the historical record that seemed to weaken the author's initial conclusion; Bowers e-mail to author, January 9, 2006.

51. Ibid., emphasis in original. See Robert T. Jones, "The Spanwise Distribution of Lift for Minimum Induced Drag of Wings Having a Given Lift and a Given Bending

Moment," Technical Note 2249 (Washington: National Advisory Committee for Aeronautics, December 1950).

52. Edward Uden, "Das Entwurfskonzept der Horten-Nurflügel" [The Design Concept of Horten Only-Wing Aircraft], Nurflügel Symposium, Oskar Ursinus Society, September 30, 1995, Scheidegg, Germany, U7–U17; Bowers personal communication, March 22, 2002.

53. Ibid., March 22, 2002; Bowers e-mail, January 9, 2006.

54. Bowers e-mail to nurflugel@yahoogroups.com, March 11, 2002, Digest Number 856; quote from Bowers e-mail to author, January 9, 2006.

55. Bowers e-mail to nurflugel@yahoogroups.com, March 11, 2002, Digest Number 856.

56. Bowers e-mail to author, March 22, 2002; Hinote e-mail to glidertech@yahoogroups.com, December 13, 2003. Bowers attendance at the July 17, 1997, Flying Wing Symposium organized by the National Soaring Museum near Elmira, New York, marked the beginning of his work to understand Horten aerodynamics.

57. Stadler e-mail to author, July 17, 2005.

58. Bowers e-mail to nurflugel@yahoogroups.com, March 11, 2002, Digest Number 856, 3. Horten wrote little on the minimum spar bending moment, see Horten and Selinger, *Nurflügel*, 98, author trans.

59. Jan Scott, "Reimar Horten Visited," 6; RHTT, 5; Reimar Horten, "Lift Distribution on Flying Wing Aircraft," *Soaring*, June 1981, 40, Jan Scott, trans.

60. KGNI 1, October 4–5, 2005; Rudi Opitz, author interview notes on paper, July 22, 1995, NASM author files.

61. Günter Brinkmann and Hans Zacher, *Die Evolution der Segelflugzeuge* (Bonn: Bernard & Graefe, 1992), Heinz Gerstle, trans., 93.

62. Chris Bryant letter to author, April 4, 1994; Eric Brown, *Wings of the Weird and Wonderful* (Shrewsbury, England: Air Life Publications, 1983), 145–149.

63. Walter G. Vincenti, *What Engineers Know and How They Know It: Analytical Studies from Aeronautical History* (Baltimore: Johns Hopkins, 1994), 78.

64. RHTT, 317–318.

65. OVI.

66. Donald J. Bush, *The Streamlined Decade* (New York: George Braziller, 1975), 3n4.

67. N.a. "Horten IV and VI—High Performance Glider Aircraft," *LUFTFAHRT International*, No. 12, November–December 1975, 1847, Lucian Heichler, trans.

68. Bernhard Mattlener effused about the H 3000 sport wing in "The Technology," 4, posted December 10, 2005, to www.nurflugel.com. I am indebted to Dr. Alan Meyer for suggesting this line of thinking. Meyer explores the relationship between pilot identity and the types of machines they choose to fly in his work on postwar private aviation. See Alan D. Meyer, "Why Fly? A Social and Cultural History of Private Aviation in Post-World War II America, 1945–1985" (PhD dissertation, University of Delaware, forthcoming), especially chapters 2 and 4.

Glossary of Terms, Phrases, and Abbreviations

A

AAF—U.S. Army Air Forces.

Adverse yaw—Tendency for an aircraft to swing sideways away from the direction of a turn as the pilot applies the ailerons to roll the aircraft into the turn, sometimes called aileron yaw.

Aerodynamic twist—Altering the distribution of lift by employing different airfoil profiles at intervals along the wingspan.

Aileron—Hinged control surface mounted near each wing tip and deflected by the pilot to maneuver an aircraft about its roll axis.

Airfoil—Curved surface designed to generate lift when moving through the air.

Airfoil profile—Outline of an airfoil section, see also *Airfoil section*.

Airfoil section—Two-dimensional cross-section of an airfoil, see *Airfoil profile*.

Airfoil thickness—Distance between the top and bottom of a wing at its thickest point.

Airframe—Underlying structural framework of an aircraft.

Akaflieg—Academic flying group, a group of aeronautical engineering students led by one or more university professors, who designed, built, and tested experimental aircraft with financial backing from the German government and the full support of a major university, which provided classrooms, workshops, and laboratories. Three were formed in 1920 at Aachen, Darmstadt, and Berlin-Charlottenburg. In 2009, there were ten.

Altimeter—Cockpit instrument that measures changes in barometric pressure to determine height above sea level.

Angle of attack—Acute angle between the chord of an airfoil and its direction of motion relative to the air.

Angle of incidence—Angle made by the wing chord relative to the long axis of the fuselage.

Aspect ratio—Wingspan divided by the average chord. Long-span wings with short chords, such as those fitted to modern sailplanes, have high-aspect ratios, while short wings with broad chords have low-aspect ratios.

AVA—Aerodynamische Versuchsanstalt, Aerodynamic Experimental Center at Göttingen, Germany.

B

Bank—Moving control surfaces such as *elevons* at the wing tips to cause differential lift that pushes one wing tip down and the other up, usually during a turn, to rotate an aircraft about its *longitudinal axis*, see *Roll.*

Bell lift distribution—Reimar Horten's name for the unique distribution of lift across the span of a wing, which he applied to the design of all-wing aircraft to improve control and stability.

Bungee cord—Rubber cable composed of individual strands, from 30 m to 70 m (100 to 200 ft.) long and covered with waterproof fabric, stretched by a launch crew to catapult a sailplane into the air.

C

Camber—Curve of an airfoil section from the *leading edge* to the *trailing edge.*

Canopy—Streamlined hatch that allows the pilot to enter and exit the cockpit, made of transparent material for maximum pilot visibility.

Cantilever wing—Wing supported only at the root without external wires or struts.

Centering a thermal—Adjusting aircraft turn radius to remain within the strongest region of lift in a thermal.

Center of gravity—Abbreviated CG, the point about which an aircraft at rest balances.

Center of pressure—Sometimes called the center of lift, a point about which a wing's lift forces balance. The relationship between the center of lift and the *center of gravity* is important to stabilize and control an aircraft.

Center section—Central portion of an all-wing aircraft housing the cockpit and main landing gear and to which are attached the left and right wing sections.

Chord—Pronounced "kord," wing width from forward leading edge to rear trailing edge.

Clen Antú—Argentine Indian term meaning "sunbeam."

Colibri—Spanish for "hummingbird."

Control surface—A device with relatively flat surface hinged to an aircraft. When deflected by the pilot, it controls the flow of air moving around it.

Coordinated turn—When pilots flying conventional airplanes use the rudder mounted at the end of the fuselage for leverage to overcome *adverse yaw*, and execute smooth curving turns, the airplane skidding neither left nor right, they are said to be executing coordinated turns. Considered an important design goal to render an all-wing aircraft capable of performing coordinated turns.

D

D-tube—Leading edge section of a wing that approximates the shape of the letter *D*, where the vertical back of the D meets the front of the main wing *spar* and the curved front forms the streamlined leading edge.

DFS—Deutsche Forschungsinstitut für Segelflug, or German Research Institute for Gliding, created by Hermann Göring after the Nazis came to power in 1933 by reorganizing, renaming, and relocating the previous organization called the Rhön-Rossitten-Gesellschaft. Göring named Professor Walter Georgii to head the new institute and moved it from atop the Wasserkuppe 115 km (72 miles) southwest to Darmstadt.

Differential gearing—Setting up the aileron control surfaces to lessen *adverse yaw* by using less aileron deflected down to roll the aircraft and more aileron deflected up. See *Frise aileron*.

Dihedral—Wing layout that resembles a flattened *V* and inclines the wing tips slightly in relation to the wing roots.

Directional stability—Tendency of an aircraft in flight to resist disturbances about the yaw axis.

DLV—Deutscher Luftsportverband, or German Airsports League, founded by Nazi decree on March 25, 1933, by combining several similar organizations, survived until 1937. See *NSFK*.

Dope—Nitro-cellulose liquid solution applied with a brush to the fabric covering an aircraft to tighten and strengthen the fabric by shrinking and to make it waterproof by filling the pores in the weave.

Drag—Friction and mass of the air that resists the movement of an aircraft in flight, often broken down further according to cause, such as induced drag formed as the byproduct of lift and parasite drag produced by various protuberances and intersections of components on an aircraft.

Drag rudder—Control surfaces mounted near each wing tip to control *yaw* when deployed one at a time by the pilot, or control aircraft speed and descent rate when deployed in unison, see *Spoiler*. Unlike an *aileron* or *elevon*, a drag rudder rests flush with the top and bottom surface of the wing until it protrudes directly into the airstream when activated by the pilot.

DTM—Deutsches Technikmuseum (German Museum of Technology) Berlin, was called the Museum für Verkehr und Technik (Museum for Transport and Technology) in 1994 when NASM lent four Horten sailplanes for restoration and preservation.

Dutch roll—Simultaneous roll and yaw of an aircraft in opposite directions caused when the pilot displaces the controls. Phenomenon resembles the technique for competitive ice, or "speed" skating, when the skater rolls his or her body in one direction while simultaneously skating in the opposite direction.

DVL—Deutsche Versuchsanstalt für Luftfahrt e. V., the German Experimental Institute for Aviation, headquartered in the Adlershof district of Berlin, equipped with laboratories, engine test stands, hangers, and wind tunnels.

Dynamic stability—Tendency of a displaced body to return to equilibrium with decreasing amplitude of motion.

E

Elevator—A pilot deflects this aircraft control surface to change pitch attitude.

Elevon—Control surface often fitted to an all-wing aircraft near each wing tip because the absence of a tail prevents use of a conventional elevator control, so named because elevons combine the effects of both the elevator and ailerons to control pitch and roll.

Empennage—Group of control surfaces clustered at the tail of a conventional aircraft, usually consisting of a rudder hinged to a vertical stabilizer for *yaw* control and an elevator hinged to a horizontal stabilizer for pitch control.

F

Favorable yaw—Tendency of an aircraft to swing sideways in the direction of a turn as a pilot rolls the aircraft, see *Adverse yaw*.

Flight ratings—Awarded to sailplane pilots in training, the A rating was earned after one flight covering 300 m (980 ft.) or lasting thirty seconds; the B rating for two flights of forty-five seconds each straight ahead, plus one sixty-second flight that described an *S* course through the air; and the C rating for flying at least five minutes without altitude loss using slope lift.

Flutter—Rapid bending and twisting vibration of a wing, control surface, or other airframe part, singly or in unison, occurring at such frequencies and phasing as to reinforce one another due to sympathetic resonance between components; often occurs at high speed and can lead to structural disintegration.

FMA—Fuerza Militar de Aviones (Fábrica Militar de Aviones), the Military Airplane Factory in Argentina.

FOD—Foreign Object Damage, any material that a jet turbine engine can suck through its intake to damage or destroy the engine.

Frise aileron—"Freeze" aileron, an aileron designed with a flattened and lowered leading edge, and hinged to protrude into the air stream when the pilot deflects the trailing edge up in order to increase drag and reduce *Adverse yaw*. Used often in conjunction with *differential gearing*.

Fuselage—Central component of conventional aircraft to which are mounted the wings and tail. Often cylindrical, and oriented with its long axis parallel to direction of flight, the fuselage usually contains the engine, cockpit, passenger cabin, and other pieces of equipment, but it produces little or no lift.

G

G—Acceleration force measurement, 1G (or Gravity) is equivalent to the pull, or force, of gravity on a stationary mass at the earth's surface. A sailplane pilot can experience multiple G-forces in turbulent air, or while maneuvering the aircraft; for example, while turning and simultaneously climbing inside a thermal.

Geometric twist—During construction, installing each wing rib on the main *spar* at varying angles of incidence to twist the wing a small amount up or down along its length, thereby imparting *washout* or *washin* and altering a wing's *lift distribution*, see *Aerodynamic twist.*

Glide—Sustained forward flight of an aircraft from an elevated point to a lower point and whose speed is maintained by descending through the air, see *Soar*.

Glide ratio—Ratio of distance travelled to altitude lost, a handy measure of sailplane performance, see *L/D*.

Glider—Motorless aircraft designed to remain airborne solely through the energy of gravity. As the term "sailplane" came to be applied more and more to gliders capable of gaining altitude, or soaring, after launch, the term "glider" was applied to motorless aircraft that could only steadily descend to a landing after launch, see *Sailplane*.

H

Habicht—"Hawk" in German.

I

IAC—Instituto Aerotécnico Córdoba, the Aeronautical Engineering Institute of Córdoba, Argentina.

I. Ae.—Instituto Aerotécnico Córdoba, or Aeronautical Engineering Institute of Córdoba in Argentina, the research arm of the state-run Military Aircraft Factory, or Fábrica Militar de Aviones, see *FMA*.

IMC—Instrument Meteorological Conditions, cloud or other weather phenomenon that prevents a pilot from using visual cues outside the cockpit, such as geographic landmarks or the horizon, to maneuver and navigate an aircraft. When flying in IMC, the pilot must rely on special instruments to prevent disorientation and loss of control and to navigate a course.

INAV—Instituto Argentino de Vuelo a Vela, the Argentine Institute of Soaring created to study the meteorology of soaring flight and to develop new sailplane designs.

Induced drag—Drag resulting from wing lift.

Inspektion der Jagdflieger—Unit within the RLM responsible for readiness, training, and tactics of the Luftwaffe fighter aircraft forces.

J

JG—Abbreviation for *Jagdgeschwader*, literally "hunting wing" in German, the wing a grouping of approximately 100 to 120 fighter aircraft.

Jungflieger—"Young Flier," the young aviation enthusiasts who joined the DLV to learn to fly and to learn aeronautics, see *DLV*.

K

KGNI—Author interviews with Karl Nickel and his wife, Gunilde (nee Horten), recorded October 4–8, 2005, on eight cassette tapes, located in Smithsonian National Air and Space Museum, Aeronautics Division, office files.

L

L'alita—"The little one" in Spanish, name for the Horten sailplane built by three men in Argentina from 1949 to 1953.

Laminar flow—Smooth airflow characterized by layers of moving air that become slower as they get closer to a wing's surface. A typical drawing depicts the different speeds as layers, or laminations. An airfoil with extensive laminar flow exhibits less drag and better performance than an airfoil without it.

Lateral axis—Also called the pitch axis about which an aircraft's nose moves up and down.

Lateral instability—Once disturbed by an outside force such as turbulence, the tendency for an aircraft to continue to rotate about its longitudinal, or roll, axis.

L/D—"L over D," the ratio of lift-to-drag of an aircraft where L (lift) is distance traveled horizontally for each unit of D (drag), or altitude lost. A handy and useful measure of sailplane performance. Both *glide ratio* and L/D refer to the same values.

Leading edge—Edge of wing or control surface closest to the normal direction of flight, opposite of *trailing edge*.

Lift coefficient—Measure of how efficiently a wing, at a certain angle of attack, converts the air flowing around it into lift.

Lift coefficient, maximum—The *lift coefficient* at the maximum angle of attack beyond which a wing stalls and loses lift.

Lift distribution—Intensity and location of lift across the surface of a wing, governed by such variables as *aerodynamic twist*, *geometric twist*, *washout*, and *washin*.

Load—Force applied to an aircraft, or aircraft component, from wind gusts, G-force, and the like, see *G*.

Longitudinal axis—Also called the *roll* axis, the aircraft rotates around this axis when the pilot *banks*, as for a turn.

Luftwaffe—German air force from 1933 to 1945.

M

Mach—When followed by a number, Mach compares aircraft speed to the speed of sound, which varies with altitude. Mach 1 is 1,124 km/h (698 mph) at 6,080 m (20,000 ft.).

Mass balance—Weight (usually lead cast into a streamlined shape) added to a control surface ahead of its hinged axis to prevent *flutter*, see *Static balance*.
Mitteneffekt—"Middle effect," term coined by Reimar Horten for the lift he mistakenly believed was lost along the center section of many of his wings. Only after World War II did he realize that Mitteneffekt was a false indication caused by the method he used to calculate the *lift distribution*, which did not account for wing sweep.
Moment—Tendency, or measurement, of a body to move about an axis or point.
Moment arm—Perpendicular distance from the point of application of force to the axis or point of rotation.
Museum für Verkehr und Technik Berlin—Museum for Transport and Technology, in 1994 borrowed four Horten sailplanes from NASM to restore and preserve, see *DTM*.

N

NACA—National Advisory Committee for Aeronautics, the aeronautical research establishment operated by the USA, renamed the National Aeronautics and Space Administration, or NASA, in April 1958.
NACP—National Archives of the USA located at College Park, Maryland.
NASM—Smithsonian Institution's National Air and Space Museum. Main building located on the National Mall in Washington, D.C., branch facilities located in Maryland and Virginia.
NSFK—Nationalsozialistisches Fliegerkorps, the National Socialist Flier Corps, created in April 1937 after the Nazis dissolved the *DLV*. The NSFK absorbed all sport flying groups in Germany.

O

Off-field landing—Also called "landing out," the term is used when a sailplane pilot is not close to an airport and has to land on the nearest open ground, whether in a farm field, along a road, or other reasonably flat surface.
OSTIV—Organisation Scientifique et Technique du Vol á Voile, the international group that encourages and coordinates developing "the science and technology of soaring and the development and use of the sailplane in pure and applied research," as stated in article 2 of the OSTIV constitution.
OVI—Videotaped author interview with Rudi Opitz, National Soaring Museum, Elmira, New York, July 18, 1997, Al Robinson videographer, preserved in the NASM Archives.

OVL—Videotaped Rudi Opitz lecture during Flying Wing Symposium, National Soaring Museum at Elmira, New York, July 17, 1997, Al Robinson videographer, preserved in the NASM Archives.

P

P—*Planeador*, Spanish for "glider," as in the *I. Ae. 37-P*, manned delta wing glider built by Horten's team in 1954 entirely of wood with a semiprone cockpit.

Parabel—German for "Parabola," name given by Reimar Horten to a wing he built in 1939 with a continuously curved leading edge; did not fly.

Piernifero—Spanish term meaning "leg-trained," refers to several very light Horten sailplanes built in Argentina, designed for maneuver on the ground and launched using the pilot's legs.

PUL—Panek-Ultra-Light, aircraft design that resulted from collaboration that began in 1987 between Reimar Horten, Siegfried Panek, and his partner, Bernhard Mattlener, with the goal to design a powered sport all-wing aircraft. Three examples of the PUL-10 followed the PUL-9 that first flew in 1990.

R

RCS—Radar cross-section, the detectable radar energy reflected by an object; objects that reflect a small amount of detectable energy are said to have a small RCS.

Reflex airfoil—Airfoil, or profile, with trailing edge curved, or reflexed, upward in order to increase stability and reduce a wing's tendency to pitch down; for example, while accelerating in a dive.

Rhön Mountains—Range of low mountains in central Germany running southwest to northeast and intersecting the borders of the states of Hesse, Bavaria, and Thuringia. *Wasserkuppe* is the highest peak.

RHTT—Reimar Horten Oral History Tape Transcript of the first nineteen cassette tapes, David Myhra interviews with Reimar Horten, 388 typewritten pages but not dated.

Rib—A wooden, airfoil-shaped framework fixed transversely to the wing *spar* to maintain the wing's shape and distribute aerodynamic loads to the spar.

RLM—Reichsluftfahrtministerium, the German Air Ministry, established by Hitler in April 1933.

Roll—See *Bank*.

Root—See *Wing Root*.

Rotate—To raise the nose of an aircraft by manipulating the control surfaces, usually during takeoff.

Royal Aircraft Establishment—Leading aeronautical research center in Britain, located at the Farnborough Airfield about 50 km (30 miles) southwest of London.

S

Sailplane—Motorless aircraft designed to use various forms of lift to maintain or increase altitude after launch, see *Glider*.

Ship—Slang for sailplane, derived from the notion that sailplanes and marine sailing vessels both derive their motive power by "sailing" on the wind.

Side Slip—Motion of an aircraft opposite of skidding, the pilot yaws aircraft opposite the direction of bank; useful for losing altitude without increasing speed or when landing in a crosswind requires the aircraft to track straight down the runway while crabbing into the wind.

Sitzung Nurflügelflugzeuge—Meeting of Only-the-Wing Aircraft held at Berlin in April 1943, sponsored by the Lilienthal-Gesellschaft für Luftfahrtforschung (Lilienthal Society for Aviation Research), attended by scientists, engineers, and mathematicians to assess the current state of all-wing aircraft design; Horten brothers presented two papers.

Skid-roll, also *Slide-roll*—Roll of an aircraft coupled with simultaneous *yaw* in the same direction, as if to turn, often the desirable result of building *dihedral* and *sweepback* into the wing.

Slide-roll moment—Aerodynamic force that simultaneously yaws and rolls a wing.

Soar—Whereas a glide is a steadily descending flight, to soar requires a pilot to use various forms of lift to maintain altitude or climb. Soaring makes it possible to attain great heights and distances. See *Glide*.

Sonderkommando L In 3—Special Command attached to Luftwaffen-Inspektion 3, or Air Force unit Inspection 3 (abbreviated L In 3 on the *RLM* organization chart). L In 3 also called Inspektion der Jagdflieger, the unit within the RLM responsible for readiness, training, and tactics of the Luftwaffe fighter aircraft forces.

Spar—Often extends from wing tip to wing tip, primary internal structure supporting a wing much as the keel supports a ship. Usually rectangular cross-section and made solid of wood or metal, but it is sometimes hollow and referred to as a box spar.

Spin—Loss of aircraft control following a *stall* when half the wing remains stalled. Resulting dissymmetry causes the aircraft to fall in a spinning dive.

Spoiler—Control surfaces fitted to a wing that reduce, or spoil, lift. Deployed in unison by a pilot, they adjust aircraft speed or descent rate, but deployed singly, they cause an aircraft to *yaw*, see *Drag rudder*.

Stall—Occurs when a pilot has reduced the speed of an aircraft until there is not enough air flowing over the wing to produce lift. Many aircraft must be stalled to touch down during landing; therefore, the pilot must be able to maintain control during a stall.

Stall speed—Airspeed at which a wing *stalls*.

Static balance—Control surface that balances about its hinge axis when not subjected to aerodynamic forces.

Static longitudinal stability—Tendency for an aircraft to restore itself to steady level flight after a disturbance along its pitch axis, whether from a pilot moving the controls, encountering turbulence, or shifting weight. Important for all-wing aircraft lacking the tail and fuselage that stabilizes conventional aircraft.

Sweepback—Backward slant of the leading or trailing edge of an airfoil.

Swept wing—Wing with leading and trailing edges slanted backward.

Symmetrical airfoil—An *airfoil* whose top and bottom surfaces have identical shapes.

T

Taper ratio—Ratio of decrease in wing chord along the entire wingspan, measured at intervals starting at the wing root and ending at the wing tip.

Task—Problem or challenge assigned to pilots competing in a soaring contest, from simple duration aloft and distance flown to speed around a designated triangle, or speed and distance flown within a prescribed area. Prevailing weather conditions usually determine the task selected.

Trailing edge—Edge of a wing or control surface that is farthest from the normal direction of flight, see *Leading edge*.

Trim—Adjusting the aircraft controls or installing weights to adjust the *center of gravity* in order to hold a desired airspeed or flight attitude.

Tuft-testing—Method of visualizing airflow by gluing up to several hundred tufts of wool, cloth, or similar material at spaced intervals onto some portion of the surface of an aircraft, then flying and observing how airflow moves the tufts. Tufts sometimes mounted on short masts to reveal flow pattern in different regions.

Turnbuckle—Device connecting the threaded ends of two cables. Rotating the turnbuckle draws the cables together or pushes them apart to adjust length in fine increments, yet the joint remains solid.

U

Urubú—"Vulture" in the language of an Indian tribe in Argentina.

V

Variometer—Instrument installed in sailplanes to detect minute changes in rate of climb or descent; aids pilot in finding lift.

Vertical axis—Axis of control passing through the top and bottom of an aircraft around which an aircraft is said to *yaw* either left or right. See *Lateral axis*, *Longitudinal axis*, and *Yaw axis*.

Vertical fin—Device mounted at the tail of a conventional aircraft to stabilize it about the vertical, or yaw, axis.

W

Washin—Wing twist that increases the angle of attack at the wing tip, causing the wing to stall at the wing tip and rendering the *elevons* useless to keep the wings level.

Washout—Wing twist that deceases the angle of attack at the wing tip, causing the wing to stall first at the wing root rather than the wing tip, where the *elevons* can continue to provide roll control.

Wasserkuppe—"Water peak" in German for the fog and rains that regularly soak the area, at 950 m (3,100 ft.) the highest peak in the *Rhön Mountains* in central Germany. The Wasserkuppe is the geographic and cultural epicenter of the German soaring movement between World Wars I and II and site of the German national soaring contest from 1920 to 1938.

Werk Nummer—Pronounced "vairk noomer," manufacturer's serial number, abbreviated W. Nr.

Wing root—Base of a wing where it joins the fuselage.

WROH—Walter and Reimar Horten Oral History Interviews conducted by David Myhra from 1982–1988 and recorded on sixty original cassette tapes, converted to CDs and preserved in NASM Archives, Washington, D.C.

Y

Yaw—Aircraft movement side-to-side about the *vertical axis*, as when the pilot changes the direction of flight.

Yaw axis—Another name for *vertical axis*.

References

UNLESS NOTED otherwise, all original source documents, personal communications, and personal interviews are located in the author's files, Aeronautics Division, National Air and Space Museum (NASM), Smithsonian Institution, Washington, D.C. Note that the author read many of the German documents in translation.

Unpublished Sources

Draft manuscript, Horten, Reimar, and Peter F. Selinger, *Nurflügel: Die Geschichte der Horten-Flugzeuge 1933–1960.*

Horten, Walter. "Eidesstattliche Erklärung," (Personal Statement). December 8, 1961.

Nickel, Karl. "Memorandum, Die Horten-Nurflügel-Flugzeuge, Erinnerungen von Karl Nickel" (The Horten All-Wing Aircraft, Memories of Karl Nickel), February 1, 2001, access restricted.

———. "Statements about some claims of Reimar Horten in his book Horten/Selinger: *Nurflügel*," September 2005.

———. "Statements about some claims of Reimar Horten in his book Horten/Selinger: *Nurflügel*, Addendum 1, 29 September 2005."

———. Comments on draft manuscript, *Only the Wing: Reimar Horten's Epic Quest to Stabilize and Control the All-Wing Aircraft*, October 8, 2005.

Scheidhauer, Heinz. Flight logbook, copies of pages relevant to Horten *H VI V1* sailplane, anonymous source.

Stadler, Reinhold. *Horten IV, Vergleich der Leistungsdaten* (Comparison of Performance Data), January 25, 1998.

———. "Die Auslegung der Horten-Nurflügelflugzeug" (Interpretation of the Horten Only-Wing Aircraft), presented at the twelth Nurflügelsymposium, Scheidegg, Germany, October 2–3, 1999.

Uden, Edward. "Das Entwurfskonzept der Horten-Nurflügel" [The Design Concept of Horten Only-Wing Aircraft], Nurflügel Symposium, Oskar Ursinus Society, Scheidegg, Germany, September 30, 1995.

Public Records

Arcier to Commanding General, Air Material Command, June 6, 1946, File "452.1—Airplanes—German (General) 1933–1948," Box 4035, RD 3888, RG 342, NACP.

"Daily Activity Report—Technical Intelligence—1 June 1945." Reel 5720, Frame 0819, Air Force History Office collections, Bolling Air Force Base, MD.

Foreign Records Seized Collection, Berlin Document Center Microfilm, NSDAP Zentralkartei, A3340-MFKL-J082 and NSDAP Ortsgruppenkartei, A3340-MFOK-I066, RG 242, NACP.

Harris, L. M., and J. T. Grey. "A Survey of German Chemical and Wood Industries Producing Plastic and Wood Materials for Aircraft Tooling and Fabrication," FIAT Final Report No. 446, U.S. Technical Industrial Intelligence Committee, Aeronautics Subcommittee, July 27–August 1945, NASM Archives Division, file "S0003510 Germany/WW II Industry #2."

Gebruder Horten (Horten brothers, Reimar, and Walter). "Stabilitätsbetrachtungen an Pfeilflügeln," 164/Vorbericht: *Flugzeugkonstruktion. Bericht über die Sitzung Nurflügelflugzeuge*, Lilienthal-Gesellschaft für Luftfahrtforschung, am 14 April 1943, Berlin (Considerations on the Stability of Arrow Wings, Preliminary Report 164: Airplane Construction, Report of the Flying Wings Meeting, sponsored by the Lilienthal Society for Aeronautics Research, April 14, 1943, Berlin), Reel 2630, Frame 126–139, Captured German and Japanese Technical Documents on Microfilm, NASM Archives Division.

———. "Zehn Jahre Entwicklung am Nurflügel-Schnellkampfflugzeug" (Ten Years Development of the Flying Wing High-Speed Fighter), Reel 2630, Frame 144–155, Captured German and Japanese Technical Documents on Microfilm, NASM Archives Division.

Lehmann, "H IV Flight Test, Brandenburg, 21 May 1943, Lehmann," Reel 3743, Frame 650, Captured German and Japanese Technical Documents on Microfilm, NASM Archives Division.

Museum für Verkehr und Technik—Berlin. Inventarisations-und Restaurierungszeichnungen Flugzeugtypen (Inventory of Aircraft Drawings): Horten II, Zeichn.-Nr.: 2.00–B11, June 11, 1996, and Nr.: 4.08.009–020 dated July 8, 1996, NASM Aeronautics Division.

Peabody, P. E. "Information of development on 'Flying Wing' type planes, obtained from the brothers Horten, German pioneers in that field. British source. Received in Britain 28 June 1945." July 10, 1945, file "Captured Personnel Material B-800–826," 2, Box 431, G-2 Library "P" file, RG 165, NACP.

Reichert, Carl E. "Summation of All-Wing Aircraft," 4, Aircraft Laboratory, Engineering Division, U.S. Army Air Forces Materiel Command, March 22, 1948, File "452.1—All-Wing Aircraft, 1948," Box 3720, R and D Case Files, RG 342, NACP.

TSNAC-4/BLE/emn/Intelligence to AAF OIC, Northrop Aircraft Inc., Hawthorne, CA, August 29, 1947, Cross Reference and Suspense Record, File "452.1—Gliders—Sailplanes—German 42–48," Box 4025, RD 3886, RG 342, NACP.

Published Sources

Books

Aird, Forbes. *Racer's Encyclopedia of Metals, Fibers & Materials*. Osceola, WI: Motorbooks, 1994.

Albrecht, Ulrich. "Military Technology and Nation Socialist Ideology." *Science, Technology and National Socialism*, eds. Monika Renneberg and Mark Walker. Cambridge: University Press, 1994.

Amtmann, Hans. *The Vanishing Paperclips*. Boylston, MA: Monogram, 1988.

Anderson, John D. *A History of Aerodynamics and Its Impact on Flying Machines*. Cambridge: University Press, 1998.

Barnaby, Ralph S. *Gliders and Gliding: Design Principles, Structural Features, and Operation of Gliders and Soaring Planes*. New York: Ronald Press, 1930.

Barnes, C. H. *Shorts Aircraft since 1900*. London: Putnam, 1989.

Blenk, Hermann, and Werner Schulz. *Jahrbuch 1956: Der Wissenschaftlichen Gesellschaft für Luftfahrt E. V.* Braunschweig, Germany: Friedrich Vieweg & Sohn, 1957.

Bower, Tom. *The Paperclip Conspiracy: The Battle for the Spoils and Secrets of Nazi Germany*. London: Michael Joseph, 1987.

Bowers, Peter. *Modern Soaring Guide*. Blue Ridge Summit, PA: TAB Books, 1979.

Bridges, David H. "Of Aeronautics, Aerophysics, and Aerospace: A History of Aerospace Engineering at Mississippi State University," *Aerospace Engineering Education during the First Century of Flight*, eds. McCormick, Barnes, Conrad Newberry, and Eric Jumper, Reston, VA: AIAA, 2004.

Brinkmann, Günter, and Hans Zacher. *Die Evolution der Segelflugzeug*. Bonn: Bernard & Graefe, 1992.

Brown, Eric. *Wings on My Sleeve*. Shrewsbury, England: Airlife, 1961.

Brütting, G. "Hofmann's Flight through Four Local Thunderstorms." Ed. Wolf Hirth, *The Art of Soaring Flight*, trans. Naomi Heron-Maxwell. Stuttgart: Vaihingen-Stuttgart, 1938.

Burzaco, Ricardo. *Las Alas de Perón-Aeronautica Argentina 1945/1960*. Buenos Aires: Artes Gráficas Morello, 1995.

Bush, Donald J. *The Streamlined Decade*. New York: George Braziller, 1975.

Central Intelligence Agency. *The World Factbook 2003*.

Chanute, Octave. *Progress in Flying Machines*. New York: The American Engineer, 1894.

Corn, Joseph J. *The Winged Gospel: America's Romance with Aviation, 1900–1950*. New York: Oxford University Press, 1983.

Development and Procurement of Gliders in the Army Air Forces 1941–1944, No. 47 of the USAF Historical Studies (Kansas State University: MA/AH Publishing).

de Vries, John A. *Taube, Dove of War*. Temple City, CA: Historical Aviation Album, 1978.

Evans, Richard J. *The Coming of the Third Reich*. New York: Penguin, 2004.

———. *The Third Reich in Power 1933–1939*. New York: Penguin, 2005.

Ferguson, Eugene S. *Engineering and the Mind's Eye*. Cambridge, MA: MIT Press, 1994.

Forest Products Laboratory. *Wood Handbook: Wood as an Engineering Material*. Reprint ed. Madison, WI: Forest Products Society, 1999.

Francillon, René. *Lockheed Aircraft since 1913*. Annapolis: Naval Institute, 1987.

Fritzsche, Peter. *A Nation of Fliers: German Aviation and the Popular Imagination*. Cambridge, MA: Harvard, 1992.

Galland, Adolf. *The First and the Last*. New York: Ballantine, 1954.

Goñi, Uki. *The Real Odessa: Smuggling the Nazis to Perón's Argentina*. London: Granta Publications, 2002.

Green, William. *Warplanes of the Third Reich*. Garden City, NY: Doubleday, 1972.

Horten, Reimar, and Peter F. Selinger. *Nurflügel: Die Geschichte der Horten-Flugzeuge 1933–1960*. Graz, Germany: H. Weishaupt, 1983.

Horton, V. R. *The Wing, (Selected Articles)* FTD-ID(RS)T-0870–83, trans. Leo Tanner Association, Foreign Technology Division, Translation Division, Wright-Patterson Air Force Base, OH: U.S. Air Force Systems Command, July 25, 1983.

Hunt, Linda. *Secret Agenda: The United States Government, Nazi Scientists, and Project Paperclip, 1945 to 1990*. New York: St. Martin's, 1991.

Hurt, H. H. *Aerodynamics for Naval Aviators*. Revised ed. Aviation Maintenance Publishers, 1965.

Krauss, Serge. *Tailless Aircraft: An Extensive Bibliography for Subsonic Types*. Sixth ed. Cleveland Heights, OH: Serge Krauss, 1999.

Launius, Roger D. *Innovation and the Development of Flight*. College Station: Texas A & M, 1999.

Lippisch, Alexander. *The Delta Wing: History and Development*. Ames: Iowa State, 1981.

McFarland, Marvin W. ed., *The Papers of Wilbur and Orville Wright, Including the Chanute-Wright Letters and Other Papers of Octave Chanute, Volume One: 1899–1905*. New York: Arno, 1953.

Mosley, Leonard. *The Reich Marshall: A Biography of Hermann Goering*. Garden City, NY: Doubleday, 1974.

Munson, Kenneth. *Airliners between the Wars 1919–1939*. London: Blandford, 1972.

Myhra, David. *The Horten Brothers and Their All-Wing Aircraft*. Atglen, PA: Schiffer, 1998.

———. *DFS 228*. Atglen, PA: Schiffer, 2000.

Nickel, Karl, and Michael Wohlfahrt. *Tailless Aircraft in Theory and Practice*. Wright Patterson Air Force Base, OH: Air Force Institute of Technology, and Reston, VA: American Institute of Aeronautics and Astronautics, 1994.

North, Lisa. *Civil-Military Relations in Argentina, Chile, and Peru*. Berkeley, CA: University of California, 1966.

Pagen, Dennis. *Understanding the Sky: A Sport Pilot's Guide to Flying Conditions*. Spring Mills, PA: Sport Aviation, 1992.

Pape, Garry R., and John M. Campbell. *Northrop Flying Wings: A History of Jack Northrop's Visionary Aircraft*. Atglen, PA: Schiffer, 1995.

Penrose, Harald. *British Aviation: The Pioneer Years*. London: Putnam, 1967.

Riedel, Peter. *Über sonnige Weiten: Erlebte Rhöngeschichte 1933–1939*. Stuttgart: Motorbuch, 1985.

Rudolph, James D., ed., *Argentina, A Country Study*. Foreign Area Studies, area handbook series, Washington, DC: American University, 1985.

Schabel, Ralf. *Die Illusion der Wunderwaffen: Die Rolle der Düsenflugzeuge und Flugabwehrraketen in der Rüstungspolitik des Dritten Reiches*. Munich: R. Oldenbourg, 1994.

Schatzberg, Eric. *Wings of Wood, Wings of Metal: Culture and Technical Choice in American Airplane Materials, 1914–1945*. Princeton, NJ: Princeton, 1999.

Schmitt, Günter. *Hugo Junkers and His Aircraft*. Berlin: VEB Verlag für Verkehrswesen, 1988.

Shepelev, Andrei, and Huib Ottens. *Horten Ho 229, Spirit of Thuringia: The Horten All-Wing Jet Fighter*. Hersham, Surrey, England: Classic Publications, 2006.

Simons, Martin. *The World's Vintage Sailplanes 1908–1945*. Melbourne: Kookaburra Technical Publications, 1986.

———. *Sailplanes 1920–1945*. Königsburg, Germany: Eqip, 2001.

———. *Sailplanes 1945–1965*. Königsburg, Germany: Eqip, 2002.

Smith, J. Richard, and Eddie J. Creek. *Me 262*. Vol. 2. West Sussex, England: Classic Publications, 1998.

Solingen, Etel. *Industrial Policy, Technology, and International Bargaining: Designing Nuclear Industries in Argentina and Brazil*. Stanford, CA: Stanford University, 1996.

Stewart, Ken. *The Glider Pilot's Manual*. Shrewsbury, England: Airlife, 2001.

Sweetman, Bill. *Stealth Bomber: Invisible Warplane, Black Budget*. Osceola, WI: Motorbooks, 1989.

Taylor, H. A. *Fairey Aircraft since 1915*. Annapolis: Naval Institute, 1974.

Teale, Edwin W. *The Book of Gliders*. New York: E. P. Dutton, 1930.

The 21st Century World Atlas Millenium Edition. Naples, FL: Trident, 2000.

United States Army Air Forces. *Pilot Training Manual for the C-47*. Rep. Len Morgan and G. G. O'Rourke, *Morgan's Aviation Hall of Fame*. New York: Arco Publishing, 1970.

Vajda, Ferenc, and Peter Dancey. *German Aircraft Industry and Production 1933–1945*. Warrendale, PA: Society of Automotive Engineers, 1998.

Vincenti, Walter G. *What Engineers Know and How They Know It: Analytical Studies from Aeronautical History*. Baltimore: Johns Hopkins, 1994.

Wegg, John. *General Dynamics Aircraft and Their Predecessors*. Annapolis: Naval Institute, 1990.

Welch, Ann. *The Story of Gliding*, 2nd ed. London: John Murray, 1980.

Wilkinson, Kenneth G. *The Horten Tailless Aircraft*. Royal Aircraft Establishment, Farnborough Hants, Class number 629.13.014.48 (43) Horten, "R. A. E. Report No. F. A. 259/1, Technical Note No. Aero. 1703," October 1945.

Woodward, Betsy, and Alex Stirnemann. *The World's Sailplanes*. OSTIV & Schweizer Aero-Revue, 1958.

Wooldridge, E. T. *Winged Wonders: The Story of the Flying Wings*. Washington, D.C.: Smithsonian Institution, 1983.

Zanrosso, Giorgio. *History and Evolution of the Glider, Volume I (1773–1914)*. Vicenza, Italy: Egida, 1996.

———. *History and Evolution of the Glider, Volume III (1945–1998)*. Venezia, Italy: Il Cardo, 1998.

———. *History and Evolution of the Glider, Volume IV (1957–2000)*. Vicenza, Italy: La Serenissima, 2000.

Journal and Magazine Articles, Papers, and Reports

"A Memorial to Robert Kronfeld." *Aeroplane*, April 30, 1948, 494.

"All-Wing Sailplane." *Flying*, November 1952, 50.

Appleyard, D. C., and M. A. Biot. "The Horten Tailless Aircraft—CIOS Target Number 25/157." London: His Majesty's Stationary Office, May 1945.

Backstrom, Al. "Small Is Beautiful—Midget and Submidget Sailplanes." *Soaring*, June 1980, 20.

Baker, David. "In Valleys of Shadow: The Black World of Stealth (Part One)." *Air International*, September 1995, 147.

Bartolini, Rogelio. "The Flying Winglet." *Sailplane and Glider*, September/October 1954, 20.

———. "Test Flights of the Horten-X." *Sailplane and Glider*, March/April 1955, 15.

Blunt, Vernon. "Editorial." *Sailplane and Glider*, Vol. XX, No. 4, April 1952, 1.

———. "Another Viewpoint." *Sailplane and Glider*, August 1952, 7.

———. "International Contests: Another Viewpoint—continued." *Sailplane and Glider*, September 1952, 2.

———. "Editorial." *Sailplane and Glider*, December 1952, 1.

Bowers, Peter M., "Interesting Gliders." *Soaring*, January–February 1959, 23.

Cuadrado, José. "Argentine News." *Sailplane and Glider*, May–June 1954, 5.

Ewald, Bernd. "Will We Have a Horten Flying Wing in the Air Again?" *VGC News*, Winter 2001, 21.

Ewald, Jochen. "News from the Research Students: Idaflieg Summer Meeting 2005." *Quiet Flyer Magazine*, August 2006, 76.

"Fifth Mid-West Soaring Contest." *Soaring*, July–August 1952, 8.

"Full-Sized Wind Tunnels." *Aeroplane*, January 29, 1930, 167.

Gallagher, Bill. "Charlie Spratt and Rudolph Opitz Honored during 40th Annual United States Soaring Hall of Fame Induction Ceremonies at the National Soaring Museum." *NSM Historical Journal* 16 (1994): 4–5.

George-Falvy, Dezsö. "Performance Analysis of the 'Horten IV' Flying Wing." Presented at VIII Congress of Organisation Scientifique et Technique Internationale du Vol a Voile, Cologne, Germany, June 1960.

"Gliding Reorganized in Argentine—Research Institute Founded." *Sailplane and Glider*, November 1951, 253.

Hall, Stanley H. "Horten Tailless Sailplanes—Types III and VI," Report No. NM-14. Northrop Aircraft Inc., Hawthorne, CA, August 30, 1948.

Handiwerk, Bertrand. "The Modified Horten IV of the AHQ BAFO Club." *Soaring*, November–December 1950, 9.

Harris, L. M., and J. T. Grey. "A Survey of German Chemical and Wood Industries Producing Plastic and Wood Materials for Aircraft Tooling and Fabrication," FIAT Final Report No. 466. U.S. Technical Industrial Intelligence Committee, Aeronautics Subcommittee, July 27–August 1945.

Helbig, Hans. "Zwei Preise." *Luftfahrt und Schule*, August 1938, 245.

"Here's Your Complete Contest at a Glance." *Soaring*, September–October 1952, 3.

Hoinville, Fred. "Castillian Skies through Friendly Australian Eyes." *Sailplane and Glider*, September 1952, 16.

———. "The 19th American National Contest." *Sailplane and Glider*, October 1952, 10, 17.

Hoinville, Grace. "Midget Sailplanes—Why Not?" *Sailplane and Glider*, November–December 1954, 10–11.

"Horten IV and VI High Performance Glider Aircraft." *LUFTFAHRT International*, 1847–1870.

Horten, Reimar. "Das Nurflügelflugzeug." *Luftfahrt und Schule*, August 1938, 244–246.

———. "Nurflügel-Jagdflugzeug 'Horten IX.'" Part 1. *Der Flieger*, 1950.

———. "The 'Horten-Parabola,'" trans. G. S. Neumann. *Sailplane and Glider*, August 1951, 186.

———. "Von der 'Ho-IV' zur 'Ho-VI.'" *Thermik*, March 1952, 33.

———. "The 'INAV-1.'" *Sailplane and Glider*, August 1953.

———. "Lift Distribution on Flying Wing Aircraft." *Soaring*, June 1981, 42.

Horten, Reimer, and Walter Horten. "Zehn Jahre Entwicklung am Nurflügel-Schnellkampfflugzeug." Lilienthal-Gesellschaft für Luftfahrtforschung, Berlin, April 14, 1943.

———. "Plastic, All-Wing Horten Aircraft." *Flugsport*, trans. H. I. Lewenz, in R. T. P. 3 Translation No. 2337, issued by the Ministry of Aircraft Production, London, December 15, 1943, 249–251.

"International Gliding Competition, 1952." *Sailplane and Glider*, June 1952, 8.

Jones, Robert T. "The Spanwise Distribution of Lift for Minimum Induced Drag of Wings Having a Given Lift and a Given Bending Moment." Technical Note 2249, Washington, DC, National Advisory Committee for Aeronautics, December 1950.

Lange, E. T. "Weather at the 19th." *Soaring*, 22.

Lippisch, Alexander. "Bestimmung der Auftriebsverteilung längs der Spannweite," (Definition of the Lift Distribution along the Wingspan), part 1. *Flugsport*, December 20, 1933, 549.

———. "Bestimmung der Auftriebsverteilung längs der Spannweite," part 3. *Flugsport*, February 7, 1934, 49.

Mattlener, Bernhard. "The PUL-10." *T.W.I.T.T. Newsletter*, June 1997, 7.

McCallum, Miles. "Fly the Wing." *Kitplanes*, February 1993, 75.

"Model Sailplanes in Competition." *Sailplane and Glider*, June 1936, 101.

Müller, Heiko. "PUL-10: Hortens letzter Enwurf—Renaissance für den Nurflügel als UL" (Horten's Last Design—Renaissance for the Only-Wing as an Ultralight). *Aerokurier*, March 1994, 64.

O. W. N. "Editorial." *Sailplane and Glider*, April 1953, 1.

Platt, Veronica. "Olympic Impressions." *Sailplane and Glider*, August 1952, 3.

Platz, Reinhold. "A Novel Sailplane." *Sailplane and Glider*, April 1954, 17.

Prandtl, Ludwig. "Über Tragflügel kleinsten induzierten Widerstandes," (About Wings of Smallest Induced Drag). *Zeitschrift für Flugtechnik und Motorluftschiffahrt*, December 28, 1932, 305.

PUL-10 aircraft sales brochure produced by Nurflügel Flugzeugbau G.m.b.H. (aircraft construction company) ca. 1992, NASM Archives Division.

Raspet, August. "The Horten IV—Another Sailplane Catalogue Item." *Soaring*, November–December 1950, 8.

———, and György-Falvy, Dezsö. "Boundary Layer Studies on the Phoenix Sailplane." *Schweizer Aero-Revue*, March 1961, 151.

Reeves, E. J. *Texas Spirals: Bulletin of the Texas Soaring Association*. September 1952, 4.

———. "Soaring Surprise." *Flight Magazine*, October 1952, 20.

"Reimar Horten." *VGC News*, Winter 1993/4, 23–24.

Rosenlof, Kim. "Horten Flying Wing: Germany's Tailless Gliders Led the Way." *Flight Journal*, June 2000, 62.

Said, Bob, ed. "Ross RJK-5." *Soaring 1983 Sailplane Directory*, November 1983, 60.

Scott, Jan. "Reimar Horten Visited." *Bungee Cord*, Vol. VI, No. 2, June 1980, 6.

———. "In Memoriam." *Bungee Cord*, Fall 1993, 4.

"Se jubila el inventor del avión a chorro sin cola," *Los Principios*, Friday, September 1, 1978.

Siepen, Howard. "On the Wings of the Wind." *National Geographic*, June 1929, 759.

Simpson, J. A. "The Evolution of the Sailplane." *Soaring*, March–April 1946, 11.

Slater, A. E. "Gliding Notes." *Aeroplane*, May 19, 1950, 596.

"The 19th National." *Soaring*, September–October 1952, 3.

"The Horten IV: An Amazing German Tailless All-wing Glider." *Flight*, August 5, 1943.

Ursinus, Oskar. "Russicher Segelflug-Wettbewerb in der Krim 7–25 September 1924." *Flugsport*, November 29, 1924.

———. "Vom Krim-Segelflug-Wettbewerb." *Flugsport*, November 25, 1925.

———. "Modelle." *Flugsport*, June 25, 1930.

Von Beckh, Harald J. "The Development and Airborne Testing of the PALE Seat." Report No. NADC-81200-60, Naval Air Development Center: Warminster, PA, June 20, 1981.

Von Hermann, Fritz. "Prone Flying." *Sailplane and Glider*, June 1952.

Walther, Helmut. "Die vergessenen Horten-Deltas: Vom Nurflüler zum Deltajäger (Teil 1)." *Flugzeug Classic*, July 2004.

Weyl, A. R. "Tailless Aircraft and Flying Wings—A Study of Their Evolution and Their Problems," part 1. *Aircraft Engineering and Aerospace Technology*, December–January 1944.

———. "Tailless Aircraft and Flying Wings—A Study of Their Evolution and Their Problems," part 2. *Aircraft Engineering and Aerospace Technology*, December–January 1944.

———. "Stability of Tailless Aeroplanes," part 1. *Aircraft Engineering and Aerospace Technology*, March 1945.

———. "Stability of Tailless Aeroplanes," part 2. *Aircraft Engineering and Aerospace Technology*, April 1945.

———. "Tailless Aeroplane Control Systems," part 1. *Aircraft Engineering and Aerospace Technology*, May 1945.

———. "Wing Tips for Tailless Aeroplanes." *Aircraft Engineering and Aerospace Technology*, September 1945.

———. "German Views on Auxiliary Powered Sailplanes. *Sailplane and Glider*, October 1949.

Zacher, Hans. "Ergebnisse der Leistungsmessung und Flugeigenschaftsprüfung des Segelflugzeuges D.30 'Cirrus.'" *Mitteilungen der Flugtechnischen Fachgruppen und Arbeitsgemeinschaften*, vol. 6, no. 1, 1944.

Internet Resources

Discussion group message traffic and articles at www.nurflugel.com.

George-Falvy, Dezsö. "Memoirs—I flew the Horten IV." www.dezso.org/writings/horten.html (website no longer exists).

———. "A Pilot's Notes on Flying the Horten IVa," September 28, 2005, www.dezso.org/writings/horten.html (website no longer exists).

Kuhlman, Bill and Bunny Kuhlman. "Twist Distributions for Swept Wings, Part 1." On the Wing . . . #161, www.B2Streamlines.com.

———. "Twist Distributions for Swept Wings, Part 4." On The Wing . . . #164, www.glidedyndns.org/on-the-wing4/index.html (website no longer exists).

Loewer, Peter. *Seeds: The Definitive Guide to Growing, History, and Lore*. Quoted in Flying Wings: Aviation and Aeromodelling—Interdependent Evolutions and Histories, at www.ctie.monash.edu.au/hargrave/wings_home.html.

Mattlener, Bernhard. "The Company." December 10, 2005, www.nurflugel.com.

———. "Products." December 10, 2005, www.nurflugel.com.

———. "The Technology." December 10, 2005, www.nurflugel.com.

Short, Simine. "Stamps That Tell a Story." *Gliding and Motorgliding International*, December 2002, February 2004, www.glidingmagazine.com/.

Index

Aerodynamische Versuchsanstalt (AVA), 26, 115; on all-wing designs, 119; *H III d* demonstration for, 120–121; *H V c* tested by, 118
Ahlborn, Friedrich, 14
ailerons: and *H I,* 37, 40–44, 240n83; and *H II,* 49–50; and *H II L,* 65; and *H III,* 75–76; and *H IX V2,* 128; and *H X a Piernifero,* 173; and *H XIII a,* 134; and PUL-10, 186; and yaw, 4, 41
Akaflieg Darmstadt, 99–100
Albrecht, Ulrich, 132
Allen, Michael, 189, 191
Allies, 141–144, 252n1
all-wing aircraft, 34–35, 37, 192; defined, 10–11; influences on, 27, 28; origins of, 14–16
Amtmann, Hans, 92
Arado 234, 128
Aramburu, Pedro Eugenio, 167, 180, 182
Argentina: and Horten aircrafts, 174–175, x; Nickels in, 152; Reimar Horten in, 28, 145–146, 182–184
Arnold, Henry H., 18, 143
Astralon, 68–69
AVA. *See* Aerodynamische Versuchsanstalt (AVA)
Avion Model 1, 17–18
B-2, 7, 22, 185, 191
B-47 Stratojet, 22
B-52, 22, 99
Bach, Liesel, 28
Backstrom, Al, 257n1
Balado, Rogelio, 177, 181
Bartolini, Rogelio, 170, 172, 173–174
Bazet, Eduardo, 153
Berger, Franz J., 142, 185
Berlin Technische Hochschule, 82
Blech, Werner, 78–80, 244n11
Bowers, Albion, 57; on effects on yaw, 138–139, 189–191; on lift distribution, 51, 53; and Prandtl-D Ship 1, 1–4
Bowlus Baby Albatross, 83
Brown, Eric, 192–193
Bryant, Chris, 192
BSLD. *See* lift distribution: bell-shaped (BSLD)
Bullot, Frank, 4
Burgess, W. Starling, 15–16
Burgess-Dunne AH-7, 16
Burzaco, Ricardo, 154, 165, 177, 178
Bush, Donald, 14
Button, Hollis, 155, 159

canopy: and *H I,* 36; and *H III,* 79; and *H III f,* 104; and *H X a,* 171;

canopy (*continued*)
and *H XVI Colibri,* 175; and *Ho 33,* 195
Capper, John E., 15
Cayley, George, 14
center of gravity (CG), 32; and *H 1,* 40, 43; and *H 1I,* 50, 51, 59, 65; and *H 1II,* 76; and *H III d,* 120; and *H IV a,* 90, 98, 157, 162–163; and *H V a,* 71, 90; and *H IX,* 138; and *H IX V2,* 125; and *H XIII a,* 134; and *I. Ae. 34 a,* 165; and *I. Ae. 38,* 181–182; and *I. Ae. 41 Urubú,* 165; and lift distribution, 67; and *Parabel,* 86; and PUL-10, 187; and transport carrier, 180
Cerny, Walter, 18
CG. *See* center of gravity (CG)
Chanute, Octave, 57
Christiansen, Friedrich, 61–62
Clen Antú. See *I. Ae. 34 a*
Club Argentino de Planeadores Albatros, 152
Club Cóndor, 175, 207
cockpit: and *H I,* 36; and *H IV a,* 104–105, 107, 121, 158, 192, 193; and *H VI,* 107, 193; and *H X a,* 170–171; and *H XIII b,* 135; and *I. Ae. 37-P,* 177; as prone, 88–89, 102–104, 130; as semiprone, 90–92, 192, 246n42; as supine, 192
Collum, Malcolm, 5
Concept Note SWS No. VII, Horten Ho 229, 9
Condor III, 98
Convair B-36 Peacemaker, 20
Convair XF-92A, 177
Cuadrado, José, 153, 154

D 30, 99–101, 104, 105–106, 108, 109, 163
Dailey, John R., 5
De Havilland Mosquito, 117
Delta I, 28, 237n27; influence of on *H 1,* 35, 36, 40, 45, 88
delta wing airplane, 177–179
Deutsche Forschungsinstitut für Segelflug [German Research Institute for Gliding] (DFS), 46, 64, 130, 138, 151–152
Deutschen Luftfahrtverband (DLV), 26, 34, 44, 62
Deutsche Versuchsanstalt für Luftfahrt e. V. (DVL), 26, 82, 118, 123
Dewoitine, Emile, 145
DFS 228, 130
D-Hangwind, 45, 46, 47
Diesing, Ulrich, 118
Dinort, Oskar, 62–64, 75, 77, 79
Dirección de Fomento de la Aviación Civil, 166
Dirección Nacional de Fabricaciones e Investigaciones Aeronáuticas (DINFIA), 183
Dittmar, Heini, 122
DLV. *See* Deutschen Luftfahrtverband (DLV)
Dori, Claudio, 152, 166
Döring, Kurt-Bertram von, 87
Doyle, Jorge, 177
drag rudders: and *H I,* 37, 41–43, 239n68; and *H II,* 50, 56, 57, 65–67; and *H III f,* 104; and *H IV a,* 91, 95, 98, 157, 162; and *H VI,* 106; and *H IX V1,* 123; and *H IX V2,* 125, 127–129; and *H XVI Colibri,* 175; and *I. Ae. 34 a,* 148, 150; and *I. Ae. 34 m,* 153; and *I. Ae. 38,* 180; and *I. Ae. 41,* 164; and *Parabel,* 86
Dunne, John W., 15–16
Dutch roll, 118, 123, 138, 189–190
DVL. *See* Deutsche Versuchsanstalt für Luftfahrt e. V. (DVL)
Dynamit A. G., 62–63, 68–72, 196

Earhart, Amelia, 16
Edwards, Glen W., 20–21
Edwards, Noah, 3
Eichmann, Adolf, 145
Electronic Pressure Measurement (EPM) device, 3–4
elevators, 10, 15, 27; and *H 1,* 37, 39–44; and *H II,* 49–50; and *H III,* 75–76; and *H IV a,* 95; and *H XIII a,* 134
elevons, 15, 67, 96; and delta wing airplane, 177–178; and *H II,* 49–50, 56, 57–58, 76; and *H II L,* 64–65, 240n7; and *H III,* 75–76; and *H IV a,* 93, 95, 98, 162–163; and *H VI,* 106–107; and *H IX,* 185; and *H IX V2,* 126, 129; and *H X a,* 170–174; and *H XVI,* 175; and *I. Ae. 34 a,* 148; and *I. Ae. 41,* 164–165; and *Parabel,* 86; and transport carrier, 180; and yaw, 189, 191
Elias, Willi, 155
Eschenauer, Arthur, 114
Etrich, Igo, 14–16
Evans, Richard, 13
Everett, Lew, 160
Ewald, Jochen, 102

F-104, 178
Fábrica Militar de Aviones (FMA), 146, 147, 176, 178, 183
Fairey, Richard, 144
Fentanes, Manuel, 177
Flieger Hitlerjugend, 132
Flight, 16, 27, 167
Flinsch, Bernhard, 99
Flinsch, Harold, 159
Flugsport, 11, 27, 54, 61, 157, 190
Fokker, Anthony, 170
foot-launched sailplane, 169–170
Frantz, Anselm, 124
Frise elevon, 95, 96, 98
Frise, Leslie George, 95
Frise noses, 171
Fritzsche, Peter, 13
Frondizi, Arturo, 178, 181, 182
fs 17, 88–89

G.38, 27, 28, 35, 51
Galland, Adolph, 86
George-Falvy, Dezsö, 160, 162–164
Georgii, Walter, 151–152
Göppingen 4 Gövier, 83
Göring, Hermann, 61, 64, 117–119, 188; and Hortens, 121–124, 128–130, 137–139
Gosslau, Fritz, 28
Göthert, Rudolf, 119, 120
Goupil, Alexandre, 57
Grafton, Sue, 3
Groenhoff, Günther, 28, 237n27
Grunau Baby, 172–175
Grupo de Desarrollo de Aviones, 183

H 3000, 188
Habicht, 60
Hallion, Richard P., 254n43
Hantsche, Lydia, 3
Harte, Richard, 15, 16
Hawkins, Victoria, 3
Hawks, Frank, 16
Heinkel, Ernst, 83–84
Heinkel He 111, 123
Heinkel He 177, 125
Heinkel He S 011, 135
Heinzel, Karl, 5–7
Helbig, Hans, 82, 245n16
Hieckmann, Kurt, 64, 79, 80
Hilger, Gisela. *See* Horton, Gisela Hilger
Himmler, Heinrich, 145
Hinote, Bill, 190
Hirth, Wolf, 33
Hitler, Adolf, 13, 64, 87, 88, 132
Ho 229 V2. See *Horten IX V2*

Ho 229 V3. See *Horten IX V3*
Hoinville, Fred, 169, 170, 257n1
Hol's der Teufel, 68, 69, 243n64
Horelick, Lauren, 5–7, 9
Horten, Gisela Hilger, 146–147
Horten, Dieter, 147
Horten, Elizabeth, 23–24, 28, 141–142; encouragement of children by, 31, 38, 236n5
Horten, Gunilde. *See* Nickel, Gunilde Horten
Horten, Karen, 147
Horten, Max, 23–24, 141–142, 236n2; death of, 252n3; encouragement of children by, 31, 38, 239n71; scholarship of, 236n2
Horten, Reimar, 12–13, 23, 193–195, 221; and aeronautics, 23; on aircraft being undetectable by radar, 6, 184–186; and all-wing aircraft, 9–10, 12, 16, 44–45; all-wing jet fighter report by, 119; in Argentina, 145–146, 182–184; avoiding conscription, 131; barriers to work of, ix–x; and bell-shaped lift distribution (BSLD), 1–4, 12, 13, 57–59, 188–190; and bomber jet, 136–137; captured by Allies, 141, 252n1; in competitions, 29–30, 82, 99, 101, 152–154; and delta wing airplane, 177–179; designations for aircraft designed by, 11; developing composite materials for aircraft construction, 68–70, 72; education of, 26–28, 114, 117, 248n88; experimenting with plywood on models, 239n72; fighter planes designed by, 111–112; flying distance of models by, 33, 238n49; on Göring, 121–122; and Heinkel, 83–84; and high-aspect ratio wings, 87–88; and *H I,* 34–36, 38–41, 46–47, 62, 174–175, 240n79; and *H II,* 49, 60–61; and *H II L,* 62–63; and *H III,* 18, 75–78, 82; and *H III d,* 250n40; and *H IV a,* 93, 95–96, 98, 102, 160, 161, 163–164; and *H V a,* 19, 69–72, 196; and *H V b,* 72–73; and *H V c,* 73; and *H VI,* 106; and *H VI V1,* 108, 109; and *H VI V2,* 105; and *H VII,* 113–115; and *H VIII,* 119–120; and *H IX,* 129–130, 137–138; and *H IX V1,* 122–123; and *H IX V2,* 123–128, 251n76; and *H X,* 169; and *H X a,* 170–174; and *H X Volksjäger,* 132–133, 252n91; on *H XIII a,* 134; and *H XVI Colibri,* 175–176; and *I. Ae. 34 a,* 147–149; and *I. Ae. 41 Urubú*, 164–167; and improvements to stability and control of aircraft, 190–192; and INAV, 151–152; influences on, 16, 28; interrogated by Allies, 141–142; and "kniendliegende" cockpit, 89–90; and Lippisch, 46–47; in Luftwaffe, 112; marriage of, 146–147; and model planes, 238n50, 238n52; and music lessons, 236n5; and Nazi Party, 13, 34, 238n59; *Nurflügel,* 12; and *Parabel,* 85–86; and parabolic wing, 241n13; as pilot, 24, 26, 29, 33–34, 236n7; and *Professor,* 37–38; and prone cockpit, 103; and proverse yaw, 2–3; and PUL, 186–187; safe-handling qualities added to planes by, 67; and sailplanes, 142–143; and semiprone cockpit, 90–92, 246n42; and supersonic aircraft, 133–135, 144; support from Nazis, 139; testing models, 30–33; transport plane designed by, 179–182; and wing loading design, 81, 84; workshops of, 115–117
Horten, Sabina von der Gröben, 112–113, 114, 248n7

Horten, Walter, x, 12, 23, 24–25, 27, 248n7; acquiring materials for planes, 112–114; and aeronautics, 24; and all-wing aircraft, 12, 34–35; attending Sitzung Nurflügelflugzeuge, 119; on aviation workers at end of war, 131; and bomber jet, 136–137; captured by Allies, 141; in competition, 29–30, 82; on Dinort, 63; education of, 28; flying distance of models by, 33, 238n49; on Göring, 121–122; and Heinkel, 83; and *H I,* 38–40, 43–44; and *H II,* 61; and *H III,* 18, 76–78, 195; and *H III d,* 120–121, 250n40; and *H IV a,* 96; and *H V a,* 19, 71, 196; and *H V b,* 72–73, 112–113; and *H V c,* 73; and *H VI V1,* 108; and *H VI V2,* 105; and *H IX,* 129–130, 138; and *H IX V1,* 123; and *H IX V2,* 124–125, 129, 251n76; interrogated by Allies, 141–142; in Luftwaffe, 62, 86–87; and model planes, 238n50; and music lessons, 236n5; as pilot, 25, 26, 29, 33–34, 192; refusing to send design for *H VII,* 183; on replacement fighter planes, 111–112; and semiprone cockpit, 90; and supersonic flight, 135; and support from Nazis, 139; testing models, 30–33

Horten, Wolfram, 23–25, 46, 55, 87, 236n5

Horten 229 V3. See *Horten IX V3*

Horten aircrafts: history of, 222–229; materials used in, 204–207; production of, 202–204, 229–232; speed comparison of, 213–215; speed notes on, 216–219; speed performance of, 210–212; weight comparison of, 199–201; wing comparison of, 196–199

The Horten Brothers and Their All-Wing Aircraft (Myhra), 12–13

Horten G.m.b.H., 122, 127–128, 136

Horten Ho 229 (Rys and Murawski), 9

Horten Ho 229, Spirit of Thuringia (Shepelev and Ottens), 13

Horten I, 24, 38–44, 192, 240n79, 240n83; in Argentina, 174–175; demonstration flight of, 44–46; design of, 34–36; design prize for, 62; influences on, 35, 36, 40, 45, 88; lower-aspect-ratio wings of, 87–88; and precision, 36; size of, 35–36, 239n62; and yaw, 37, 41–42

Horten I b, 175, 176

Horten II, 24, 49–50, 241n19; clear plastic panels for, 57, 62, 70; criticism of, 76; and drag rudders, 50, 56, 57, 65–67; engine of, 59, 61; and *H IX,* 127; lift distribution of, 55, 188; metal framework of, 55–56; Reimar Horten piloting, 57, 60–61; serial production of, 61; shape of, 50–51, 53; size of, 51; and supine cockpit, 56–57, 59–60, 89, 192; Walter Horten piloting, 59–61; wing design of, 50, 51, 88; wing modification of to deliver ammunition, 88; and yaw, 50, 57, 190–191

Horten II f, 127

Horten II L, 61–68, 79–82, 151, 240n4, 240n7

Horten III, 18, 75–76, 78–83, 88, 96, 133, 188, 195

Horten III b, 82, 84

Horten III d, 250n40

Horten III f, 65, 102–104, 127, 143, 151

Horten III g, 136

Horten III h, 143, 151

Horten IV, 188, 193

Horten IV a, 192–193, 196, 221; in competition, 99, 101; construction of, 117; control surfaces of, 93,

Horten IV a (*continued*)
95–96; design of, 93–94; and flutter, 98, 109, 157; interest in by Allies, 143–144; and lift distribution, 190; lift-to-drag ratio of, 98–99, 102, 160–161, 163; prone cockpit of, 88; Scheidhauer piloting, 96–99; semiprone cockpit of, 104–105, 107, 121, 158, 192, 193; in U.S. competition, 154–158
Horten IV b, 136, 192
Horten V a, 19, 69–72, 75, 84, 192
Horten V b, 18, 72–73, 112–113
Horten V c, 73, 113, 118
Horten VI, 105–109, 143, 188, 193
Horten VI V1, 107–109
Horten VI V2, 105, 151
Horten VII, 113–115, 117
Horten VIII, 119–120
Horten IX, 105, 119, 121–122, 185–186, 190
Horten IX V1, 122–123, 138
Horten IX V2, 13, 123–129, 137, 192, 251n76
Horten IX V3, 5–9, 13, 127, 144
Horten IX V6, 127
Horten X, 144
Horten X a, 170–174
Horten X c Piernifero 3, 191
Horten XIII, 144
Horten XIII a, 132–134, 252n86
Horten XIII b, 134–135
Horten XV a, 254n31
Horten XVI Colibri, 175–176
Horten XVIII, 144
Hurt, Hugh H., Jr., 54

I. Ae. *See* Instituto Aerotécnico Córdoba (I. Ae.)
I. Ae. 27 Pulqui I, 149
I. Ae. 33B, 146
I. Ae. 34 a, 147–149, 154, 157, 165, 167, 176, 254n31
I. Ae. 34 m, 153–154
I. Ae. 37-P, 177–178
I. Ae. 38, 176, 180–182
I. Ae. 41 Urubú, 164–167
I. Ae. 48, 178
INAV-1A, 221
Instituto Aerotécnico Córdoba (I. Ae.), 11, 146, 150, 166
Instituto Aerotécnico Córdoba (IAC), 183–184
Instituto Argentino de Vuelo a Vela (INAV), 151–152, 154
Instituto de Investigaciones Aeronáuticas y Espaciales, 183
International Plastic Modelers Society, 9

Jackson, Deborah, 3
Jacobs, Hans, 45, 240n86
Jakab, Peter, 5
Jensen, Chris, 3
Johnson, George, 9
Johnson, Richard H. “Dick,” 135, 158–161, 164
Jones, Robert T., 2, 188–189
Jungflieger, 26, 30
Junkers, Hugo, 16, 17, 27, 28, 35, 51
Junkers F.13, 27–28

Kegel, Max, 66–67
Klein, Armin, 2
Klingberg Wing, 189
Kracht, Felix, 130, 138, 139
Kranich, 98
Krasinski, José, 177
Krauss, Serge, 10
Kronfeld, Robert, 155
Kunz, Seff, 63

laminar airflow, 135–136, 160
laminar airfoils, 135, 158, 164, 192
Lampros, Alex, 250n61
landing gear, 33, 88; of all-wing aircraft, 10; of *H I,* 36; of *H II,* 56, 61; of

H IV a, 91, 107, 163; of *Ho 33,* 195; of *H V a,* 70, 72; of *H VI,* 107; of *H X a,* 172; of *H XVI Colibri,* 175; of *I. Ae. 34 m,* 153; of *I. Ae. 41,* 164; of PUL-10, 187; of Pulqui II, 146; of transport carrier, 179
lift distribution, 186, 188–189; bell-shaped (BSLD), 1–4, 12–13, 57–58, 67, 126, 164, 188–191; and delta wings, 177; elliptical model of, 53–55; and *H II,* 51, 57, 59; and *H III,* 78–79; and *H IX V2,* 126; and wing sweep, 59
Light Sport Aircraft (LSA), 188
Lignofol, 71, 106, 109
Lilienthal-Gesellschaft für Luftfahrtforschung (Lilienthal Society for Aviation Research), 119
Lilienthal, Otto, 56–57, 88
Lippisch, Alexander, 18, 28–29; and all-wing aircraft, 61; attending meeting on *H IX,* 121; attending USA National Soaring Competition, 155; attending Sitzung Nurflügelflugzeuge, 119; on aviation workers at end of war, 131; and Delta I, 40, 88; and *Hol's der Teufel,* 68; influence of on Reimar Horten, 16, 33, 35, 36, 45; and lift distribution, 54–55, 190; and parabolic wing, 241n13; and Reimar Horten, 46–47; and tailless aircraft, 50, 240n86
Lippisch Storch tailless glider, 157
Lonardi, Eduardo, 167, 180
Lower Rhine Flying Club, 26, 108
Luftwaffe, 34; and Horten aircrafts, 62–63, 69–70, 82, 123; Hortens in, 71, 86–87; and loss of aircrafts and crew, 121; and need for planes, 111, 117
Lyon, Dick, 135, 158

Mantelli, Adriano, 151–152
Mattlener, Bernhard, 186, 188, 260n68
Maxim, Hiram, 15
McElhinney, Peter, 5–7
Mengele, Josef, 145
Messerschmitt [aircraft company], 112, 122, 188
Messerschmitt, Willy, 141
Messerschmitt Bf, 111
Messerschmitt Me 163, 28
Messerschmitt Me 163A, 122
Messerschmitt Me 262, 123–124, 126, 128, 129, 137
Messerschmitt Me 262A-1a, 124
Meyer, Alan, 260n68
Mipolan, 68–69
Mississippi State College/University (MSC/MSU), 158–163
Mitteneffekt, 59, 70, 84–85, 105
Multhopp, Hans, 146
Murawski, Marek: *Horten Ho 229,* 9
Museum für Verkehr und Technik (MVT), ix
Myhra, David, 12–13; *The Horten Brothers and Their All-Wing Aircraft,* 12–13

N-9M, 19, 20, 21
NASM. *See* Smithsonian National Air and Space Museum (NASM)
Nationalsozialistisches Fliegerkorps (NSFK), 61–62, 80
Nazi Party, 13, 34, 62
Nickel, Gunilde Horten, 23, 24; in Argentina, 152; giving location of planes to Allies, 143; on *H I,* 47; on Reimar Horten, 28, 83, 142
Nickel, Karl, 142; in Argentina, 152; on bell-shaped lift distribution, 188; and *H I,* 47, 239n68; and *H II,* 50, 65–66; and *H III,* 79; and *H III d,* 120, 250n40; and *H III f,* 65, 67; and *H IV a,* 102; and *H XIII a,* 133; and *I. Ae. 34,* 165; model experiments by, 176–177; on Northrop, 150; as

Nickel, Karl (*continued*)
pilot, 136; piloting in prone position, 102–104; on Reimar Horten, 28, 67; *Tailless Aircraft in Theory and Practice,* 12, 188; and transport plane, 180
Nimbus 4D, 99
Northrop, John K. "Jack," 13–15, 150–151; and all-wing aircrafts, 16–22, 116, 138
Northrop Aircraft Corporation, 18, 21, 150–151
Northrop Grumman Corporation, 7–8
Northrop N-1M, 18–19
Nurflügel (Horten and Selinger), 12
Nurflügel G.m.b.H., 187

Ojeda, Cesar Raul, 151, 153, 178, 179
Opitz, Rudolf (Rudi), 154–163, 193
Ortner, José, 152, 153
Oskar-Ursinus-Vereinigung Wintertagung (Oskar Ursinus Conference), 190
Ottens, Huib: *Horten Ho 229, Spirit of Thuringia,* 13
Otto Lilienthal Prize, 82

P-51 Mustang, 135
Panek, Siegfried, 186–187
Panek Ultra-Light (PUL-9), 186–187
Panek Ultra-Light (PUL-10), 186–187
Parabel, 81, 84–86
Parker, Ray, 160
Perón, Evita, 146
Perón, Juan D., 13, 145–146, 149–150, 167, 178, 180
Piernifero. See *Horten X*
Platz, Reinhold, 170
Post, Wiley, 16
Prandtl, Ludwig, 26, 144, 241n14; and bell-shaped lift distribution, 1–2, 13, 188–189; and lift distribution, 190; observing Horten aircraft, 120–121; and parabolic wing, 241n13; on supersonic flight, 135; on wing design, 51, 54
Prandtl-D 3, 3
Prandtl-D 3c, 3
Prandtl-D Ship 1, 1–2, 4
Professor, 37–38, 239n71
Pulqui I. *See* I. Ae. 27 Pulqui I
Pulqui II, 146, 176

Quick, A. W., 119, 120

radar, 8, 10–11, 184–186, 194
Raspet, August, 155, 158–162, 163, 164
Rechlin Flight Test Center, 129
Reeves, E. J., 158
Reichert, Carl, 21
Reichsluftfahrtministerium (RLM), 34, 96, 121, 123, 188; on all-wing aircraft, 118; Reimar Horten hiding work from, 114–116; and support of Horten aircrafts, 61, 64, 73, 87, 95, 105, 112, 139
Reiher, 98
Reitsch, Hanna, 64–68, 76, 82
Reynolds, Osborne, 36
RJ-5, 158–161, 164
RLM. *See* Reichsluftfahrtministerium (RLM)
Roosevelt, Franklin, 18
Ross, Harland, 135, 158, 159
Rossi, Guido, 166
Rovesti, Plinio, 151–152
Ruff, Siegfried, 135
Rys, Marek: *Horten Ho 229,* 9

Sanchez, Rubén, 166
Schatzberg, Eric, 122
Scheidhauer, Heinz, 77, 153, 257n107; in Argentina, 152; cockpit designed for, 90, 105; and foot-launched sailplane, 170; and *H I,* 175; and *H II,* 78–79; and *H II L,* 63, 64; and *H III,* 65, 79–80, 120–121; and

H IV a, 96–99, 101–102, 104, 105, 163; and *H IV b,* 136; and *H VI VI,* 108; and *H VII,* 129; and *H IX,* 122–123; and *H X,* 170, 174; and *H XIII a,* 133; and *H XVI Colibri,* 176; and *I. Ae. 34 m,* 154; and *I. Ae. 37-P,* 178; and *I. Ae. 41 Urubú,* 165–167; as test pilot, 84, 88, 193
Schempp, Martin, 33
Schmitt, Günter, 27
Schmitt, Jupp, 46
Schmitz, Franz Wilhelm, 27–29, 35, 51
Schrödter, Rolf, 38, 55–56, 59, 241n19
Schweizer 1–23, 158
Schweizer SGS 2-8, 83
Scott, Jan, ix, 184, 190, 191
Selinger, Peter F., 11; *Nurflügel,* 12
Shepelev, Andrei: *Horten Ho 229, Spirit of Thuringia,* 13
Siarez, Fernando, 151, 167, 179
Sitzung Nurflügelflugzeuge, 118–119, 120
Slater, A. E., 161
Slingsby Kirby Gull 1, 83
Slingsby Sky, 154, 158
Slingsby Type 34 Sky, 166
Smithsonian National Air and Space Museum (NASM), ix, 4–9, 240n4
Soldenhoff, Alexander, 37
Spitfire, 112
spoilers: of *H II L,* 65; of *H III,* 96; of *H IV a,* 93, 95–96, 101–102, 162; of *H VI,* 106; of *H IX V2,* 125; of *H XIII a,* 134
Stackhouse, Don, 189
Stadler, Reinhold, 51, 54, 190, 241n13, 252n91; evaluation of Horten aircraft by, 59, 126, 163, 186
Stadlman, Anthony, 16
Stamenovic, Marko, 4
Storch I, 28
Strebel, Hermann, 79, 104, 108, 133–134, 136
Stüper, Josef, 118
Sturm, Waldemar, 176
Suitor, Rachel, 3
supersonic flight, 120, 133–135, 144, 178

tailless aircraft, 16, 28, 37; defined, 11; and *H I,* 33; and *H IV a,* 104, 158, 162; and *H V c,* 118; and *I. Ae. 38,* 179; landing of, 125; and Lippisch, 50, 240n86; and Northrop, 19; stability of, 14–16, 44–45, 57, 119, 138; and U.S. Army Air Forces, 143; and Walter Horten, 111–112
Tailless Aircraft in Theory and Practice (Nickel and Wohlfahrt), 12, 188
Tank, Kurt, 141, 145, 146, 150, 176, 182
Taube, 15
Texas Soaring Association, 158
Tscheranovsky, B. I., 37

U-2R, 99
Uden, Edward, 57, 186, 189, 191
Udet, Ernest, 64, 112–114; support of Hortens by, 61–63, 82, 83
Ursinus, Oscar, 11–12
Urubú. See *I. Ae. 41 Urubú*
U.S. Air Force, 7, 14, 20, 21, 22, 92, 155, 177
USA National Soaring Competition, 154–159
U.S. Army, 15–16, 19, 92, 142, 151, 162
U.S. Army Air Corps, 18–19
U.S. Army Air Forces, 5, 20, 143–144
U.S. Navy, 15–16, 142, 162

Vega, 16–17
vertical/short takeoff and landing (VSTOL), 162
vertigo, 67
Vintage Sailplane Association, ix, 184
Viswanathan, Sathy, 2

Volksjäger, 132
Volkssturm, 131–132
von Kármán, Theodore, 18, 142

Waddell, Abigail, 3
Wasserkuppe, Germany, 12, 37, 116, 175; as home of sport soaring movement, 25–26, 29; Hortens competing at, 30, 33, 45–47, 62, 79–82, 84, 88, 104–105
Weihe, 98
Weiss, Anna, 5–7
Weiss, Edmundo Osvald, 149
Wells, H. G., 15
Wels, Franz, 14–16
Wenk, Fredrich, 37
Weyl, A. R., 57
Whitcomb, Richard T., 2
Wilkinson, Kenneth G., 65, 142–144, 161
Wills, Chris, 130
Wills, Philip, 130
wing loading, 81, 88, 196–199; of *H I*, 36; of *H II*, 78, 80; of *H III*, 78–80, 83, 84; of *H IV a*, 99, 101, 147; of *H V b*, 18; of *H VI*, 106; of *H IX V2*, 124; of *H X a*, 170, 175; of *H XVI Colibri*, 175–176; of *Parabel*, 85–86
wing sweep, 28, 32, 119; and Horten aircrafts, 86, 108, 157, 190
wing tips, 32–33, 36; and *H 1*, 37, 42, 44; and *H 1I*, 56; and *H 1II*, 76; and *H II L*, 66; and *H IV a*, 90–91, 98, 101, 163; of *H VI*, 106; and *I. Ae. 38*, 180; and *I. Ae. 41*, 163, 164; and lift distribution, 52, 55, 189
Wohlfahrt, Michael: *Tailless Aircraft in Theory and Practice*, 12, 13, 188
Wright, Orville, 14–15, 56–57, 88
Wright, Wilbur, 14–15, 56–57, 88

XB-35, 19–20, 138
XP-56, 19
XP-79, 19

yaw, 1–4, 123; and all-wing aircraft, 14, 20–21; and bell-shaped lift distribution, 57–58; and center of gravity, 32; and Delta I, 28; effects on, 138–139, 189–190; and *H I*, 37, 41–45; and *H II*, 50, 57, 190–191; and *H II L*, 65–66; and *H III*, 75–76, 157; and *H IV a*, 95, 98, 162, 164; and *H V c*, 118; and *H IX V1*, 138; and *H IX V2*, 128–129; and *H X a*, 170–173; and *H XIII a*, 134; and *H XIII b*, 135; and *I. Ae. 37-P*, 177; and *I. Ae. 38*, 180; and *I. Ae. 41 Urubú*, 164; and PUL-10, 186, 187
YB-49, 20, 21–22, 116, 138–139

Zacher, Hans, 99, 101, 102
Ziller, Erwin, 123, 127–130, 137–138, 193, 251n76
Zindel, Ernst, 27
Zoukei-Mura, Inc., 9
Zuárez, Alfonso, 166

About the Author

RUSSELL E. LEE is a curator in the Aeronautics Division at the Smithsonian Institution's National Air and Space Museum, Washington, D.C. He specializes in soaring flight and sailplane technology as well as other types of sport aircraft. Lee graduated in 1992 from George Mason University with a master's degree in American history, and he has flown a variety of single-engine and motorless aircraft since earning a private pilot's license in 1979.